Uncertainty and Forecasting of Water Quality

Editors:
M. B. Beck and G. van Straten

With 143 Figures

Springer-Verlag
Berlin Heidelberg New York Tokyo 1983

M. B. BECK
Department of Civil Engineering
Imperial College of Science and Technology
London SW 7 2 BU
England

G. van STRATEN
Department of Chemical Technology
Twente University of Technology
P. O. Box 217
7500 AE Enschede
The Netherlands

IIASA
International Institute for Applied
Systems Analysis
A-2361 Laxenburg
Austria

ISBN 3-540-12419-5 Springer-Verlag Berlin Heidelberg New York
ISBN 0-387-12419-5 Springer-Verlag New York Heidelberg Berlin

Library of Congress Cataloging in Publication Data
Main entry under title:
Uncertainty and forecasting of water quality.
Bibliography: p.
Includes index.
1. Water quality--Mathematical models--Adresses, essays, lectures.
2. Lakes--Mathematical models--Adresses, essays, lectures.
I. Beck, M. B.
II. Straten, G. van (Gerrit van).
TD370.U44 1983   363.7'394'0724   83-4679

This work is subject to copyright. All rights are reserved, whether the whole or part of the material is concerned, specifically those of translation, reprinting, re-use of illustrations, broadcasting, reproduction by photocopying machine or similar means, and storage in data banks.

© International Institute for Applied Systems Analysis, Laxenburg/Austria 1983

Printed in Germany

The use of registered names, trademarks, etc. in this publication does not imply, even in the absence of a specific statement that such names are exempt from the relevant protective laws and regulations and therefore free for general use.

Offsetprinting: Weihert-Druck GmbH, Darmstadt. Bookbinding: Graphischer Betrieb Konrad Triltsch, Würzburg
2061/3020 - 543210

PREFACE

Since the International Institute for Applied Systems Analysis began its study of water quality modeling and management in 1977, it has been interested in the relations between uncertainty and the problems of model calibration and prediction. The work has focused on the theme of modeling poorly defined environmental systems, a principal topic of the effort devoted to environmental quality control and management.

Accounting for the effects of uncertainty was also of central concern to our two case studies of lake eutrophication management, one dealing with Lake Balaton in Hungary and the other with several Austrian lake systems.

Thus, in November 1979 we held a meeting at Laxenburg to discuss recent methodological developments in addressing problems associated with uncertainty and forecasting of water quality. This book is based on the proceedings of that meeting.

The last few years have seen an increase in awareness of the issue of uncertainty in water quality and ecological modeling. This book is relevant not only to contemporary issues but also to those of the future. A lack of field data will not always be the dominant problem for water quality modeling and management; more sophisticated measuring techniques and more comprehensive monitoring networks will come to be more widely applied. Rather, the important problems of the future are much more likely to emerge from the enhanced facility of data processing and to concern the meaningful interpretation, assimilation, and use of the information thus obtained.

JANUSZ KINDLER
Chairman
Resources and Environment Area
IIASA

FOREWORD

From the point of view of analysis and prediction, environmental systems can be said to lie midway between the two extremes of electrical circuit systems and social systems. This situation presents special problems in the analysis of environmental and, more specifically, water-quality and ecological systems. On the one hand, *a priori* theory, with its basis in the physical and biological sciences, would seem to be capable of predicting observed behavior relatively accurately. On the other hand, however, it is especially difficult to conduct planned experiments against which *a priori* theory can be tested. Under these somewhat ambivalent conditions there has arisen a growing incompatibility between what can be simulated in principle with a model and what can be observed in practice. To a great extent this accounts for the gap that has developed between the "larger" simulation models, with which there is little hope of conducting rigorous calibration exercises given currently available field data, and the much "smaller" models that have been so calibrated. There is, in short, a pressing need to reconcile such incompatibility in the current developments in water quality modeling.

The purpose of these proceedings, as with the meeting on which they are based, is therefore to present and discuss *applicable methods* both for identifying (calibrating) water quality models from uncertain experimental data and for analyzing prediction-error propagation. There are three themes on which the book places special emphasis:

- *Identification*. This has special emphasis because many previous publications that have dealt with the subject have not addressed some of the more challenging methodological questions and because many methods that might appear attractive in theory are neither easy nor useful to apply in practice.
- *Relation Between Identification and Prediction*. Few publications consider this relation in any detail, although it is clear that the uncertainties in a model and its predictions are a function of how the model has been identified and calibrated.
- *Interaction Between Case Studies and Methodological Development*. There are many problems generated in the course of a case study, so much so that it becomes almost redundant to demonstrate the applicability of a technique by means of an abstract example. It is the accumulation of experience from case studies that is the genesis of new methodological advances.

The book is divided into four parts: an introduction (consisting of two papers); a part on uncertainty and model identification or calibration (consisting of eight papers); a part on uncertainty, forecasting, and control (consisting of seven papers); and a final commentary. Throughout the book there is an emphasis on the interaction between developing approaches and methods and applying them to cases.

In Part One, two papers (by Beck and by Young) provide an historical perspective and the synthesis of a methodological framework for the book. Both papers draw upon the use of recursive estimation algorithms in the context of time-series analysis as the basis for much of their discussion. Quantitative, theoretical aspects of these algorithms, however, are not introduced in detail; rather, it is the conceptual representation of the problems, their solution, and the interpretation of case-study results that are of primary importance. Broadly speaking, therefore, Parts Two and Three of the book include the case studies, and the problems of analysis thereby generated, which Part One uses here as the raw material for methodological developments and the synthesis of a framework for modeling and forecasting environmental systems behavior.

Part Two contains eight papers. Although there is an intimate relationship between identification (calibration) and prediction, as argued in Part One, these papers have been collected together in one section because they all focus principally on the problem of calibration. Three of the papers (by Hornberger and Spear, by Halfon and Maguire, and by Chahuneau et al.) discuss the application of techniques of Monte Carlo simulation for the purposes of model calibration. There are also two papers in Part Three that deal with Monte Carlo simulation methods and they are closely related to these three papers. Hornberger and Spear present further results for a case study of eutrophication in Peel-Harvey Inlet (Australia) using a novel approach to the generation of hypotheses under conditions of sparse field data. Their work plays an important role in the discussion given by Young in Part One. The paper by Chahuneau et al. is likewise closely related to Spear and Hornberger's work, exploring questions of uncertainty in the fluid mixing and transport properties of a lake. Somlyódy's paper considers the hitherto little-discussed problem of uncertainty in the input (forcing function) observations and compares the importance to parameter variability of such uncertainty with the uncertainty of model calibration results. He uses a hydrodynamical model for Lake Balaton (Hungary) as an illustrative case study. The papers by van Straten and by Mejer and Jørgensen are concerned with specific problems generated in applying off-line (as opposed to recursive) parameter-estimation algorithms in lake water quality modeling. van Straten, for example, reports results for maximum likelihood estimation of the parameters in a fairly complex phytoplankton model for Lake Ontario. The last two papers in this part of the book, by Ikeda and Itakura and by Tamura and Kondo, both discuss the identification of statistical relationships in a set of field data using as few *a priori* assumptions as possible about the structure of these relations. Ikeda and Itakura's paper deals with a study of Lake Biwa (Japan), whereas Tamura and Kondo concentrate more on the theoretical development of their algorithm, with some illustrative results for the Bormida River in Italy.

There are seven papers in Part Three. The first two, as already mentioned, discuss the application of Monte Carlo simulation methods in analyzing the effects of uncertainty on the confidence attached to model predictions. Gardner and O'Neill's paper reviews results for the effects of parameter uncertainty and covers part of the substantial contribution that they and their colleagues have made to this field. Fedra adds an important extension of the earlier work by Spear and Hornberger by emphasizing the significant relationship between calibration and prediction (see also Beck's paper in Part One); his work, focusing mainly on the ecological aspects of lake behavior, is applied to a case study of Attersee (Austria).

The next two papers, by Reckhow and Chapra and by McLaughlin, are complementary to each other. Both apply a first-order analysis of prediction errors, but, whereas Reckhow and Chapra treat the specific cases of simple nutrient loading models, McLaughlin presents an analysis of error propagation for a general class of distributed-parameter models. McLaughlin's results are in fact stated in a form equivalent to that of recursive estimation and filtering algorithms, which has important parallels with the discussion of Beck's paper.

The contributions by Whitehead and by Koivo and Tanttu also draw upon methods of recursive parameter estimation and prediction, but broaden the discussion of the book toward management issues. Both papers are concerned with case studies of UK river systems, the Cam and the Bedford-Ouse (both of which are central to the discussion of Beck's paper), although Whitehead introduces a wider, and more practice-oriented perspective on planning and operational management problems, including the application of Monte Carlo simulation. Koivo and Tanttu restrict their paper to more theoretical aspects of operational (real-time) forecasting of water quality.

Fisher's paper, the last in Part Three, continues with and expands the theme of uncertainty in control and management studies. As opposed to Whitehead's interest in point-source discharges of waste material, Fisher deals with the problem of managing non-point-source nutrient inputs for the control of eutrophication in lakes, a problem background common to nearly all the other papers in the book.

The short Part Four contains a contribution stimulated by the discussion at the meeting. Sharefkin was prompted to prepare a paper giving a Bayesian interpretation to certain problems of identification. In particular, he is concerned with how *"a priori* information" and "data information" are combined and mapped into *"a posteriori* information", after the identification exercise; he constructs his arguments around the papers presented by Hornberger and Spear and by Chahuneau et al.

In closing this introduction, a brief comment can be made on the lively discussion of the meeting. It is typical of IIASA meetings that a large part of the available time is allocated to the informal exchange of views, experiences, and opinions; in our case it was generally felt that these deliberations were truly instrumental in the success of the meeting. It is not an easy task to convey an impression of the discussion to the reader who did not participate himself. However, many of the themes touched upon then return in a digested form in the chapter immediately following this introduction. Thus, here we wish to attempt merely to draw a few conclusions and to evaluate briefly "where we are" and "where we go" in water quality modeling and prediction.

Not surprisingly a substantial amount of time was devoted to the technical details of both the models and the methods and there is little purpose in repeating such details here. Of greater interest is the observation that the debate about which models to build and which methods of analysis to apply was sometimes obscured by unspoken differences in the objectives of modeling and analysis. There is no doubt that the objective of using modeling as a tool for structuring information, with the ultimate aim of gaining an improved understanding of the complexity of a system, demands an attitude different from that required when the goal is to apply models for control, or, even more ambitious, for design and management. Whereas for operational-control purposes it may be completely satisfactory to have a purely data-based, simple model, even a black-box model that is well-calibrated within the range of presently observed states, an application for design and management requires a different approach because in these cases answers are expected for situations often deviating very significantly from the presently observed conditions. This, then, can be our first conclusion: that prior to any technical analysis it is important to ask "why is it all being done?", and "what do we wish to do with it?".

Following the organization of this book, the next issue for consideration is model structure identification, which was discussed largely in relation to model calibration and parameter estimation. Especially when management applications are the ultimate aim, a preference can be observed among modelers to construct their models on the basis of some assumed *a priori* physical, chemical, and biological theories. However, this perhaps necessary step may lead to problems of over-parameterization and surplus content when the model is calibrated against field data, although such a condition is by no means easy to recognize. While techniques have been developed for determining the order of a model for certain classes of linear models, any general theory seems to be lacking for the type of commonly-met nonlinear models derived from *a priori* theory. And although analysis of the parameter variance—covariance structure and, in association, a sensitivity analysis may provide some insight, there are numerous pitfalls in such analyses. Thus, we may perhaps conclude that a definite need exists for developing techniques to detect which components of the models can be considered to be "hard" and which "soft", when confronting the model response with the field data.

The problems encountered in model structure identification are closely related to the problems of prediction and forecasting. Propagation of uncertainty is an important issue for study, and there was quick agreement among the participants that any model-based prediction should be accompanied by error-budget calculations. The controversy between models based largely on empirical observations and models based on *a priori* theory leads to the dilemma pointed out by Beck in the first paper: that with the former we may predict the "wrong" future with great confidence, and with the latter the "correct" future might be predicted but without any confidence whatsoever. Or, as one of the participating biologists tersely put it: "model predictions are either false or trivial". Although certainly excessive, this statement may serve as a warning that one should be aware of the limitations of modeling in the field of water quality systems. It should not be overlooked, however, that this awareness can only arise from the development and application of methods for quantifying the extent of these limitations, examples of which are amply supplied in the remainder of this book. Indeed, the issue of model credibility is perhaps the most important challenge for future work on water quality modeling. Therefore, the final conclusion to be drawn from the stimulating discussion of the meeting is that improvement of predictive power is to be expected only if we succeed in finding proper methods to bring together the rigor of data-based analysis with adequately designed experimentation and the achievements of *a priori* theory derived from past experience.

BRUCE BECK
GERRIT VAN STRATEN

CONTENTS

PART ONE: INTRODUCTION

Uncertainty, system identification, and the prediction of water quality   3
    M.B. Beck
The validity and credibility of models for badly defined systems 69
    P. Young

PART TWO: UNCERTAINTY AND MODEL IDENTIFICATION

An approach to the analysis of behavior and sensitivity in environmental systems . . 101
    G.M. Hornberger and R.C. Spear
Distribution and transformation of fenitrothion sprayed on a pond: modeling
    under uncertainty 117
    E. Halfon and R.J. Maguire
Input data uncertainty and parameter sensitivity in a lake hydrodynamic model . . . 129
    L. Somlyódy
Maximum likelihood estimation of parameters and uncertainty in phytoplankton
    models 157
    G. van Straten
Model identification methods applied to two Danish lakes 173
    H. Mejer and L. Jørgensen
Analysis of prediction uncertainty: Monte Carlo simulation and nonlinear
    least-squares estimation of a vertical transport submodel for Lake Nantua . . . 183
    F. Chahuneau, S. des Clers, and J.A. Meyer
Multidimensional scaling approach to clustering multivariate data for
    water-quality modeling. 205
    S. Ikeda and H. Itakura
Nonlinear steady-state modeling of river quality by a revised group method of
    data handling 225
    H. Tamura and T. Kondo

PART THREE: UNCERTAINTY, FORECASTING, AND CONTROL

Parameter uncertainty and model predictions: a review of Monte Carlo results 245
    R.H. Gardner and R.V. O'Neill
A Monte Carlo approach to estimation and prediction 259
    K. Fedra
The need for simple approaches for the estimation of lake model prediction
    uncertainty 293
    K.H. Reckhow and S.C. Chapra

Statistical analysis of uncertainty propagation and model accuracy 305
    D.B. McLaughlin
Modeling and forecasting water quality in nontidal rivers: the Bedford Ouse study . . 321
    P.G. Whitehead
Adaptive prediction of water quality in the River Cam 339
    H.N. Koivo and J.T. Tanttu
Uncertainty and dynamic policies for the control of nutrient inputs to lakes 357
    I.H. Fisher

PART FOUR: COMMENTARY

Uncertainty and forecasting of water quality: reflections of an ignorant Bayesian . . 373
    M. Sharefkin

Author index 381
Subject index 383

Part One

Introduction

UNCERTAINTY, SYSTEM IDENTIFICATION, AND THE PREDICTION OF WATER QUALITY

M.B. Beck

International Institute for Applied Systems Analysis, Laxenburg (Austria)

1 INTRODUCTION

There would be little disagreement among water quality modelers with the opinion of Orlob (1983a) that virtually all the significant developments since the (now) classical work of Streeter and Phelps (1925) have occurred within the past two decades. During the 1960s and early 1970s there was a very substantial investment in model-building associated with water quality management projects, particularly in the United States. The main legacy of this initial investment is a well-established interest in the development of progressively larger and more complex simulation models. "Large" is admittedly a rather imprecise description of a model, although a glance at some of the recent literature on water quality modeling will give some impression of the intended meaning (for example, Russell, 1975; Patten, 1975, 1976; Jørgensen and Harleman, 1978; Scavia and Robertson, 1979). There is no doubt that the immense scope for complex system simulation created by the advent of electronic computers has fostered the rapid growth of "large" water quality models.

Relatively little attention, however, has been given to the problems of uncertainty and errors in the field data, of inadequate numbers of data, of uncertainty in the relationships between the important system variables, and of uncertainty in the model parameter estimates. It is only during the last seven or eight years, for example, that an increasing but still comparatively small number of detailed studies in system identification (model calibration and verification) have been reported. These later developments might be summarized by the statement that only "small" models have so far been calibrated rigorously against *in situ* field data (see Beck, 1980, for a survey of the literature). The reasons for this are naturally of concern in this study, as are the intimated distinctions between "large" and "small" models and the emerging recognition of "uncertainty"; these topics will be discussed in more detail later in this paper.

The history of water quality modeling has been shaped by a number of quite separate and almost independent contributions from various different scientific and engineering disciplines. It is instructive, therefore, to review some of the major individual trends of the past, since this will help us to define the gaps remaining in the subject today.

Hydrological sciences. The hydrological sciences have only recently become more involved with problems of water quality. It is significant that the International Association of Hydrological Sciences convened its first meeting on water quality in 1978 (IAHS, 1978) and that a subsequent meeting on hydrological forecasting (IAHS, 1980) attracted only six papers on water quality from a total of more than 70 contributed. There is, nevertheless, a strong tradition of research in stochastic hydrology. Calibration, parameter estimation, and uncertainty in rainfall–runoff and streamflow-routing models have all been studied extensively. However, as will become apparent later, system identification in quantitative hydrological modeling is substantially different from system identification in water quality modeling.

Systems ecology. Developments in systems ecology display a strong concern with the stability and structural properties of a model *once it has been developed* (see, for example, Adachi and Ikeda, 1978; Goh, 1979; Siljak, 1979). The problems of "uncertainty" are now also widely recognized (Argentesi and Olivi, 1976; O'Neill and Gardner, 1979; Reckhow, 1979; Scavia et al., 1981), although research is generally concentrated on prediction error analyses of models which are once again assumed to have been developed and calibrated previously. There appears to be a distinct separation of the problems of calibration and prediction, and a lack of detailed studies in data analysis and the identification of models by reference to *in situ* field observations. This point of view seems to be held by others: in 1975 Eberhardt responded to the question "whither systems ecology?" with the remark that "we should be very careful to avoid letting our computers run too far ahead of what statistical methods and data tell us about the real world" (Eberhardt, 1977). Ulanowicz (1979) would also appear to be in agreement when he speaks favorably of "... modeling in an *a posteriori* fashion, allowing the data to define interactions [between variables]".

Sanitary engineering. Unlike both the ecological and the hydrological sciences, sanitary engineering has until recently been reluctant to adopt the techniques of mathematical modeling (Andrews, 1977; Andrews and Stenstrom, 1978; Olsson, 1980). This is partly a function of different objectives — the interest is in new process designs and process control rather than "scientific understanding" — and partly a function of different traditions — sanitary engineers do not generally present their knowledge and hypotheses in a mathematical format. Consequently, while routine operating data from wastewater treatment are readily available, the majority of studies in time-series analysis have been confined to "black box" modeling approaches (see also Beck, 1980). This is quite the opposite of the situation in systems ecology, where models have flourished despite a relative dearth of field observations.

System identification. This topic is usually associated with control theory, statistics, and econometrics, and may perhaps be considered an intruder in water quality modeling (Åström and Eykhoff, 1971; Eykhoff, 1974). An impressive array of techniques has been developed for the analysis of relatively conventional, well-posed problems, such as those encountered in identifying models for aircraft dynamics, chemical unit processes, and physiological systems (see Isermann, 1980), but water quality model identification involves problems that are not at all well-posed so that these techniques may not be applied in any straightforward way.

This survey suggests that one of the major areas which needs investigation is the problem of uncertainty. There are two key issues: first, the problem of uncertainty in the structure of the mathematical relationships hypothesized for a particular model; and second, the uncertainty associated with the predictions obtained from models. These two issues are discussed briefly and qualitatively in a companion paper (Beck, 1981); a more detailed and quantitative statement of the same arguments is given here.

The questions posed by the issue of uncertainty are clearly questions about the reliability of models and their forecasts. In particular, this paper is concerned with the intimate relationship between the two aspects of uncertainty introduced above. In other words, it emphasizes the importance of the relationship between calibration and prediction, a relationship that is largely ignored in systems ecology. It is therefore assumed that, in addition to providing a concise representation of existing knowledge about a system's behavior, mathematical models are intended for forecasting, which in turn implies a concern with management. If decisions are to be made on the basis of the model's forecast, how confident can one be that this forecast is correct, and what is the risk of making a wrong decision? These have become familiar questions in water quality modeling and management.

Two subsidiary topics underpin the central theme of uncertainty in calibration and prediction; they are both related predominantly to problems of calibration (system identification). First (in Section 3.1), we characterize some limiting properties of the field data available for calibration of water quality models, compare our results with quantitative hydrological system identification, and examine why there has been little progress in the identification of wastewater treatment process dynamics. This leads, second (in Sections 3.2 and 3.3), to an extensive discussion of the identification of model structures using experimental field data. It is then possible to identify the desirable properties and some probable limitations of algorithms for ill-posed problems in system identification. It will also become apparent why time-series analysis, calibration, and "curve-fitting" — perhaps contrary to popular opinion — are concerned with the derivation of models that are more than simple black-box descriptions.

Overlying the main theme of uncertainty, however, is the question of "large" and "small" models mentioned briefly above. We shall interpret this as the larger question of "dominant" and "not-so-dominant" approaches to the subject of modeling in general, and the whole of Section 3 is therefore a statement of the "not-so-dominant" approach. However, since we are trying to make a fair assessment of the advantages and limitations of different approaches to modeling, we should consider both calibration and prediction, and not calibration alone; Section 4 therefore concentrates on the predictive aspects of modeling.

First, however, it is necessary to explain what is meant by the "dominant" approach to water quality modeling, and why there is a need to question it.

2   A DOMINANT APPROACH AND SOME CONCERNS

An obvious barrier in the identification of water quality models has been the lack of field data suitable for analysis. We suggest that it is partly the vacuum created by this absence of adequate field data that has led to the predominance of the large simulation model (broadly speaking, one which contains numerous state variables at many spatial locations in the system). Large models of this type grew out of the double assumption that there are few constraints on numerical complexity and that more detail necessarily means a better model. The checks and balances provided by readily available data have not generally seemed to restrain the growth of model complexity. There are, nevertheless, those who would argue against large models, believing that "small is beautiful" on a number of grounds: because it is not possible to verify larger models against the available *in situ* field data; because the responses generated by large models are not readily intelligible; and because techniques for optimal management and policy design cannot accommodate large models. Indeed, the discussions at recent workshops would have been much less lively had it not been for these differences in opinion (see, for example, Russell, 1975; Vansteenkiste, 1975, 1978).

However, it is unwise to rely entirely on the labels "large" and "small" in the discussion that follows; the use of the terms "dominant" (conventional) and "not-so-dominant" (unconventional) to describe the two approaches to modeling is to be preferred, although this again is not altogether satisfactory. In the present case, the more established conventional approach* involves the (conceptual) subdivision of the field system into smaller, individual components, whose (conceptual) behavior can usually be approximated by laboratory-scale replicas (for example, chemostat and open-channel flow experiments). It is assumed that the submodels describing these components can be verified against the experimentally observed behavior of the replica; and that the model for the field system can be assembled by linking together the submodels. Such models tend to be large. But largeness and the inclusion of great detail do not necessarily imply accuracy and reliability. It seems obvious that accuracy and reliability can only be assessed by *rigorous* tests of the model's hypotheses against *in situ* field observations. It is at this point that the problems of calibrating large models arise, although they are rarely adequately recognized (see also Thomann et al., 1979) and almost never adequately resolved. The systematic recognition and resolution of these problems presents major, and possibly insurmountable, difficulties given current methods of analysis. These difficulties are fundamental to the discussion of this paper.

The above description merely depicts the archetypal form of the dominant approach to modeling. However, those who follow this approach also subscribe, in effect, to a school of thought that works principally from what may be called *a priori* theory. Thus the model is supported by arguments that allow extrapolations from laboratory systems and equivalent or similar field systems. At the calibration stage it is assumed that *a priori* theory is correct unless *demonstrably* inadequate, and the difficulty clearly lies in demonstrating inadequacy. For as long as there is a lack of *in situ* field data, the need to question the original extrapolations will remain in doubt. Here, then, lies the chief distinction between the dominant and unconventional (not-so-dominant) approaches to water quality modeling. With the unconventional approach the analyst works much more from the *in situ* field data. It is assumed that the underlying mechanisms of system behavior can be identified directly from these data. The model is supported by what is identifiable from the *in situ* observations and, if these observations are few, the resulting models would tend to be small. (It would be a mistake, however, to assume that the field data will always be few, as we shall see later.) Thus, in the unconventional approach, *a priori*

---

* Referred to as "reductionist" by Young (1978, 1983).

theory is considered, at best, as a fertile source of more or less speculative hypotheses. Moreover, the legitimacy of the extrapolations inherent in the dominant approach would appear to remain unproven without the directly evaluative properties of this unconventional approach.

2.1 Some Concerns about the Dominant Approach

The items below reflect a number of major concerns about the dominant approach to modeling. They are quoted as points with which to open a discussion, in which inevitably there will be differences of opinion, not because of any intention to detract from past achievements.

Item 1. In the proceedings of a recent seminar on water quality management Fleming (1979) discusses the topic of assessing water resources problems that link sediment, hydrological, and water quality processes. It is accordingly proposed that each of these three components should be treated as a deterministic system for which mass and energy balances in time and space can be calculated. Superficially there is nothing particularly disconcerting about this recommendation, although a more careful examination of the proposed mathematical model reveals features that are indeed somewhat more disturbing. The hydrological catchment model is divided among three phases: in each phase there are three interacting submodels that deal individually with the movement of sediment, the quantity of flow, and the changes in water quality. Together with a groundwater model, these submodels are used to assess the quantity and quality of water resources; this assessment interacts with an optimization model, which in turn is linked to regional, national, and international models. These are then apparently to be subordinated to sociopolitical–economic models. Notwithstanding that this is presumably intended to be a linked set of models (and not one single, enormous, economic model that embraces all the other models), the catchment model has impressive dimensions, for it is not to be restricted to a small stretch of river. On the contrary, the proposal was made specifically for a major international river system.

Concern 1. In the discussion referred to above, Fleming states that the computer era has produced (through modeling techniques) an "acceleration in our understanding of natural processes". This may be so; but alternatively it may be that the computer era has merely fostered the growth and popularity of large simulation models with little accompanying increase in understanding. In addition the only constraint on mathematical modeling is identified as the "ability of the planner to grasp the potential of the method". One could argue conversely that such over-enthusiastic attitudes toward modeling, with their emphasis on determinism, tend to mask the undoubted difficulties of accounting for uncertainty. Complexity and completeness cannot necessarily be equated with accuracy; given the limitations of the data available, a "complete" model containing a large number of parameters is more likely to produce predictions with serious hidden ambiguities than anything else.

Item 2. Figure 1 appears in a book on ecological modeling from the mid-1970s (Park et al., 1975). The authors, referring to the diagram, state that their results for modeling and calibration are "relatively good". There are other "equally good" results provided.

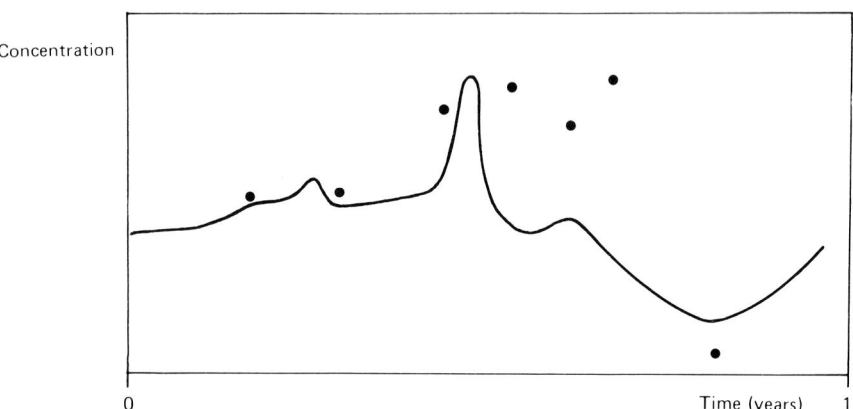

FIGURE 1   An example comparison of the model response with the field observations.

Concern 2. The more sceptical reader looking at Figure 1 might well protest that this diagram shows nothing more than a rather whimsical fluctuation peppered with a few dots. Our concern here is that there ought to be a less subjective method of assessing the performance of models. However, as Lewandowski (1981) points out, while there are several definitions of model validity, there are very few applicable methods for model validation. Thus, while such an analytical "vacuum" exists, there can be little more than a somewhat fruitless exchange of opinions about the "goodness" (or otherwise) of models.

Item 3. In a case study of lake eutrophication management, a relatively complex microbiological model was developed to describe nutrient transformation processes in the lake (Leonov, 1980). The model responses were fitted to the data by adjusting the values of the parameters by trial and error; some of the parameters were assumed to vary with time in order to improve the calibration results. In particular, the values obtained for one of the parameters imply that a more or less "regular" behavior pattern prevails for eleven months of the year but that behavior during the remaining month is highly "irregular".

Concern 3. There are two main causes for concern in this item. First, suppose that a model contains ten parameters for which values are to be estimated by a trial and error comparison of the deterministic model responses with the field data. Most modelers would admit to having preconceived notions about how a system *should* behave, and during the fitting of the model to the data will probably have a preference for adjusting the values of perhaps three or four of the ten parameters. Thus preconceived notions may dictate the outcome of the calibration — clearly excluding any serious questioning of prior assumptions and extrapolations — and preconceived notions are not always correct. The second cause for concern is that the results obtained from the calibration might be accepted without, in this case, any rigorous argument or evidence to support a hypothesis of highly irregular behavior. In fact, far from being dismissed out of hand, the peculiarities of the calibration results might suggest how the structure of the model can be improved.

Our main concerns about the dominant approach to water quality modeling may therefore be summarized as follows:

1. Attitudes toward water quality modeling have been largely characterized by a strong sense that the behavior of the systems being studied is deterministic. This has obscured the fact that the values of model parameters are rarely known accurately, and that the larger the model, the less likely it is that a *unique* set of parameter values can be estimated using a limited number of field data. Such (unjustified) confidence in *a priori* theory engenders both a reluctance to put aside "classical" assumptions and a readiness to reject the unconventional without due consideration.
2. There is no universal model for solving all manner of water pollution problems. However, once a water quality model has been constructed the costs of development (if large) may make it necessary to present the model as "universal" in order to justify these costs. If a government agency confers its "seal of approval" on a model, this lends a sense of authority to the use of that model and at the same time may undermine confidence in an alternative model. "Complexity" and "completeness" can be misrepresented (and misunderstood) as "truth" and "accuracy". There is a temptation to believe that a large, comprehensive model must be correct, for how can it possibly fail to be so if every detail of conceivable relevance has been included?
3. Confidence in a model is perhaps unavoidably subjective. Yet surely one should doubt any model that requires the use of an absurd hypothesis to make the model responses fit the data. This would merely be a sterile calibration exercise in which curve-fitting is an end in itself. It would ignore the obvious inadequacy of the model, excluding the closer examination that could actually be a source of new ideas, and would suggest that calibration is something of a backwater compared with the mainstream of model development.

These, then, are the concerns that have provoked this discussion. It would be naive to imagine, however, that these issues can be resolved as easily as they have been raised – a precise resolution of the issues requires first a precise definition of the issues, and our concerns are themselves borne of a vague sense that "all is not well with water quality modeling".

3   FIELD DATA, UNCERTAINTY, AND SYSTEM IDENTIFICATION

Most analysts and decision-makers would wish to be reassured that the patterns of behavior simulated by a model do in fact resemble observed patterns of behavior. There is thus a need for system identification, or more specifically, for model calibration, an exercise typically associated with curve-fitting and parameter estimation. However, the word "calibration" is misleading. It suggests an instrument (here, the model) whose design has been completed and whose structure is fixed; it is only necessary to make a few minor adjustments to the parameter values. This is *not* what is meant by "calibration" in the discussion of water quality–ecological modeling that follows. There are two main reasons for this: the nature of water quality field data and the nature of contemporary *a priori* theory. This section, therefore, begins with a discussion of these two factors, which form the basis for a thorough exploration of the key problem of model structure identification. An important objective of this exploration is to map the "topography" of the problem by examining it from several different perspectives; we then consider briefly the philosophical foundations upon which we hope to construct a solution. We must also dispel any expectation that some "universal" approach to modeling will be presented here. To do this, it is necessary to realize that model development may take place under various conditions: with few or no data; with some adequate data; or with too many data (a situation which could become increasingly common in the future). This discussion will consider only the second of these categories; however, a universal approach to modeling should of course be able to deal with all three.

3.1 Field Data and *A Priori* Theory

At present, field data from water quality systems are generally scarce. However, when data are available they are subject to high levels of error and uncertainty. Halfon (1979a), for instance, shows just how many sources of error can affect the data obtained from large lakes. These errors, however, are not the only problems encountered in the calibration of water quality models. It is not simply a matter of large errors, too low a sampling frequency, or too short a record of time-series data.

Young (1978) suggests that the inability to perform planned experiments (see below) is a distinctive feature of the modeling of badly defined systems: a category which clearly includes water quality–ecological systems. Successful calibration is hindered by the conditions under which field observations are obtained. Planned experiments may be defined as experiments in which the responses of some of the system variables (outputs or effects) are recorded and are assumed to be unambiguously related to changes in other (input, causative) variables. In such planned experiments all variables (except the input deliberately manipulated and any response variables thereby disturbed) are maintained at constant values. In other words, the "environment" of the system is held constant, the causative variables can be manipulated to conform with a desired pattern of change, and the experiment is planned such that unambiguous relationships between the variables can be determined. Planned experiments of this kind are virtually impossible for water quality–ecological systems; there are merely successively less good approaches to this ideal.

3.1.1 Active, Natural, and Passive Experiments

A considerable amount of work in system identification has been devoted to the problem of experimental design (Åström and Eykhoff, 1971; Söderström et al., 1974; Gustavsson, 1975). Two interpretations of this problem are particularly relevant to our discussion since they demonstrate a number of desirable features of experimental design that are rarely possible in the collection of water quality data. The *frequency-response* interpretation has immediate relevance to this discussion; the *sensitivity-analysis* interpretation is of more general interest, but is significant here in that it links together some of the arguments of Sections 3 and 4.

Let us begin with the simplest possible situation, illustrated in Figure 2. In this case the purpose of the model is to characterize the dynamic relationship between two measured variables: $u(t)$, the input, and $y(t)$, the output. The objective is therefore to design an experiment which has an appropriate set of variations of $u(t)$ with time t. Assuming a frequency-response interpretation of system behavior, this objective can be made more

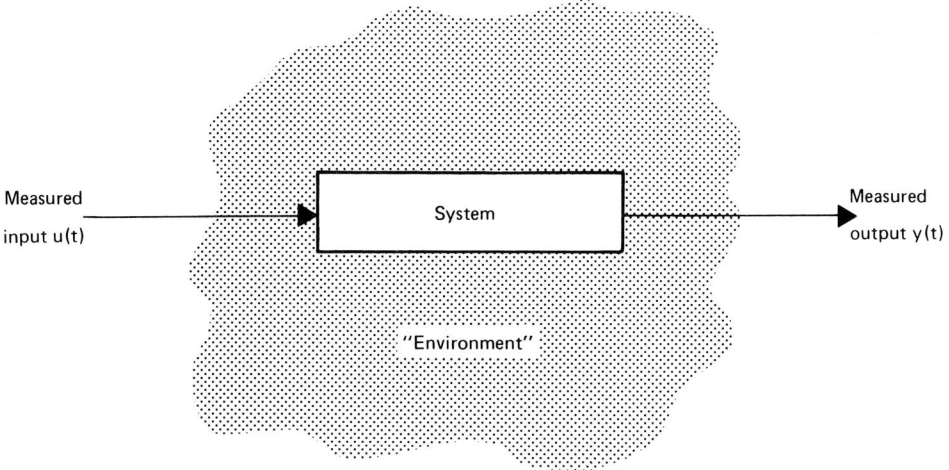

FIGURE 2   Definition of a simple system for the interpretation of data from field experiments.

specific: for example, it is desirable to design $u(t)$ such that its frequency spectrum is "matched" with the *expected* frequency-response characteristics of the system. (This "matching" does not, of course, take place with knowledge of the actual frequency-response characteristics of the system, since this is what we are trying to determine.) At the same time it is desirable to design $u(t)$ such that in the subsequent analysis it will be possible to discriminate against the effects of high-frequency random measurement errors, i.e., some of the effects of the "environment" shown in Figure 2.

Given these limited, qualitative notions of experimental design, we shall now examine the problem of identifying the dispersion mechanisms affecting the distribution of a dissolved material in a body of water. This is the classical problem of water quality modeling, and continues to attract much attention (for example, Somlyódy, 1977; Thé, 1978; Kahlig, 1979; White et al., 1980; Jakeman and Young, 1980; Young, 1983; van Straten and Golbach, 1982). The tracer experiments employed for subsequent model calibration, which we shall define as *active* experiments, are the closest field approximations to the planned experiments of laboratory science. They possess three of the most desirable properties of experiments: (i) significant, in this case impulse-like, input perturbation of the system — this is intuitively desirable because the observed effects of dispersion are most pronounced for a high-frequency input variation; (ii) relative lack of ambiguity in the assumed relationship between input and output; (iii) restriction of the experimental measurements to the quantities assumed to vary.

The problem of model calibration is thus as well-posed as it can be. However, it will swiftly degenerate should any of these three properties not hold, for example, if: (i) the frequency components of the input are not significant (except for the effects of measurement error) at the higher frequencies where the response of the system is theoretically most sensitive to the dispersion coefficient; (ii) the tracer or dissolved material is not completely conserved, i.e., there is some interaction with the environment of the experiment; (iii) other quantities, such as stream discharge, also vary with time, i.e., the environment of the experiment is not constant. Ample evidence of the limitations introduced by these compromises, especially (i) and (iii), can be found in recent analyses of dispersion mechanisms in the River Cam (Lewandowska, 1981) and the lower reaches of the River Rhine (van Straten and Golbach, 1982).

It is instructive to consider the Rhine study (van Straten and Golbach, 1982) in somewhat greater detail, partly because the problems encountered are typical of system identification in this field, and also because it provides a natural extension of the discussion to the topic of sensitivity analysis. Under the assumption that the classical advection–dispersion model would adequately describe the observed variations in the field data, van Straten and Golbach found that the loss function for calibration was much more sensitive to the value of the stream velocity than to the value of the dispersion coefficient. At the "best" value for the estimated stream velocity, both a ten-fold decrease and a two-fold increase in the value of the dispersion coefficient produced only a marginal increase (12% and 24%, respectively) in the value of the loss function. A 7% change in the estimate of the stream velocity, however, gave a 47% increase in the value of the loss function. We can therefore conclude that the gradient of the loss function surface with respect to the dispersion coefficient is relatively small close to the minimum of the loss function. In other words, for these particular experimental conditions, the output response of the model is not sensitive to the value of the dispersion coefficient, which is therefore probably not uniquely identifiable. A common consequence of this lack of sensitivity (or identifiability) is that the estimation error covariances of the associated parameters will be large, which in turn has important implications for the model predictions (see Section 4). In more general terms, however, and for more complex multiple-input/multiple-output systems, it is possible to minimize subsequent difficulties of this kind. A prior analysis of the assumed model structure may indicate a combination of input and output measurements that will allow the unique estimation of all the model parameters (see, for example, Cobelli et al., 1979). However, this merely confirms once again that a well-designed experiment must be based upon a good *a priori* model of system behavior.

Not only is the availability of a good *a priori* model unlikely, as we suggested in Section 2, but the scope for implementing active experiments is virtually confined to the classical problem of identifying dispersion mechanisms. However, this does not imply that we cannot observe *natural* experiments in complex natural systems. For example, the hydrological sciences place considerable emphasis on the identification of catchment characteristics through analysis of the response of the stream discharge to a storm (IAHS, 1980). In the light of the preceding discussion the importance of the storm is obvious: it represents a significant input disturbance of the system, and the output response can be relatively unambiguously related to the input disturbance. There are, however, significant differences between active experiments and natural (hydrological) experiments. The environment of the hydrological system is not entirely constant (the temperature will vary, for instance, thus affecting the evapotranspiration rate) and it is not possible to manipulate the input disturbance at will. This may lead to problems similar to those encountered in the Rhine dispersion study. Nevertheless, the subsequent calibration problem is likely to be as well-posed as could be expected in water quality modeling. If the assumption of a simple two-variable (input/output) relationship is not sufficient to characterize the observed behavior of the system, the comparative wealth of *a priori* theory may be used to restructure the model such that a more accurate representation is obtained (as in Whitehead et al., 1979). Thus, the field data available from natural experiments can be summarized by the stylization *"significant input perturbation: significant output perturbation"* (Figure 3).

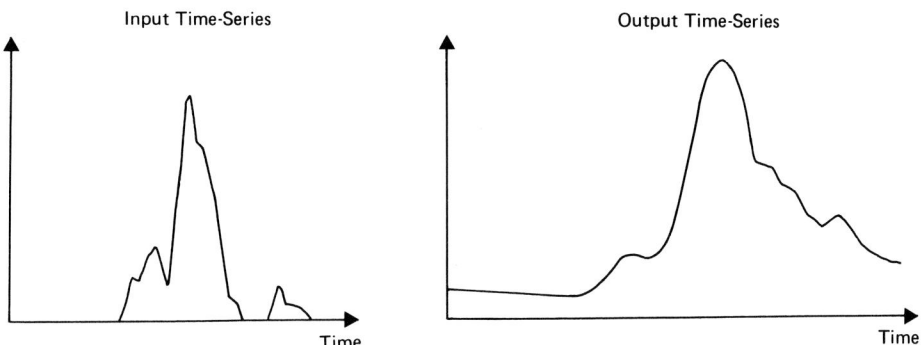

FIGURE 3 Stylized representation of the input/output data available for, e.g., rainfall/runoff model identification (natural experiments).

Natural experiments of the kind described above are quite rare in water quality–ecological systems. Consider, as a contrast to the natural hydrological experiment outlined earlier, the "extreme" output response which appears as a phytoplankton bloom in a lake. The bloom occurs because a specific but apparently commonplace sequence of environmental (input) conditions forces the system into a region of the state space in which a nonlinear mode of behavior becomes dominant. Unlike the example of the hydrological system, the response of the lake is probably not unambiguously related to an extreme input disturbance. Instead, it may be a consequence both of subtle changes in the system's environment and of a very specific combination of circumstances within the lake at the given point in time (or space). Such a situation has useful parallels with the conditions that have prompted the application of catastrophe theory to problems of water quality modeling (see, for example, Kempf and van Straten, 1980). Two points have special relevance. First, the applicability of elementary catastrophe theory depends upon the assumption that variations in the inputs are relatively slow (lower frequency) with respect to the output response variations (higher frequency). Second, the matching of this theory to observed behavior implies a quite *specific,* critically important nonlinear structure for the dynamic model of the system.

For a number of reasons, therefore, field data obtained from the predominantly *passive* experiments with water quality–ecological systems do not possess the properties required for system identification. Pictorially the data for such situations can be represented as in Figure 4 and hence stylized as *"apparently insignificant input perturbation:*

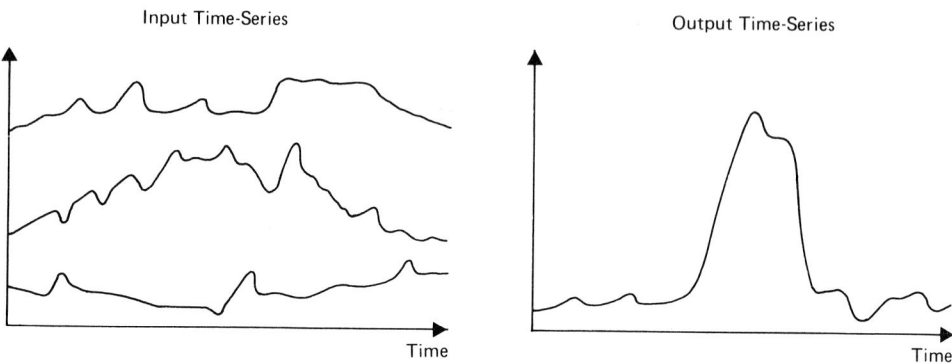

FIGURE 4 Stylized representation of the input/output data available for, e.g., lake water quality model identification (passive experiments).

significant output perturbation". If a specific model structure is needed in order to describe observed behavior, which the parallel with catastrophe theory would suggest, the probability that *a priori* hypotheses clearly embody this structure is not great, as we shall discuss in Section 3.1.2.

Finally we consider an important variation on the above theme which is in some sense the converse of the conditions depicted in Figure 4. This last class of conditions is encountered in the identification of models for microbiological wastewater treatment processes, such as, for example, an activated sludge unit (Beck et al., 1978). Figure 5 summarizes the stylized form of this category of field data, which we might define as "*apparently significant input perturbation: insignificant output perturbation*". The system, observed in this way, appears to possess remarkable stability. Typically the mean levels of the input disturbances are much higher than corresponding mean output levels (for biochemical oxygen demand (BOD), and suspended solids, for example), and

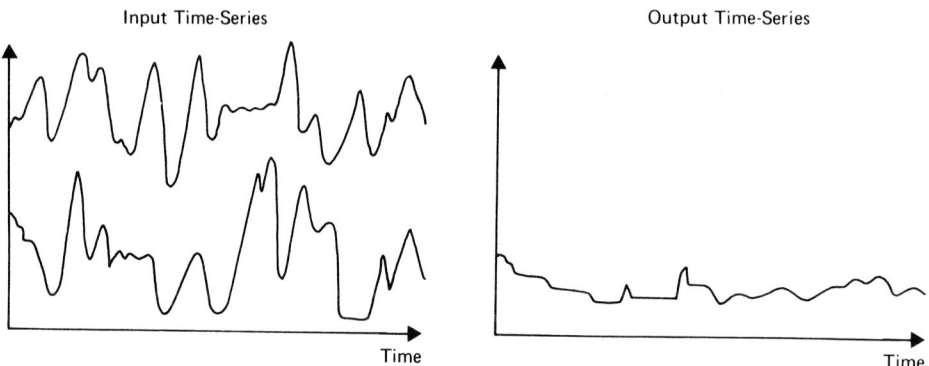

FIGURE 5 Stylized representation of the input/output data available for, e.g., wastewater treatment plant model identification (passive experiments).

some of the large-amplitude, higher-frequency variations of the inputs are attributable to random measurement error. It is not surprising, therefore, that it is difficult to relate output to input, even in a statistical sense, which in turn accounts for the tendency for strongly autoregressive models to be identified from such data. Yet in these systems conditions equivalent to those of Figure 4 do occur, and they occur for very much the same reasons as discussed above. Longer-term variations in the input disturbances can, in combination with a specific set of operating conditions, induce subtle but important changes in the microbiological state of the activated sludge. While these complex changes themselves occur slowly, they may eventually lead to what are observed as relatively rapid changes in the system's output responses (Tong et al., 1980).

3.1.2 A Priori Theory and Ambiguity

The problems of error-corrupted data and the lack of planned, active, or natural experiments (in the sense that we have used these terms) are technical, rather than fundamental problems of model calibration. Admittedly, they lead to severe difficulties in the application of methods for system identification, but they are purely technical in the sense that if the analyst knew, *a priori*, how the system behaved, then it would still be comparatively easy to distinguish the estimated patterns of behavior in the observed field data. The basic problem of calibrating water quality–ecological models is the limited degree of *a priori* knowledge about expected system behavior. In spite of very many laboratory-scale experiments and a number of major field studies, our knowledge of the relationship between the mineral, organic, and microbiological components of water quality–ecological systems is actually quite incomplete. This means that model calibration should concentrate on resolving the problem of model structure identification. Indeed, the question of accurate parameter estimation will only be touched on briefly in the following discussion.

A somewhat sophisticated, but particularly apt example of uncertainty in the structure of model relationships is given in Bierman's study of Saginaw Bay, Lake Huron (Bierman et al., 1980). Bierman noted that the output response of his model was especially sensitive to the choice of hypothesis for the growth-rate of phytoplankton. The model had originally been calibrated against field data from Saginaw Bay with phytoplankton growth expressed according to the threshold hypothesis — namely, that growth-rate is governed only by that single factor which is determined to be rate-limiting — and there was additional evidence from laboratory experiments to support the chosen hypothesis. But Bierman subsequently admits that an alternative hypothesis — the multiplicative growth hypothesis, where all factors contribute to an overall rate of growth — could probably also have been calibrated against the Saginaw Bay data. Calibration of this differently structured model with the alternative growth-rate expression would almost certainly have resulted in different estimates for all the other parameter values in the model. The significance of this example is, of course, that it demonstrates how sufficient uncertainty exists in our *a priori* theories of system behavior to allow considerable speculation about the precise structure of an appropriate mathematical model. In short, there are ambiguities in the *a priori* theory of behavior patterns in water quality–ecological systems.

There is fairly widespread recognition of the effects of such ambiguities on the process of model calibration. For instance, Halfon (1979b) presents results for a model of

Lake Ontario where an order-of-magnitude change in many of the estimated parameter values gives rise to an increase of merely 6% over the minimum of the loss function. There are many combinations of the parameter values that fit the data "equally well". It might be that the occurrence of this flat loss-function surface is dominated by errors in the field observations. If this is not the case, however, one should clearly question the appropriateness of the model structure, as does Halfon. Yet Halfon's questioning, although correctly aimed (according to the concerns outlined in Section 2) leads to answers that err on the side of accepting convention. He concludes, in effect, that all the parameters are more or less equally significant (or equally insignificant) in determining the correspondence between model and observations. There is thus no radical rethinking of the model structure; it is considered to be sufficiently aggregated and to have no redundant parameters, i.e., no *surplus content*.

The tendency, as stated in Section 1, is to accept the legitimacy of extrapolations from laboratory or similar field situations. The analyst who justifies surplus model content on the basis of these extrapolations must support one of two possible arguments: either the surplus content of the model has originally been unambiguously identified by a prior calibration for another field system, which is unlikely in the present circumstances, or his justification is founded upon a chain of similar justifications starting from an extrapolation from laboratory to field conditions. An argument such as the latter covers the possibility of a further implicit, but important extrapolation. Suppose that, in order to overcome the ambiguities of unconstrained estimation of many parameters, bounds are introduced to define "acceptable" ranges of values for the estimates. (A logical extension of this is to restrict some of the *a priori* "better known" parameters to point estimates, which are then assumed to be known without error.) But from what source of *absolute* authority are these bounds themselves derived? For example, if the bounds on the growth-rate constant of a species are drawn from laboratory-determined values, it should not be forgotten that such values are only defined *relative to the model* (kinetic expression) that was assumed and calibrated for the observed nutrient and phytoplankton concentrations in the laboratory experiment.

This is not, however, to dismiss entirely the accumulation of experience but rather to emphasize that too much confidence is frequently attached to the validity and relevance of *a priori* theory.

3.2 Model Structure Identification

Given the preceding discussion we see that calibration of models for water quality–ecological systems is unlikely to be a simple and straightforward matter of making minor adjustments to a well-designed "instrument". Instead, even before asking the question "Can I estimate the model parameters accurately?", the analyst must first ask himself whether he knows how the variables of the system are related to each other. In particular, one must ask whether information about these relationships can be identified from the *in situ* field data. Most exercises in model calibration have focused solely on the matter of parameter estimation; hence little attention has been paid to the arguably more important prior problem of model structure identification (Beck, 1979a). As a simple example, it may be a fine idea to estimate the slope and intercept of a straight line drawn through a set of data points (i.e., parameter estimation), if it has *already* been established that a straight line, and not a curve, will give the best fit to those data (i.e., model structure identification). Hence, model structure identification logically precedes parameter estimation. Undoubtedly it is a complex problem whose complete description is a function of many elements, requiring examination of the problem from several different perspectives.

Perspective 1. In order to describe the problem of model structure identification more precisely, let us begin by introducing the following general model of the system's dynamics

$$\dot{x}(t) = f\{x(t), u(t), \alpha(t)\} + \xi(t) \tag{1a}$$

$$y(t_k) = h\{x(t_k), \alpha(t_k)\} + \eta(t_k) \tag{1b}$$

where

- x = n-dimensional vector of state variables,
- u = m-dimensional vector of measured input disturbances,
- y = p-dimensional vector of (discretely sampled) measured output variables,
- α = l-dimensional vector of model parameters,
- ξ = n-dimensional vector of unmeasured (unknown) input disturbances,
- η = p-dimensional vector of output measurement errors,

and **f** and **h** are nonlinear, vector-valued functions; t is the independent variable of time, t_k is the kth discrete sampling instant in time, and $\dot{x}$ denotes the derivative of x with respect to t. From Section 3.1.1 it is apparent that system identification is generally concerned with the analysis of that which is measurable — the inputs u and the outputs y, i.e., the "external" description of the system (see also Figure 2) — in order to infer the characteristics of **f**, **h**, x, and α, i.e., the "internal" description of system behavior. This process of inference may require assumptions about the "environment" of the system, ξ and η, or it may conversely be directed to drawing conclusions about the properties of ξ and η themselves.

Perspective 2. In addition to the formalities of eqn. (1), let us introduce a complementary conceptual representation of the system, as shown in Figure 6. First note the distinction between the microscopic (block 1 in Figure 6) and macroscopic features (block 2) of the system's dynamics and between the easily and not easily measured state variables, x_m and x_u, respectively. x_u is intuitively associated with the (literally) microscopic dynamic features of the system's behavior patterns because these detailed microbiological characteristics are not directly observable. It is particularly difficult, for example, to monitor mechanisms of nutrient uptake and release by micro-organisms (a microscopic feature), but it may be supposed that such mechanisms have considerable significance for the observed variations in chlorophyll-a concentrations (a macroscopic feature). The process of inference mentioned in Perspective 1 is thus especially difficult. If the microscopic features of block 1 in Figure 6 are of central interest in determining and understanding system behavior, they must be inferred from identification of their interaction with macroscopic state variables observed in the presence of a highly uncertain "environment" (block 3) characterized by high levels of measurement error and random input disturbances.

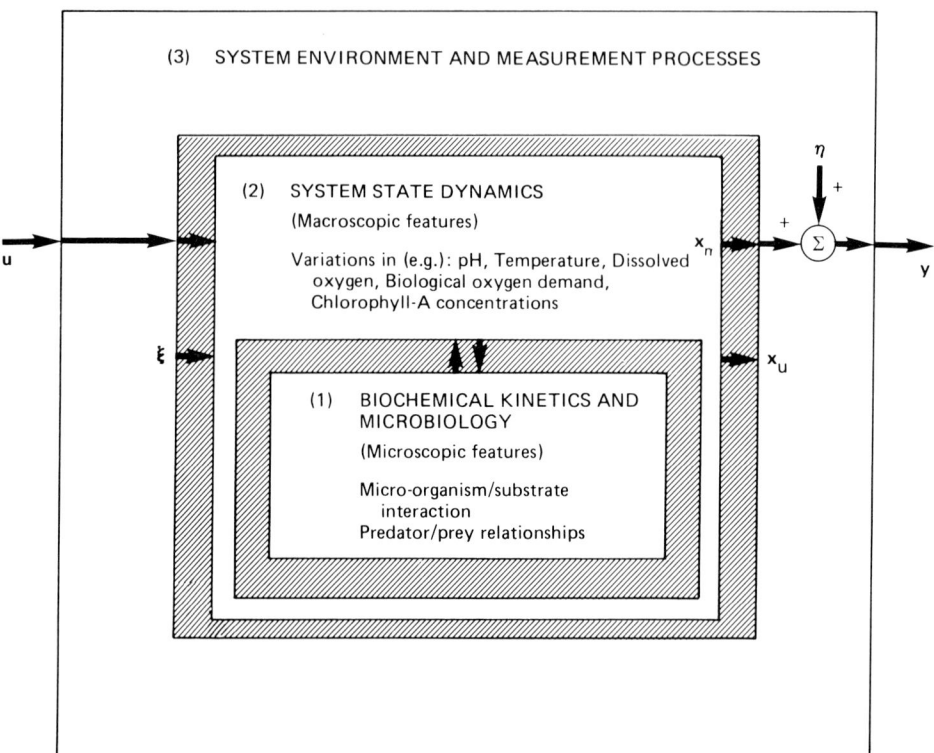

FIGURE 6 Conceptual representation of the system for perspective 2 on the problem of model structure identification.

Second, the conceptual distinctions between blocks 1, 2, and 3 can be loosely associated with three types of inadequacy in a given model structure. This requires further careful consideration. Suppose that the analyst attempts to identify a model of the system that is consistent with the "observable state" of the system; this implies a natural choice of state variables (x_m) *directly* related to the output observations, which in turn implies a strong correspondence between block 2 and "that part of the system being modeled". Thus, in the event of a demonstrable discrepancy between "that part of the system being modeled" and the observed behavior of the "system", one or more of the following underlying causes may be responsible:

(a) Interaction between x_u and x_m has not been accounted for (i.e., relationships between block 1 and block 2);
(b) The relationships among the variables of block 2 are incorrectly specified (i.e., relationships within block 2);
(c) There is significantly nonrandom interaction between x_m and the assumed "environment" of the system (i.e., relationships between block 2 and block 3).

These are important classes of causes and they will be especially useful in constructing arguments for resolving the problem of model structure identification.

Perspective 3. At this point it is appropriate to offer a working definition of model structure identification:

> Model structure identification is concerned with establishing unambiguously and by reference to the *in situ* field data how the measured system input disturbances, u, are related to the state variables, x, and how the state variables are in turn related both to each other and to the measured outputs, y.

It is thus an oversimplification to suggest that this problem is analogous to the problem of choosing the *order* of a polynomial that "best" fits the set of data. Solving the latter problem can be more specifically described as *model order estimation*, which, while it is an important component of the overall solution, nevertheless leaves some of the most challenging aspects of model structure identification unresolved. Since this may appear to be a subtle difference of definitions, it is important to make a further clarifying comment. Let us suppose, for instance, that an analysis of the input/output time-series leads to a model in which input and output are related according to the form shown in

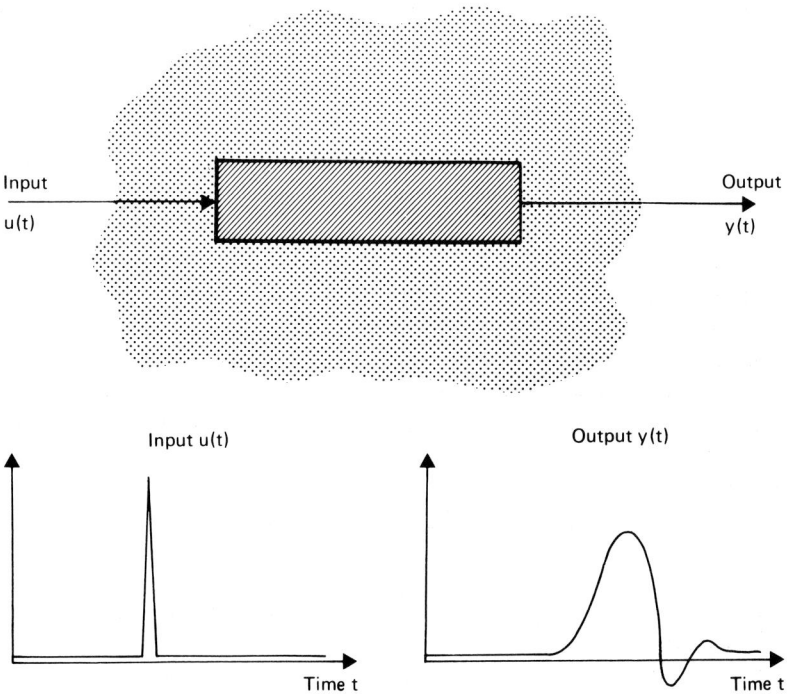

FIGURE 7 Identified input/output relationship for the observed system behavior; from the point of view of the model the system is considered to be a black box, i.e., nothing is assumed to be known about the internal mechanisms that govern the relationship between input and output.

Figure 7. Model order estimation (and parameter estimation) would constitute such an analysis; but only if the analysis were directed toward answering questions about *why* and by *what internal mechanisms* the input and output are so related would this constitute model structure identification as defined here. There is a large body of literature on methods of model order estimation (for example, Box and Jenkins, 1970; Åström and Eykhoff, 1971; Akaike, 1974; Unbehauen and Göhring, 1974; van den Boom and van den Enden, 1974; Chan et al., 1974; Söderström, 1977; Wellstead, 1976, 1978; Young et al., 1980), although it is not certain how many of these methods would be applicable under the conditions discussed in Section 3.1 (see also Maciejowski, 1979). Most of this literature is concerned with the analysis of problems in which model structure identification is either not relevant (because *a priori* theory is not fraught with ambiguities and contradictions) or not important (because just the abstract mathematical properties of the model are sufficient for the solution of the problem).

Perspective 4. Our working definition of model structure identification can be represented as in Figure 8 and hence specified in more detail. It can now be seen that:

Given the input/output data as the fixed basis for analysis, it is necessary to determine an appropriate number of state variables for the model (the "nodes" of the system description in Figure 8) and appropriate expressions for the relationships between u, x, α, and y, that is, $\mathbf{f}$ and $\mathbf{h}$ in eqn. (1).

A simple example will serve to illustrate this point. Suppose we are investigating the removal of a substrate in a closed system, and our first hypothesis is a linear model

Model I: $\quad \dot{x}_1(t) = -[\alpha_1]x_1(t)$ \hfill (2)

in which x_1, the concentration of the substrate, is the state variable, and α_1 is a parameter representing a first-order kinetic decay-rate constant. For a second hypothesis about the

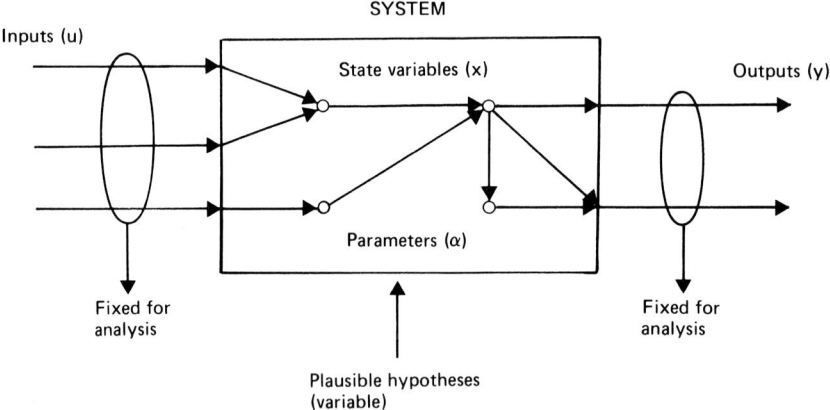

FIGURE 8 Pictorial definition of the problem of model structure identification.

observed system behavior we might propose a Monod-type kinetic expression and the presence of a mediating micro-organism in the reaction

Model II:
$$\dot{x}_1(t) = -[\alpha'_1 x_2(t)/(\alpha'_2 + x_1(t))]x_1(t)$$
$$\dot{x}_2(t) = [\alpha'_3 x_2(t)/(\alpha'_2 + x_1(t))]x_1(t) - \alpha'_4 x_2(t)$$
(3)

where the additional state variable x_2 is the micro-organism concentration and we have a vector $[\alpha'_1, \alpha'_2, \alpha'_3, \alpha'_4]$ of associated model parameters. Now recall that there are presumably some noise-corrupted measurements available from this experiment, but that we do not know which, if either, of Models I and II best describes the observed behavior. Model structure identification is thus concerned with determining whether $[x_1]$ or $[x_1, x_2]$ is the most suitable choice of state vector and with identifying an appropriate form for the expression contained in the square brackets [·] of eqns. (2) or (3). If both models are thought *a priori* to be good approximations of observed behavior, this might also be called a problem of *model discrimination* (Shastry et al., 1973; Maciejowski, 1979). But if neither hypothesis is adequate and a more complex pattern of behavior is suggested by the analysis of the data, then the generation of alternative, more appropriate hypotheses is a very difficult problem. This kind of problem will be of central concern in the following.

Perspective 5. The simple example discussed with respect to eqns. (2) and (3) can be generalized by rearranging eqn. (1a) as follows (Beck, 1979a):

$$\dot{x}(t) = \mathscr{T}\{x(t), u(t), \alpha_1(t)\} + \mathscr{U}\{x(t), u(t), \alpha_2(t)\} + \xi(t) \quad (4)$$

Here $\mathscr{T}\{\cdot\}$ includes expressions representing relationships from *a priori* theory that are considered to be *relatively well known*, for example, transport and dispersion properties; $\mathscr{U}\{\cdot\}$ accounts for all other phenomena whose significance in the observed patterns of behavior is a *matter of speculation* and for which no well-established mathematical relationships are available *a priori*. α_1 and α_2 are, respectively, the vectors of model parameters associated with $\mathscr{T}$ and $\mathscr{U}$. The distinction drawn between $\mathscr{T}$ and $\mathscr{U}$ is, of course, rather arbitrary, since there tends to be a complete spectrum of degrees of confidence in the theories incorporated in a model. Nevertheless, the ideal objective during the process of model structure identification would be to eliminate $\mathscr{U}$ from eqn. (4) by modification and/or expansion of the structure of $\mathscr{T}$.

How precisely one could approach or attain this ideal objective, by a process of both generating and assessing plausible hypotheses, is again a basic concern of this study.

Perspective 6. At this point we introduce one further conceptual representation of the problem of model structure identification. Suppose the patterns of system behavior exhibited in the (historical) field data can be represented by the set A in the set P of all possible patterns of behavior — see Figure 9. This pictorial representation has its origins in the work of Mankin et al. (1977); qualitatively, it is a powerful medium in which to express the following arguments. For reasons that will become apparent later, care must be taken to qualify P as being the set of all behavior patterns that one would expect to observe in "reality". Our first hypothesis for a model (say M_1) might be rather modest in size, allowing only a somewhat restricted type of behavior, although a reasonable

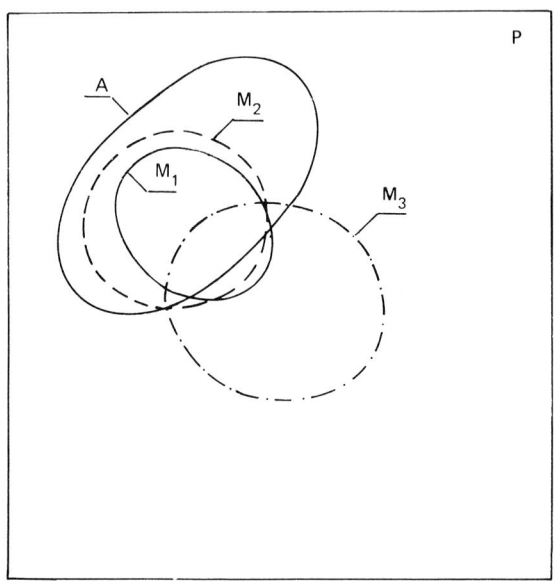

FIGURE 9 Perspective 6 on model structure identification: P is the set of all possible behavior patterns; A is the set of historically observed patterns of behavior; M_1 is the set of behavior patterns simulated by the first model hypothesis; M_2 and M_3 are alternative models hypothesized after assessing the suitability of M_1.

proportion of the set of behavior patterns simulated by the model (M_1 in Figure 1) is contained in the set A. Again, one must be careful about misinterpretation. Terms such as "small" model or "limited variety" of behavior patterns should not be equated too literally with a small number of variables, equations, or relationships. Moreover, note that, strictly speaking, A and M_1 represent observation and simulation under exactly equivalent conditions; such equivalence may not apply if M_1 contains relationships based, in effect, upon extrapolations from laboratory conditions. An example of a model typifying M_1 might be the Streeter–Phelps model of stream dissolved oxygen (DO)–biochemical oxygen demand (BOD) interaction. This model is a good starting point for analysis, although we are aware that its ability to describe system behavior is limited. Thus, given Figure 9 as a pictorial representation of the problem, what does the analyst do? His first model may not be bad, for it has captured part of the essence of reality (A and M_1 have an intersection), but it is far from being good — it does not simulate half of what was observed in practice. The crucial issue of model structure identification is this: we require a method that provides a useful feedback of diagnostic information from analysis of the first hypothesis (M_1) so that a second hypothesis (M_2) can be cast more fully within the set of observed patterns (A). It would be undesirable at an early stage of the analysis to suggest a revised model (M_3, say), probably both greater in size and with relationships different from those of M_1, that merely simulates more apparently spurious behavior.

3.2.1 An Approach Based on Recursive Estimation

At this point let us pause to assimilate the salient points of the foregoing discussion. The "problem" is model structure identification. Equation (1), the working definition, Figure 6, Figure 8, eqn. (4), and Figure 9, are different, complementary perspectives which, when assembled together, provide a broad description of this problem; some, especially eqn. (4), will subsequently become important in the development of solutions. Indeed, what is really required as a means of solution is an "intelligent" method of model structure identification — intelligent because it should indicate which parts of the structure are inadequate and how they might be corrected. Using the representation in Figure 9, such a method should maximize the probability of moving from M_1 to M_2 and minimize the probability of moving from M_1 to M_3.

One promising approach is to restate the problem of model structure identification in terms of the problem of parameter estimation (Young, 1974, 1978; Beck and Young, 1976; Beck, 1979a; Whitehead, 1979); this is assessed as "promising" simply because it generates a relative wealth of the kind of diagnostic information mentioned above, although we would not claim that this yields more than partial solutions to the problem. In fact, the very wealth of diagnostic information itself leads to other difficulties that will be discussed below. In order to develop the approach, however, it is first necessary to introduce some basic concepts underlying recursive estimation algorithms. We shall then present an illustrative case study (in Section 3.2.2), where it is reasonably straightforward to solve the problem of model structure identification, and proceed finally (in Section 3.3) to assess the prospects for further progress in this field.

For our purposes an important distinction can be made between parameter estimation algorithms that are off-line (or block data processing schemes) and algorithms that are *on-line*, or *recursive*. Figure 10 shows the essential differences between the two types of algorithm. With an off-line procedure (as in Figure 10a) the parameter estimates are assumed to be constant and equal to their *a priori* values, $\hat{\alpha}^0$, while the complete block of time-series field data — from time t_0 to t_N of the experimental period — is processed by the algorithm. Frequently all the data are processed together in one computation. A loss function, generally based on the errors between observed and model reponses, is calculated at the end of each iteration; the algorithm then attempts to minimize the loss function over the parameter space and computes an updated set of parameter values, $\hat{\alpha}^1$, for substitution into the next iteration through the data (from t_0 to t_N). A recursive algorithm, in contrast, computes revised parameter estimates, $\hat{\alpha}^0(t_k)$, at each sampling instant t_k of the field data (see Figure 10b); the minimization of the error loss function is implicitly, rather than explicitly, included in the algorithm. At the end of the block of data the estimates $\hat{\alpha}^0(t_N)$ are substituted for the *a priori* parameter values $\hat{\alpha}^1(t_0)$ of the next iteration through the data. Subsequent iterations through the set of field data are required since any initially incorrect estimates, $\hat{\alpha}^0(t_0)$, contribute larger errors to the calibration loss function than the errors contributed by initially correct estimates. (By implication, therefore, the minimum of the loss function is unlikely to have been located after the first iteration.) The ability of a recursive algorithm to estimate *time-varying* parameter values, upon which certain very useful interpretations will be placed shortly, is its greatest asset here.

In the case of a recursive estimator, therefore, the estimate $\hat{\alpha}$ of α at time t_k is given by an algorithm of the general form

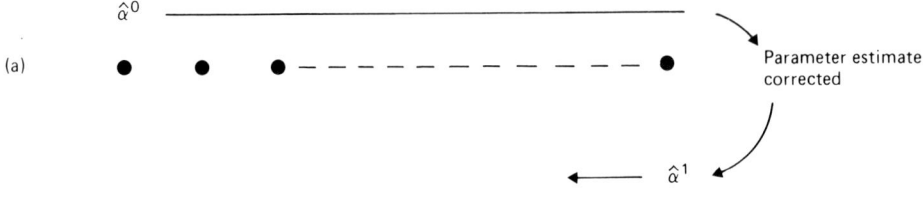

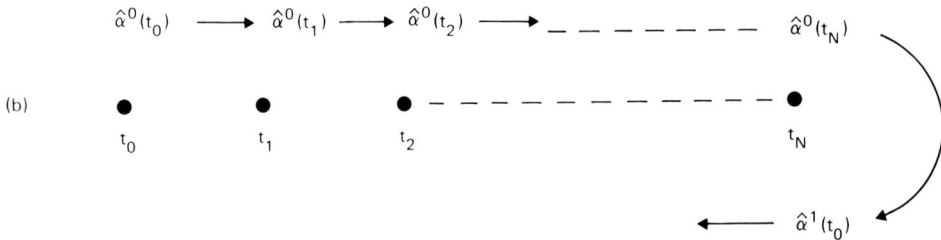

FIGURE 10 Methods of parameter estimation: (a) off-line; (b) recursive. The notation t_k denotes the kth discrete sampling instant in a time-series with N samples; the superscript i in $\hat{\alpha}^i$ denotes the estimate at the beginning of the $(i + 1)$th iteration through the data.

$$\hat{\alpha}(t_k) = \hat{\alpha}(t_{k-1}) + \mathbf{G}(t_k)[y(t_k) - \hat{y}(t_k|t_{k-1})] \qquad (5)$$

where the second term on the right-hand side is a correction term based on the error between observations $y(t_k)$ and an estimate of those output reponses, $\hat{y}(t_k|t_{k-1})$, obtained from the model using estimated values for the parameters from the previous sampling instant t_{k-1}. $\mathbf{G}(t_k)$ is a weighting matrix whose elements may be thought of as being dependent upon the levels of uncertainty (or error) specified for the model (as an approximation of "reality"), in the unmeasured input disturbances ($\boldsymbol{\xi}$), and in the output response observations ($\boldsymbol{\eta}$). For the time being our arguments will center upon the effects of the errors ($\boldsymbol{\varepsilon}$) in eqn. (5) on the performance of a recursive estimator, where

$$\boldsymbol{\varepsilon}(t_k|t_{k-1}) = y(t_k) - \hat{y}(t_k|t_{k-1}) \qquad (6)$$

Later, in Section 3.3.2, the discussion will return to a more serious consideration of the role of the matrix $\mathbf{G}(t_k)$ in solving the problem of model structure identification.

Perspective 7. Armed with a basic understanding of recursive parameter estimation algorithms let us now resume our discussion of the original problem. Imagine that the state variables x in a model may be represented conceptually by the nodes of Figure 11b (this is similar to the concept of Figure 8) and that the parameters $\boldsymbol{\alpha}$ are visualized as the "elastic" connections between the state variables. If we have assumed that all the parameters have values that are constant with time and yet a recursive algorithm yields an estimate of one or more of the parameters that is significantly time-varying, we may question the correctness of the model structure chosen. We can argue this point as follows.

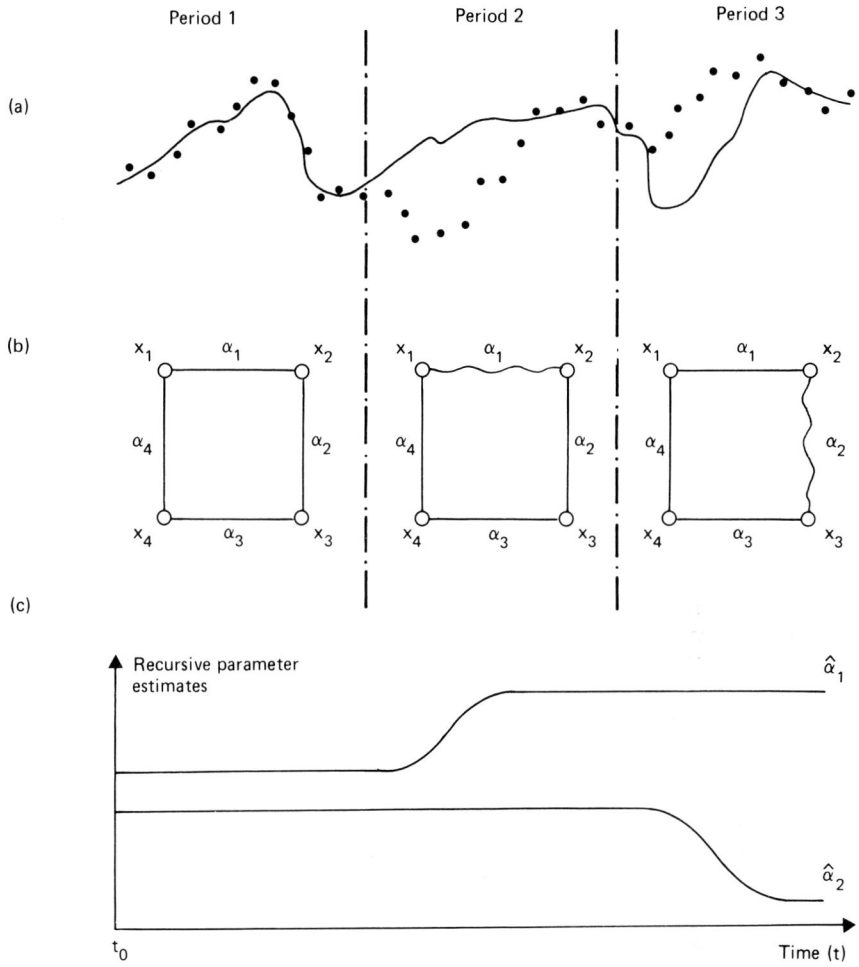

FIGURE 11 An illustrative example showing the concept of using a recursive parameter estimator in the context of model structure identification: (a) hypothetical model response and observations (dots); (b) conceptual picture of model structure; (c) recursive parameter estimates.

The general tendency of an estimation procedure is to provide estimates $\hat{x}$ of the state vector, or some function thereof, i.e., $\hat{y}$, that track the observations y. Hence, if any *persistent structural discrepancy* is detected between the model and "reality" (in other words, the errors ε exhibit a significantly nonrandom pattern), this will be revealed in terms of *significant adaptation of the estimated parameter values*. Clearly, direct adaptation of the model structure cannot occur, because the relationships between u, x, and y must be specified in a fixed manner for the purposes of implementation. Such variability of the parameter estimates with time can, of course, occur for several reasons; for instance, the parameter may be truly time-varying in accordance with a seasonal fluctuation. However, while this latter possibility is important, let us assume that it is not relevant to the illustrative example of Figure 11, which we shall now describe in more detail.

We start with Period 1 of Figure 11a. The model responses ($\hat{y}$) and output observations (y) are essentially in agreement over this period and there is no significant adaptation of the parameter estimates (according to Figure 11c). At the beginning of Period 2, however, there is a persistent discrepancy between $\hat{y}$ and y. It might be supposed, for example, that the underlying cause of the discrepancy is an inadequacy in the behavior simulated for x_1 and x_2, that α_1 is sensitive to this discrepancy (Figure 11b), and that (persistent) adaptation of the estimate $\hat{\alpha}_1$ (Figure 11c) partly compensates for the error between $\hat{y}$ and y. In Period 3 there is again disagreement between the observations and model responses, which leads to adaptation of the estimate $\hat{\alpha}_2$.

Perspective 8: A View of the Philosophical Foundations. The example of Figure 11 is clearly an idealized view of how a recursive estimation algorithm should be employed for model structure identification. Nevertheless, cast in this particular fashion such an approach has intuitively appealing interpretations. First, and by analogy with the analysis of physical structures, our aim is to expose inadequacy in terms of the "plastic deformation" (Figure 11c) of the model structure. Second, and of deeper significance, testing the model structure to the point of failure (the failure of one or more hypotheses) can be said to be consistent with Popper's view of the scientific method (Popper, 1959). An introductory text on Popper's work begins (in 1973) with the assertion that "... Popper is not, as yet anyway, a household name among the educated ..." (Magee, 1973). Judging by some of the recent literature (Holling, 1978; Young, 1978; Maciejowski, 1979, 1980), this assertion may no longer be true, at least in the present field of interest.

Especially pertinent is Holling's remark that (in discussing "model invalidation and belief") "... the model is [to be] subjected to a range of tests and comparisons designed to reveal where it fails." This remark, with emphasis on the words "range" and "designed to reveal", will be our guiding principle for solving the problem of model structure identification. But to have revealed that the model structure is inadequate is merely a part of the solution, and actually a relatively easy part. Extending the example of Figure 11 by one further step, let us suppose that the first (model) hypothesis has been identified as failing, as shown in Figure 12a. Now assume that a second hypothesis can be generated (in some way) and that it has the structure of Figure 12b with an additional state variable (x_5) and two new parameters (α_5, α_6). It may well be that calibration of the second model against the field data yields effectively invariant parameter estimates and hence that the analyst can accept the adequacy of this model structure as a conditionally good working hypothesis. These two steps of Figure 12 are consistent with the procedure of model structure identification outlined in Figure 9. The problem of how to proceed from one hypothesis to a subsequent hypothesis, however, has by no means been solved; nor can it be solved, as will become apparent later, as a matter of mere technique.

3.2.2 An Illustrative Case Study

A good example of how some of the foregoing ideas apply in practice is a study of the River Cam in eastern England (Beck, 1978a). In fact the development of a conceptual framework for model structure identification has been heavily dependent upon this case study and can be traced through three papers (Beck and Young, 1976; Beck, 1978b, 1979a). The Cam case study has therefore been an immensely fruitful prototype for the testing of ideas and methods. A "successful" case study, on the other hand, is not without disadvantages, for it creates the illusion that other case studies will be equally

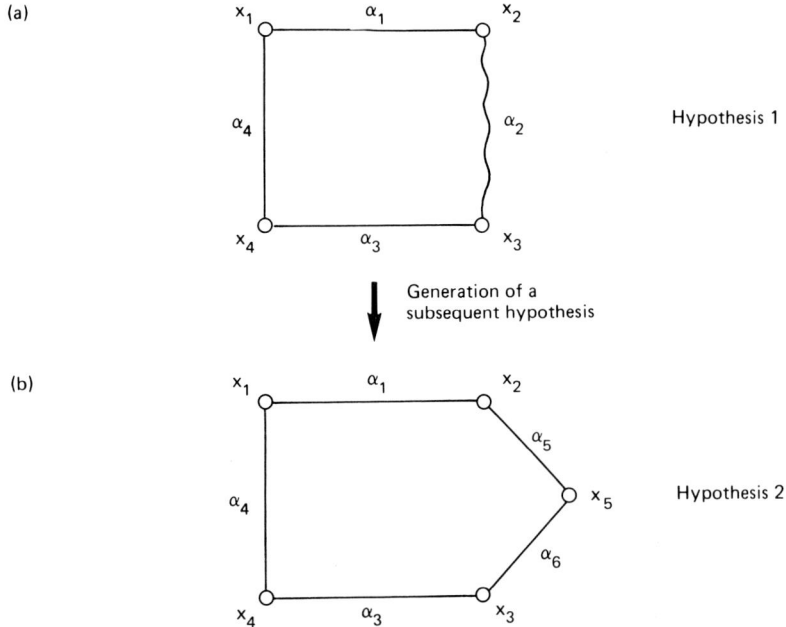

FIGURE 12 The process of model structure identification: revision of the model structure and re-estimation of the associated parameters (b) on the basis of diagnosing how the prior model structure fails (a).

successful (whereas mostly they are problematic) and it traps the analyst in dominant modes of problem-solving. However, for this discussion the interpretations of model structure identification embodied in Perspectives 2, 5, 7, and 8 will be of principal relevance.

An *a priori* model for the dynamics of DO–BOD interaction can be derived from the classical studies of Streeter and Phelps (1925). Since we wish to concentrate on model structure identification, we will state the model in the following form:

$$\begin{bmatrix} \dot{x}_1(t) \\ \dot{x}_2(t) \end{bmatrix} = \mathcal{F}'\{x(t), u(t), \alpha_1(t)\} + \begin{bmatrix} 0 & -\alpha_{1,1} \\ 0 & -\alpha_{1,1} \end{bmatrix} \begin{bmatrix} x_1(t) \\ x_1(t) \end{bmatrix} + \xi(t) \quad (7a)$$

$$y(t_k) = x(t_k) + \eta(t_k) \quad (7b)$$

where $[x_1, x_2] = x$ is the state vector comprising downstream DO and BOD concentrations, respectively, $\alpha_{1,1}$ is the first-order BOD decay-rate constant (assumed to be time-invariant), and u contains upstream DO and BOD concentrations as measured input disturbances. In terms of eqn. (4), it has been assumed for eqn. (7a) that

$$\mathcal{T}[x(t), u(t), \alpha_1(t)] = \mathcal{T}'[x(t), u(t), \alpha_1(t)] + \begin{bmatrix} 0 & -\alpha_{1,1} \\ 0 & -\alpha_{1,1} \end{bmatrix} \begin{bmatrix} x_1(t) \\ x_1(t) \end{bmatrix} \quad (8a)$$

$$\mathcal{U}[x(t), u(t), \alpha_2(t)] = 0 \quad (8b)$$

so that $\mathcal{T}'[\cdot]$ contains terms accounting for the process of re-aeration and the fluid transport and mixing properties of the reach of the river. If the discussion of eqn. (4) is recalled, it can be seen that the assumption of eqn. (8b) implies considerable confidence in the proposed model structure. This is a very deliberate use of the tactic of stressing a relatively rigid structure so that the probability of detecting a significant failure is maximized. At this early stage of the analysis it is not particularly useful to express little confidence *a priori* in the model and then to try and identify *unambiguously* where failure occurs. In such a case the postulated model structure would be, as it were, too flexible. Adaptation may or may not be significant, because one has little confidence in the model, and clear-cut answers cannot be obtained because clear-cut questions are not being asked. We may note in passing that these arguments have quantitative statistical counterparts for the use of certain recursive estimation algorithms, for example, the extended Kalman filter (Jazwinski, 1970; Beck, 1979a, b).

With the benefit of hindsight, however, it seems that a more appropriate *a posteriori* hypothesis for the model structure (in this specific instance) would be based upon the scheme of Figure 13a. An examination of the simple model structure will thus lead to a failure of the hypothesis that the observed behavior of DO–BOD interaction is represented by eqn. (7). Following Figures 11 and 12 it can be expected that there will be a deformation of the model structure as given by Figure 13b, and in fact Figure 14 shows this to be the case. Significant adaptation of the recursive estimate $\hat{\alpha}_{1,1}(t_k|t_k)$ occurs where there is a marked and persistent discrepancy between the model and observed output responses. At the same time, although this is not shown, the recursive estimate of the re-aeration rate coefficient becomes negative.

If we were now to imagine this same situation in the absence of hindsight, it would be apparent that the model structure *is* inadequate, but not necessarily *why*. The analyst would be confronted with the need to generate a second, hopefully more plausible, hypothesis. He might begin by examining the relative likelihood of four possible, generic causes of failure, for which purpose the conceptual distinctions of Perspective 2 and Figure 6 can at this point be exploited and expanded. Since the model structure of eqn. (7) is stated in terms of the macroscopic features of system behavior (DO and BOD), and since both state variables are directly measurable (i.e., $x \equiv x_m$), these four causes of failure can be classified as follows:

(i) The only reason for failure lies in an incorrect specification of the relationships among u, x, and y, which has a counterpart in α, $\mathbf{f}$, and $\mathbf{h}$ being incorrectly specified (see eqn. 1). If this is improbable, then failure is a function of interaction between "that part of the system being modeled" and its "environment", because:

(ii) ξ disturbs x in a nonrandom fashion, which has a counterpart in u being incorrectly specified;

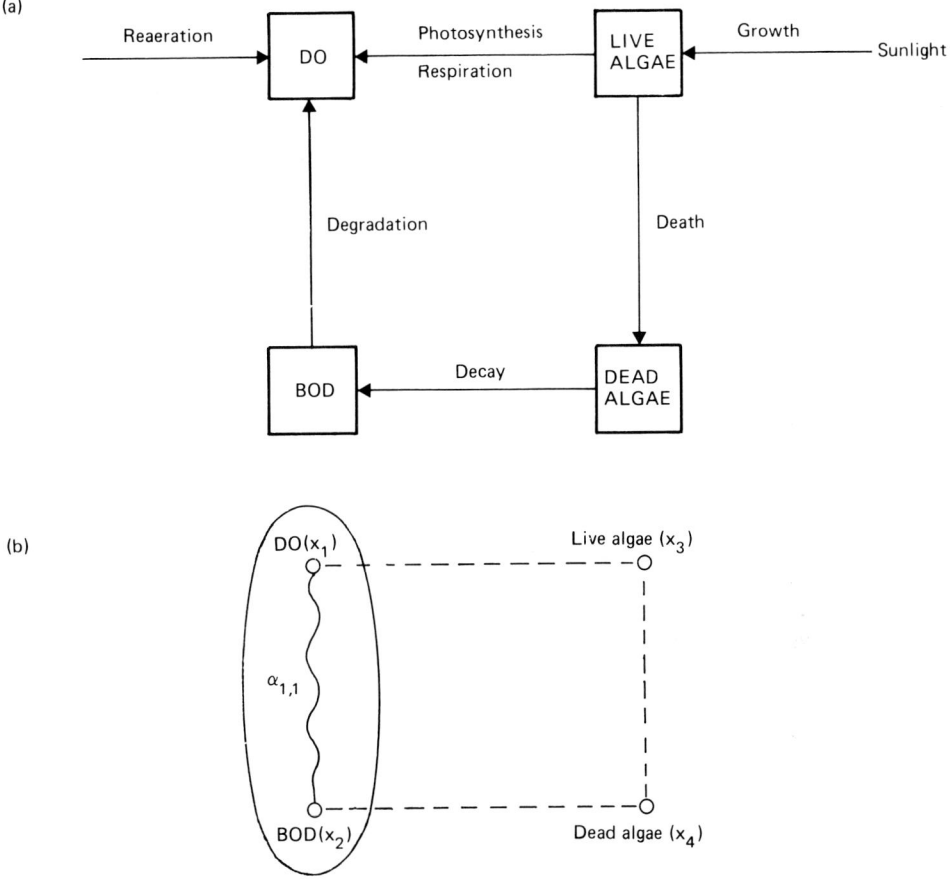

FIGURE 13 Model structure identification in the Cam case study: (a) the *a posteriori* model structure; (b) expected failure of the *a priori* model structure.

(iii) x_u interacts with x_m in a significant manner, which has a counterpart in x being incorrectly specified, i.e., $x \not\equiv x_m$;

(iv) η corrupts the relationship between x_m and y in a persistently biased fashion.

These, however, are only guidelines for the organization of one's thoughts, the beginnings of the process of diagnosis and synthesis where the analysis enters a phase in which creative speculation is necessary. It is helpful to introduce a somewhat broader organizing principle for the procedure of model structure identification. Let us simply suggest that the analyst is concerned with conducting experiments on and with the model structure. The use of $\mathcal{T}$ and $\mathcal{U}$ in eqn. (4) allows two different orientations (or objectives) for these experiments:

(i) The former ($\mathcal{T}$) in the process of *falsification*;
(ii) The latter ($\mathcal{U}$) in the process of *speculation* about alternative hypotheses.

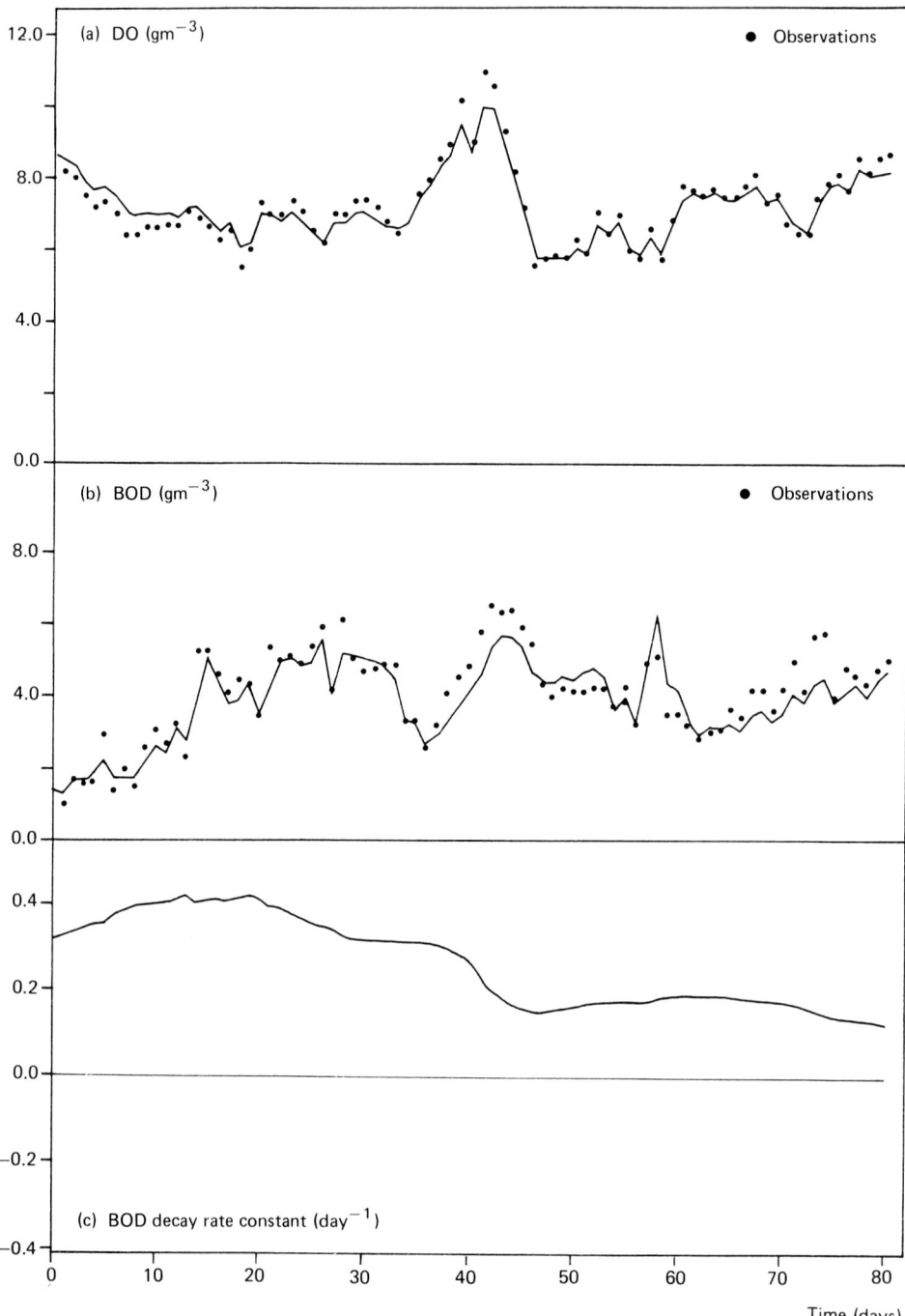

FIGURE 14 Model structure identification (*a priori* model) in the Cam case study: (a) recursive state estimates $\hat{x}_1(t_k|t_k)$ and observations $y_1(t_k)$ for dissolved oxygen concentration; (b) recursive state estimates $\hat{x}_2(t_k|t_k)$ and observations $y_2(t_k)$ for BOD concentration; (c) recursive parameter estimates $\hat{\alpha}_{1,1}(t_k|t_k)$ for the BOD decay-rate constant.

Therefore the analysis of the *a priori* model structure, as in eqns. (7) and (8) was an exercise in falsification. It merely remains to illustrate now the process of speculation.

We postulate some of the assumptions of Dobbins (1964) as suitable candidate hypotheses to be included in $\mathscr{U}$ such that

$$\mathscr{U}[x(t), u(t), \alpha_2(t)] = \begin{bmatrix} \alpha_{2,1}(t) \\ \alpha_{2,2}(t) \end{bmatrix} \tag{9}$$

in which $\alpha_{2,1}(t)$ and $\alpha_{2,2}(t)$ are, respectively, lumped variables representing all sources and sinks of DO and BOD other than those (the assumptions of Streeter and Phelps) accounted for in $\mathscr{F}$ of eqn. (8a). Relatively little confidence would be attached to the expected behavior of $\alpha_{2,1}$ and $\alpha_{2,2}$, although they would be expected to vary with time. Calibration of this revised model gives the recursive estimates $\hat{\alpha}_{2,2}(t_k | t_k)$ of Figure 15a. We now require a logical explanation of why $\hat{\alpha}_{2,2}$ varies in such a fashion, which in turn is related to the question of why the *a priori* model structure fails. From Figure 13a, it is apparent that a subsequent hypothesis is that:

(i) An additional input disturbance, the variation in sunlight, has an important indirect effect on the DO(x_1) and BOD(x_2), i.e., the *a priori* specification of u is inadequate;
(ii) This disturbance acts on x_2 through its effects on the additional states of the system, neither of which are measured, i.e., the *a priori* specification of $x \equiv x_m$ is not adequate.

When generated as a purely *deterministic* function of the day-to-day sequence of observed sunlight conditions, the estimated variations of x_4 (the concentration of "dead algae") are as shown in Figure 15b. It is but a short step from there to propose that the *apparent* rate of addition of BOD in this reach of river is proportional to the concentration of dead algal material.

It is tempting to close the issue of whether an acceptable model structure has been identified, yet at least one competing hypothesis is worthy of attention. This concerns the possibility of algal "interference" with the measurement of BOD, that is to say, the possibility of an incorrect specification of **h** (from eqn. 1) in eqn. (7b). This is quite a plausible hypothesis and it will be of relevance in a later section of the paper. It also illustrates the potential difficulty of distinguishing between system behavior and the process of observing that behavior.

3.3 Problems and Prospects

On occasion, therefore, one is fortunate and the case study described in the preceding section is just such an occasion. When calibration of a Streeter–Phelps model yields a negative-valued re-aeration rate constant, the analyst can be reasonably confident about rejection of the associated model structure. In such a situation he is forced to support an absurd hypothesis if he wishes to obtain correspondence between the given model and the data. But when eventually the diagnostic evidence favors rejection of the model, can one really hope to formalize the procedure for generating the next hypothesis?

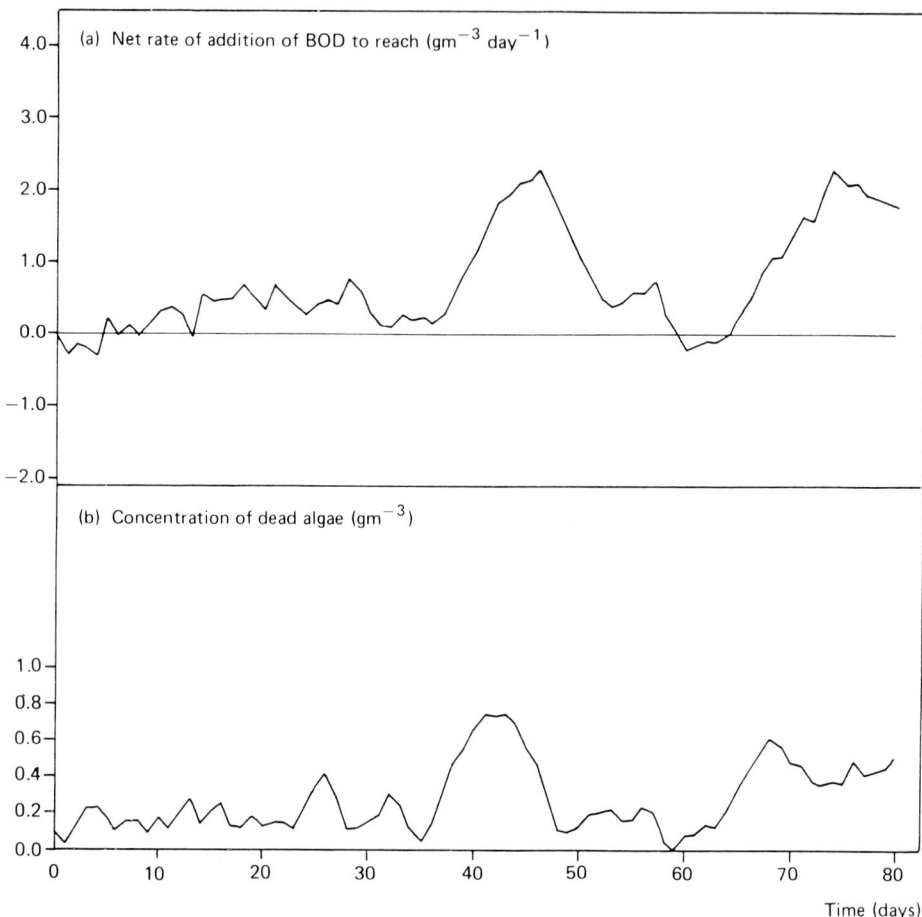

FIGURE 15 Model structure identification in the Cam case study: (a) recursive parameter estimates $\hat{\alpha}_{2,2}(t_k|t_k)$ for the net rate of addition of BOD to the reach; (b) deterministically estimated variations of the concentration of dead algae (x_4, a posteriori model).

Isn't this in fact a procedure that demands that spark of creative thought characteristic of scientific discovery? Perhaps, therefore, we should be rather modest in searching for an "intelligent" algorithm of model structure identification. While we do not believe that hypothesis generation can be reduced to a formal algorithm, it is still legitimate to ask whether there are methods that suggest feasible directions in which to cast new hypotheses, as was the intent of Figure 9.

Let us summarize the discussion so far. Because of the lack of planned experiments, because field data are highly uncertain, and because *a priori* definition of the mathematical forms of relationships among the important system variables cannot be made categorical, the calibration of water quality–ecological models is not a straightforward exercise of parameter estimation. The problem of model structure identification has to be solved before accurate estimation of the parameter values is attempted. The basic aim of model structure identification is to seek plausible hypotheses for "unexplained" relationships in a set of field data. The approach that we have outlined exploits the idea of curve-fitting as a "means-to-an-end" and not as an "end" in itself. Experience shows that approaching the problem from a variety of angles — for example, using different types of models and different estimation algorithms — can yield different clues about why a given hypothesis is incorrect and how it might subsequently be modified (Beck, 1978b). The illustrative case study of Section 3.2.2 has focused on merely a part of an overall approach from one or two angles. Falsification of the model, or components thereof, rests partly upon judgments about absurd parameter values, or about implausible variations in parameter values. Unless these variations and values can be defended by logical argument, then it must be conceded that the structure of the model does not match the structure of the observed patterns of behavior. Even in a relatively simple context, however, these kinds of solution to the problem are not easily derived. In the more complex situation to be discussed below the basic process of absorbing and interpreting all the diagnostic information generated by the analysis itself becomes very much more difficult. The evidence cannot be sharply focused in order to reveal the absurd hypothesis. But even to believe that such a sharp focus might be possible is arguably a delusion, since the field data are subject to high levels of uncertainty. Individual elements of *a priori* theory may not themselves be ambiguous. It is when many of these elements are assembled in a complex model, which is then calibrated against the *in situ* field data, that ambiguity arises. As with the examples of Bierman et al. (1980) and Halfon (1979b) quoted earlier, there is usually not a unique set of parameter values — nor is there necessarily a unique model structure — that will give a significantly superior fit between the data and the simulated responses of a complex model. The purpose of model structure identification is nevertheless to allow *a posteriori evidence* (*a posteriori* in the sense of having calibrated the model) to be brought to bear on distinguishing one or another of the possible *a priori explanations* as (conditionally) the most plausible. The difficulty lies in focusing and interpreting the *a posteriori* evidence.

3.3.1 Some Problems of Complexity

The case of the Bedford-Ouse river in central-eastern England is a natural extension of the Cam study. From 1972 to 1975 the UK Department of the Environment and the Anglian Water Authority jointly funded a major study of the Bedford-Ouse river system in order to evaluate the effects of developing a new city (Milton Keynes) in the upper part of the catchment area (Bedford-Ouse Study, Final Report, 1979). Daily data on some 16 water quality variables were collected at five locations on a 55-km stretch of the river for 14 months from late 1973 to early 1975. The character and behavior of the Bedford-Ouse system is very similar to that of the Cam (see, for example, Whitehead, 1983), but the scope of the available data base is incomparably greater.

Superficially the Bedford-Ouse study appears to offer an opportunity for *straightforward* application of those techniques that have proved so successful in the Cam study, but simply on a "larger" scale. In particular, during the spring of 1974 a substantial algal bloom occurred in the river with measured concentrations of chlorophyll-a, DO, and BOD reaching maximum levels of $300 \mu g l^{-1}$, $20 \, gm^{-3}$, and $13 \, gm^{-3}$, respectively. The modeling problem can be formulated in terms of four state variables (DO, BOD, chlorophyll-a, and suspended solids) for each reach of a three-reach system. Not only do the observed relationships among these variables appear to be significant, and therefore identifiable, but also the character of the interactions appears to vary both in time and space. In retrospect, it was probably naive to expect success, but what is important is an analysis of the problems revealed in the process of model structure identification.

Let us look first at the notion of testing the model structure to the point of failure, that is, the *process of falsification* in which $\mathscr{U}\{\cdot\} = \boldsymbol{0}$, as in the general statement of eqn. (4) and the specific example of eqns. (7) and (8). $\mathscr{T}$ thus contains various (confident) assumptions about the transport and dispersive properties of the river, re-aeration, BOD decay, and the growth, death, and photosynthetic properties of a population of algae. Six parameters are to be estimated in identical model structures for the behavior of each reach of the system (a total, therefore, of 12 state variables and 18 parameters). Figure 16 shows the recursive estimates of these six parameters for the third (downstream) reach of the river. Comparing Figure 16 with the enviable simplicity of Figure 14c, one

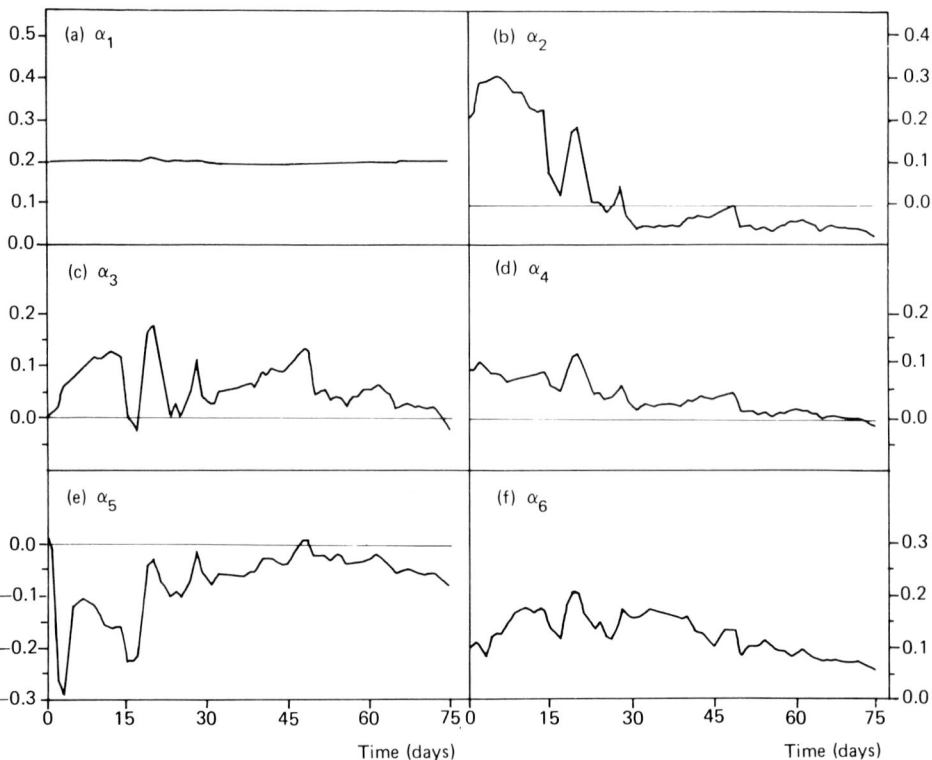

FIGURE 16 Model structure identification (the process of falsification) in the Bedford-Ouse case study (third reach): (a) re-aeration rate constant (day^{-1}); (b) maximum specific growth-rate constant for algae (day^{-1}); (c) BOD decay-rate coefficient (day^{-1}); (d) rate constant for addition of BOD to reach from suspended solid matter (day^{-1}(gm^{-3} BOD)(gm^{-3} SS)$^{-1}$); (e) death-rate constant for algae (day^{-1}); (f) rate constant for "loss" of suspended solids from the reach (day^{-1}).

would have great difficulty in answering the question "at what point does the model structure fail?" without even asking the question why it might have failed. There are clearly some apparently absurd hypotheses. For instance, the recursive estimates of both the maximum specific growth-rate (Monod-type kinetics) and death-rate constants for the algal population (Figures 16b and 16e, respectively) become negative-valued. One could argue, as a result, that the former is barely significantly different from zero and that the latter — a linear, negative, death rate — is evidence of a preferred linear growth-rate function for the algae (at least for all but the initial period of the data), but the analyst would be hard-pressed to attach great confidence to such conclusions.

There is also evidence in Figure 16 of preconceived notions dictating the outcome of this test of the model structure. For example, the remarkable stationarity of the recursive estimate for the re-aeration rate constant (Figure 16a) is a function of assuming relatively more *a priori* confidence in this particular parameter. In other words, the analyst has assumed that, if the model is to fail, its failure is unlikely to be a function of an inadequate description of the re-aeration process. Solely on the basis of these data, however, there are good reasons for arguing that the classical assumptions both of Streeter and Phelps and of dispersion in flowing media are not identifiable from the observed patterns of behavior. Figure 17 shows that with respect to the first reach of the system, which is typical of all three reaches, a classical advection–dispersion model with Streeter–Phelps assumptions produces responses such that the error between the observations and these responses is likely to be highly insensitive to the estimated values of the associated model parameters. Such a common problem of identifiability does not arise because of complexity, which we would suggest is the case in the results reported by Bierman et al. (1980) and Halfon (1979b), but because other *dominant modes of behavior* (in this instance, algal growth) almost entirely obscure these less significant modes of behavior. This is again slightly different from the situation described by van Straten and Golbach (1982), in which the character of the input disturbances is such that the system is not stimulated to respond in a manner sensitive to the dispersive properties of the river.

Since concern has been expressed in Section 2 about preconceived notions, it is important not to pass over this point without further reflection. We do not claim that our approach is without any element of subjectivity. The judgment of the analyst is not only required in specifying *a priori* confidence levels for the parameter estimates but also, of course, in deciding which hypotheses to include in the *a priori* model structure. Söderström (1977) is here in agreement, for he states that:

> "... Naturally, an objective method will produce a model structure without interference [by the analyst]. However, it is a chimera to regard this as an essential advantage. On the contrary, it may be a misleading property, since for all methods, objective as well as subjective, there is always a potential risk that a false model structure is selected."

We do claim, however, that the test procedures we have outlined are more rigorous than the approaches generally employed previously. If one adopts that view of the scientific method in which falsification of hypotheses is of fundamental importance, one might conclude that in the present example the assumptions of Streeter and Phelps cannot be expressed in a form that permits falsification. In a sense, therefore, they are not testable

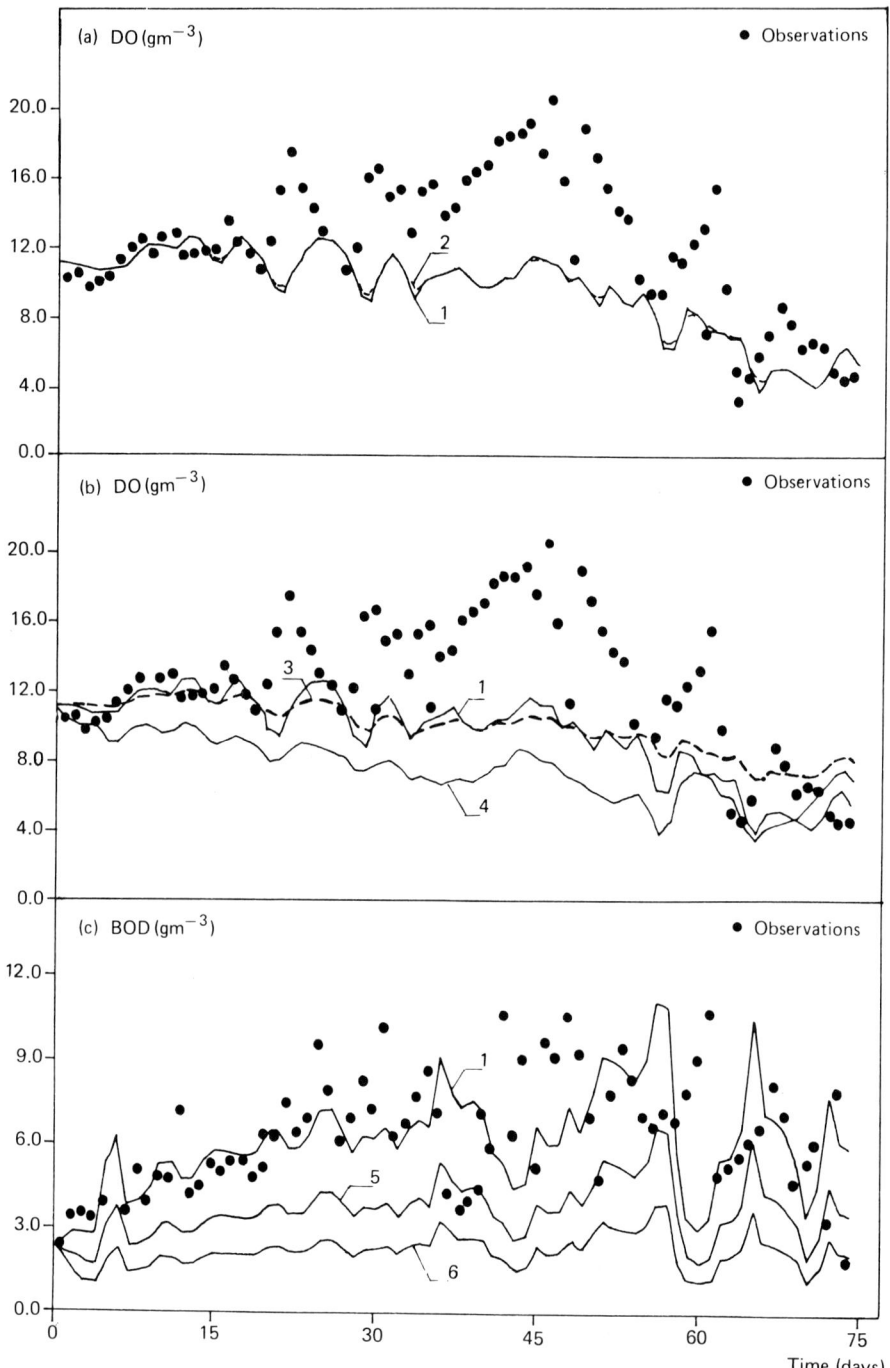

FIGURE 17 Evaluation of a classical advection–dispersion model with the assumptions of Streeter and Phelps for the first reach of the Bedford-Ouse case study. Values for the longitudinal dispersion coefficient (β_1, in km^2 day^{-1}), the reaeration rate constant (β_2, in day^{-1}), and BOD decay rate constant (β_3, in day^{-1}) for each curve are given respectively as: (1) $[\beta_1, \beta_2, \beta_3] = [0.0, 0.0, 0.0]$; (2) $[\beta_1, \beta_2, \beta_3] = [2.0, 0.0, 0.0]$; (3) $[\beta_1, \beta_2, \beta_3] = [0.0, 0.3, 0.0]$; (4) $[\beta_1, \beta_2, \beta_3] = [2.0, 0.3, 0.4]$; (5) $[\beta_1, \beta_2, \beta_3] = [0.0, 0.15, 0.2]$; (6) $[\beta_1, \beta_2, \beta_3] = [2.0, 0.3, 0.4]$.

propositions and their inclusion in any *a posteriori* model structure is tantamount to an act of faith. Such a problem may not be critically important in the context of model calibration, but could have significant consequences when the model is used for prediction, which will be of concern in Section 4.

It seems important, therefore, to question the motives for maintaining hypotheses that are not, strictly speaking, falsifiable. The reluctance to set aside convention is strong indeed, and Figure 16 illustrates well the conflict that can occur. Given prior experience that the hypothesis of BOD decay is probably not identifiable, a BOD decay-rate constant is still maintained in the model structure, but with an *a priori* estimate of zero (day^{-1}). Moreover, it would be difficult to argue that the subsequent pattern of the recursive estimates prompts the assumption of a significantly nonzero value for this parameter. This is not surprising in view of the "peculiarity" of observed conditions of simultaneously high DO and BOD concentrations and given the possibility of algal interference with the BOD test. Nevertheless, in this case there is no evidence that better hypotheses than those of Streeter and Phelps are available. In marked contrast is the discussion by Young (1983) of the classical representations of pollutant dispersion in rivers. Young clearly produces evidence that challenges conventional assumptions, a situation not without a certain irony. Originally unconventional assumptions about fluid-mixing properties, which allowed a transformation from a partial- to an ordinary-differential equation representation* (Beck and Young, 1975), were largely responsible for the developments leading to the present paper. Such developments can of course be challenged because they are unconventional, yet it is the results of precisely these assumptions that now forcefully challenge convention (Young, 1983). Moreover, subsequent reassessment of the original assumptions shows them to be fairly reasonable for that particular study (see Lewandowska, 1981).

Our reflections may then be briefly summarized as follows. The results of Figure 16 are founded upon the premise that:

(a) We have confidence in the hypotheses of Streeter and Phelps, but consider current hypotheses about mechanisms of algal growth as highly speculative.

Such a premise could be reoriented in either of two ways:

(b) We are confident about our hypotheses for algal growth, but consider the assumptions of Streeter and Phelps to be highly speculative.
(c) All hypotheses are equally speculative.

Although there is a temptation to cling to the first premise and not reject convention until it is demonstrably inadequate, this course of action may very well preclude the important possibility of revealing inadequacy.

Let us now turn from the process of falsification and look instead at the *process of speculation* (see also Section 3.2.2). If it is assumed, once again, that the Streeter–Phelps

---

* To avoid the problem of parameter estimation in partial-differential equation representations, a difficult problem to solve then as now.

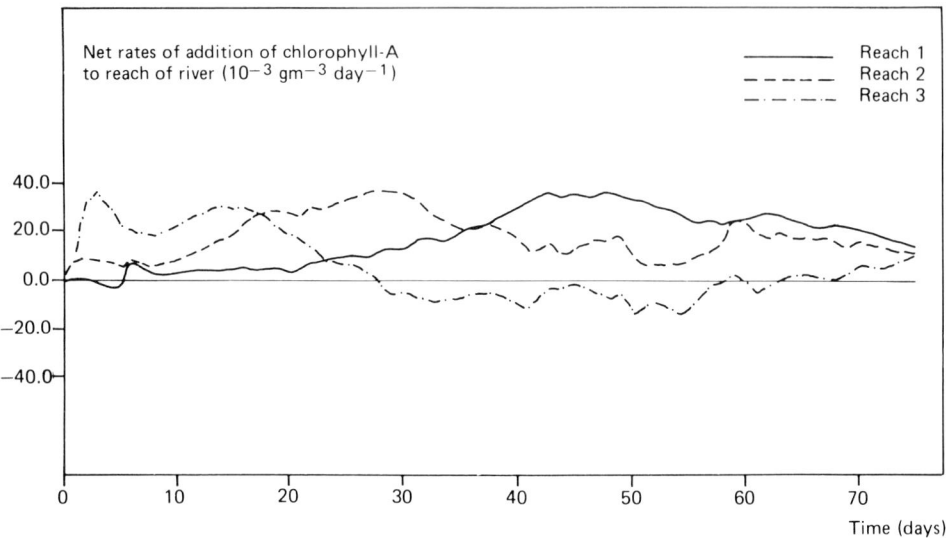

FIGURE 18 Model structure identification (the process of speculation) in the Bedford-Ouse case study: recursive estimates for the net rates of addition of chlorophyll-a to each reach of the system.

and fluid-mixing hypotheses can be included in $\mathscr{T}$, we may speculate that $\mathscr{U}$, as in eqn. (9) for the Cam study, comprises a vector of lumped parameters for all other sources and sinks of DO, BOD, and chlorophyll-a. Part of the diagnostic evidence from analysis of this speculation for the three reaches of the Bedford-Ouse system is gathered together in Figures 18, 19 and 20. One might tentatively conclude from these recursive estimates that:

(i) The rate of addition of chlorophyll-a to the system reaches a maximum first in the third (downstream) reach, then in the second, and lastly in the first (upstream) reach (see Figure 18).
(ii) The rate of addition of dissolved oxygen to the first reach is roughly proportional to the observed concentration of chlorophyll-a at the downstream boundary of that reach (see Figure 19a); the rate of addition of dissolved oxygen to the second reach is roughly proportional to the observed concentration of chlorophyll-a, except over the middle of the period recorded (see Figure 19b); the rate of addition of dissolved oxygen to the third reach is not obviously proportional to the observed chlorophyll-a concentration for most of the time (see Figure 19c).
(iii) The rate of addition of BOD in all three reaches is essentially identical over the later part of the period recorded (see Figure 20); the rate of addition of BOD to each reach roughly follows the same relative pattern as the rate of addition of chlorophyll-a to each reach over the initial part of the period recorded (compare Figures 18 and 20).

It would certainly require bold and imaginative thinking to synthesize a hypothesis from such evidence that would facilitate the step from M_1 to M_2 in the terms of Figure 9.

Uncertainty, system identification, and water quality prediction 39

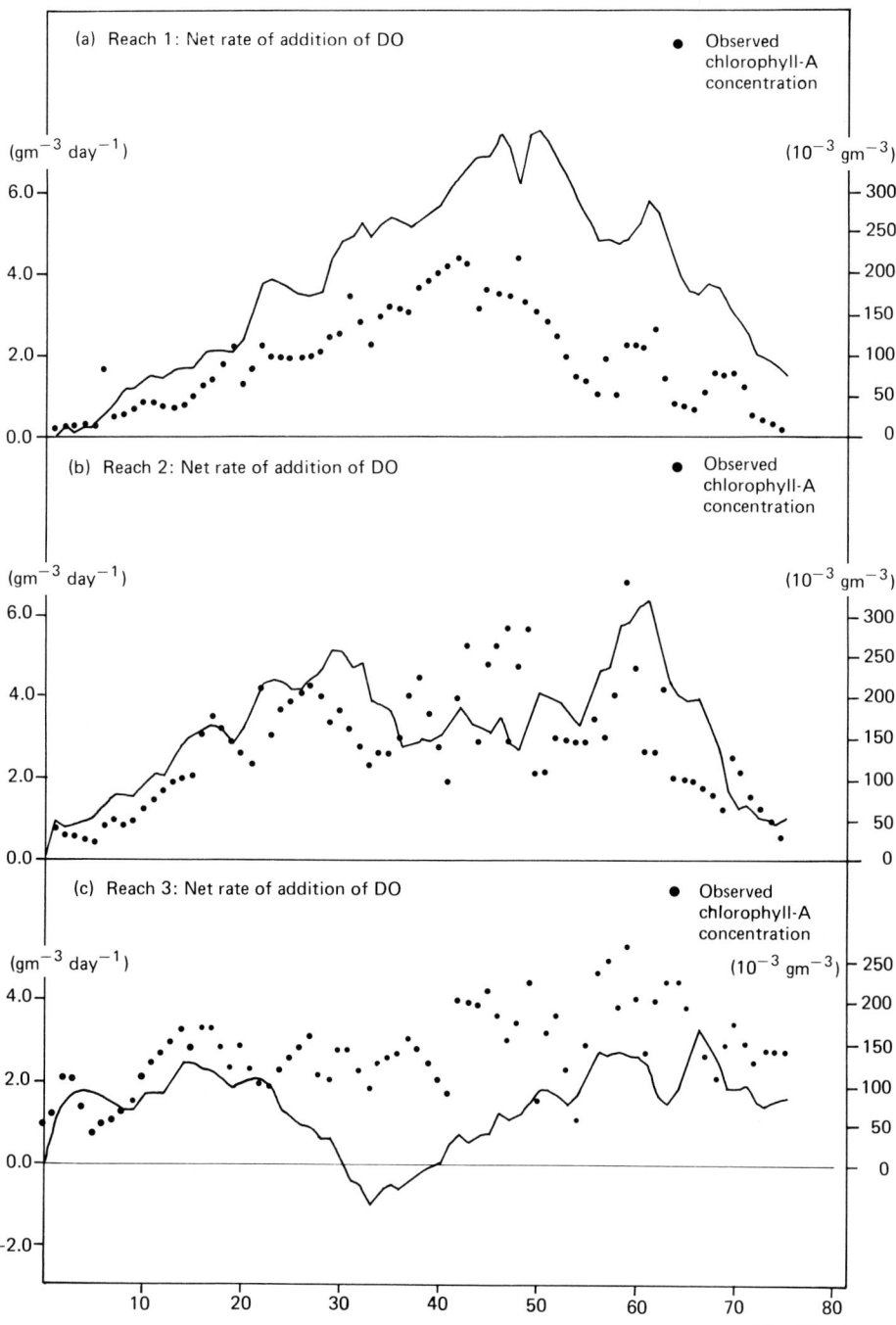

FIGURE 19 Model structure identification (the process of speculation) in the Bedford-Ouse case study: comparison of recursive estimates for the net rates of addition of DO to each reach of the system with the observed chlorophyll-a concentrations at the downstream boundary of each respective reach.

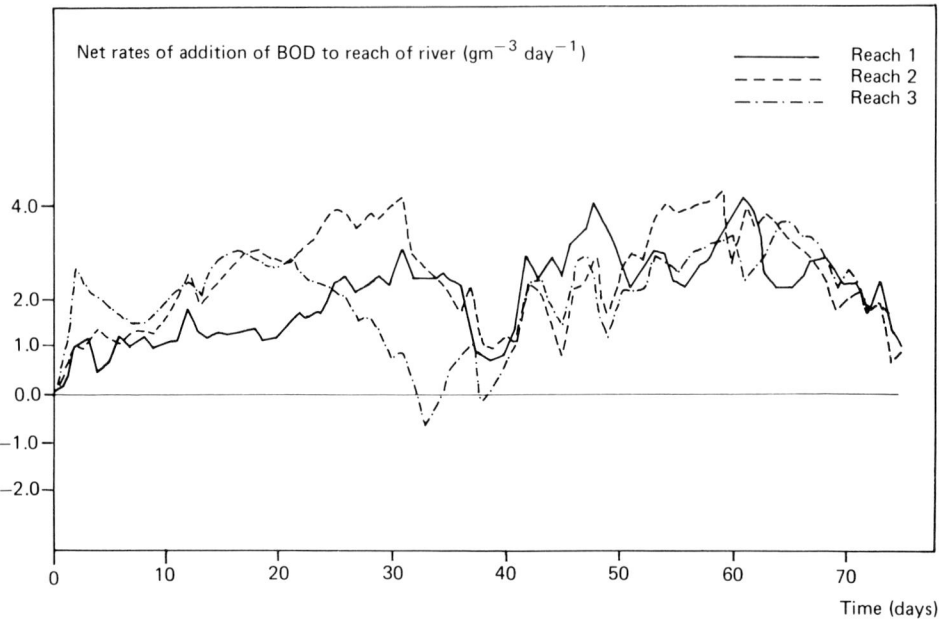

FIGURE 20 Model structure identification (the process of speculation) in the Bedford-Ouse case study: recursive estimates for the net rates of addition of BOD to each reach of the system.

How, in fact, would the analyst absorb and interpret this relative wealth of diagnostic information?

Further conjecture about model structures based on evaluation of these observations will not be presented here; we merely offer one brief speculation, principally because it introduces a problem of considerable general significance. Let us suppose that there are two distinct states of the algal population, one of which gives rise to a net addition of both chlorophyll-a and DO, while the other gives rise only to a net addition of DO. This hypothesis might be consistent with some of the diagnostic evidence and, if it were, any subsequent model structure would probably have to account for the behavior of *unobserved state variables,* i.e., x_u in Figure 6. There is, however, no really satisfactory procedure for accommodating unobserved state variables in our scheme for model structure identification. Undoubtedly, the inclusion of such variables gives additional flexibility to the model structure, in fact, so much so that the test designed to reveal structural inadequacy in terms of significant parameter adaptation would be nullified. The estimates of the unobserved state variables would be adapted rather than the parameter estimates and there would be little basis on which to judge the plausibility of such variations. The procedure adopted for the Cam study, where the dynamics of the unobserved state variables were defined as *completely deterministic* functions of the inputs and other state variables, seems to be more compatible with the idea of making bold, easily falsifiable hypotheses than procedures used elsewhere (Beck, 1979c). The identification of a time-varying parameter — or unobserved state variables, for the two concepts are barely distinguishable — is the archetypal fraud of which curve-fitting is often accused. The conclusion that the model fits the data subject to arbitrary variations in one or more of the parameters is itself of no consequence. Rather, it is the process of speculating why such variations occur that should be highly valued.

3.3.2 Design for Failure and Speculation

There are obviously lessons to be learned from the prototype case study of Section 3.2.2 and its more ambitious successor. The most important are that:

(i) The process of model structure identification proceeds as a sequence of "experiments" in which the analyst has either the objective of falsifying confident hypotheses or the objective of speculating about relatively uncertain hypotheses; in a purely technical (and perhaps more fundamental) sense these two objectives are best considered as mutually exclusive.
(ii) If speculation is the objective, then a primary concern is rearrangement of the model (utilizing the results of speculation) in such a way that its structure is likely to contain only parameters with essentially time-invariant recursive estimates when re-estimated. For complex systems and patterns of behavior it is extremely difficult to diagnose the results of speculation.
(iii) If falsification is the objective, there may be ambiguities in determining where the model fails and in distinguishing between reasonable and absurd hypotheses.

Given these limitations and inadequacies in the technical aspects of model structure identification, is it still possible to identify avenues of further progress? For it is not particularly encouraging to conclude that the analysis of an apparently simple problem leads merely to more complex unresolved problems. One might reflect, with resignation, that such is progress! Our answer to this question is nevertheless positive, although cautious: for the reasons given in the introductory paragraph of Section 3.3 it is clear that one should not expect the impossible.

Let us begin by examining point (iii) and its concern with "determining where the model fails". In the case of Figure 16, for example, it is not at all apparent which parameter is associated with the least adequate component of the model structure. It might be possible to clarify the situation by testing each hypothesis individually, assuming all but one of the parameters to be constant and known with certainty. To do this we would need to appreciate the underlying causes of variability in the recursively estimated value of a parameter. Recalling therefore the general form of eqns. (5) and (6), we have

$$\hat{\alpha}(t_k) = \hat{\alpha}(t_{k-1}) + G(t_k)\epsilon(t_k|t_{k-1}) \tag{10}$$

which for parameter α_i (when all other parameters are assumed to be known) reduces to

$$\hat{\alpha}_i(t_k) = \hat{\alpha}_i(t_{k-1}) + \sum_{j=1}^{p} g_{ij}(t_k)\epsilon_j(t_k|t_{k-1}) \tag{11}$$

It is perhaps easier to state, on the basis of eqn. (11), the following conditions required for $\hat{\alpha}_i(t_k)$ *not* to vary with time:

(a) $\varepsilon_j(t_k|t_{k-1})$ is small and random for all j,k: this is clearly the case when one has an appropriate model structure and good parameter estimates.
(b) $g_{ij}(t_k)$ is small for all j,k: this implies that the estimation algorithm computes $g_{ij}(t_k)$ such that all errors ε_j are effectively ignored, which would be correct if any hypothesis associated with α_i were truly unrelated to any persistent error ε_j.
(c) The effects of summing the products $g_{ij}\varepsilon_j$ may be self-cancelling even though both g_{ij} and ε_j may not be small; however, the implications of this condition are not immediately evident and will thus not be considered further in the present discussion.

Conditions (a) and (b) are stated as mutually exclusive extremes. A combination of the two is clearly feasible, since one simply has the condition that $\varepsilon_j(t_k|t_{k-1})$ is small when $g_{ij}(t_k)$ is not, and vice versa.

Condition (b) is the most interesting of the three because it calls into question both the role of the matrix $\mathbf{G}(t_k)$ and the way in which this matrix is computed. Conceptually, $\mathbf{G}$ is a mechanism for distributing the stress applied to the model structure (when it is to be tested for failure) among the component parameters and hypotheses. In eqn. (11), for example, this process of stress distribution has therefore been constrained to act upon a single parameter. One disadvantage with certain methods of recursive estimation — the extended Kalman filter, for example — is that $\mathbf{G}$ is largely, but not entirely, determined by the *a priori* assumptions about the model structure, parameter uncertainties, and statistical properties of the system's "environment", i.e., $\boldsymbol{\eta}$ and $\boldsymbol{\xi}$. In other words, and perhaps not surprisingly, the degree of attention paid to the *a posteriori* evidence, as represented by $\varepsilon(t_k|t_{k-1})$, is prejudiced by the *a priori* assumptions. With slight modification, however, this intuitively appealing "directional" property of $\mathbf{G}$, namely that it indicates the relative degree of significance of any ε_j with respect to a given $\hat{\alpha}_i$, might be turned to a distinct advantage. More specifically, Ljung (1979) has shown recently that it is $\mathbf{G}$ that has all the algorithmic importance for the extended Kalman filter and that, if $\mathbf{G}$ is parametrized and its elements estimated recursively, some of the *a priori* assumptions previously required become redundant. From our point of view, which is rather different from Ljung's perspective, not only does this lessen the dependence of the analysis upon *a priori* assumptions, but it also shifts the balance of the correcting mechanism $\mathbf{G}(t_k)\varepsilon(t_k|t_{k-1})$ in eqn. (10) toward greater exploitation of *a posteriori* evidence. Furthermore, this has a strong equivalence with recent developments proposed by Young (1979) for other forms of recursive estimation algorithms.

In these various suggestions for circumventing the problems of point (iii), and also the problem of point (ii) to which we now turn, there is nevertheless a disquieting element concerning complexity and computational effort. Recursive estimation of the elements of $\mathbf{G}$, in addition to estimating the model parameters, would seem an added burden in the process of diagnosing the results of model structure identification. It is therefore sensible to seek some informative, yet easily computable scalar quantity that aggregates the multivariate character of the analysis. An obvious choice would be the determinant of a matrix (or submatrix) — or a function of this and determinants of other matrices — that appears naturally within the estimation algorithm. This too could be a promising direction for further progress, as is clearly evident from parallel developments in model order estimation (for example, Woodside, 1971; Wellstead, 1978; Young et al., 1980).

Yet there is something more fundamental, and perhaps more disturbing about this problem of complexity and computational effort. Consider, for instance, the context of the whole of this section of the paper. It deals with only one or two kinds of diagnosis, within one particular approach to the problem of model structure identification, for one of three categories of data (the intermediate category of "some adequate data", as explained in the introduction to Section 3), and applied to modestly sized, even humble models. Nevertheless, the approach is sufficiently complex that it loses the attractive simplicity of the Monte Carlo approaches adopted elsewhere for the analysis of "scarce data" situations (Hornberger and Spear, 1980, 1981; Spear and Hornberger, 1980; van Straten, 1980; Fedra et al., 1981). And it is a sobering thought indeed, if one reflects upon the vast potential of telemetered, on-line water quality monitoring networks (Marsili-Libelli, 1980), that a lack of field data will not always be the critical constraint on water quality modeling. There is every possibility, therefore, that future critical constraints will involve precisely this area of absorbing and interpreting the results of data analysis.

Could one thus argue a case in favor of other equally profitable but simpler approaches? At least two alternatives come to mind: the ubiquitous "trial and error" comparisons of deterministic simulation responses with the field data (although serious concern about such an approach has already been expressed in Section 2); and approaches based on off-line methods of parameter estimation, which, while they are relatively effective (see for example, Di Toro and van Straten, 1979; van Straten, 1983), do not have the same potential for insight as the approach described here. For both of these alternative approaches there is little evidence of associated work on the problem of model structure identification and one could argue that simplicity, when dealing with complex models and large numbers of data, is a chimera. In the absence of comparable approaches, the question of the advisability of following the principles outlined here must be examined with some deeper appreciation of how consistent these principles are with the scientific method. For this reason we contend that certain aspects of Sections 3.2.1 and 3.3.1 are neither fanciful nor esoteric excursions into the realm of philosophy.

Our last consideration deals with point (i) of the lessons to be learned and also relates back to the discussion of convention, confidence, and speculation in Section 3.3.1. Let us suppose that in a given study the ultimate objective is to reconstruct *in situ* "experiments" from the observed data by *analytical methods*. In other words, as scientific endeavor moves outside the laboratory it carries with it the notion of recreating the "controlled" conditions of a laboratory experiment in the field system itself (see, for instance, Lack and Lund, 1974). The objective of recovering experiments may be worthy and it would appear to relate to the central issue of extrapolation from laboratory systems which was raised in Section 1. It seems reasonable to attempt to design the analysis of model structure identification so that it compensates for the variable environmental conditions of the "experiment" (recall here the discussion of Section 3.1.1). An apt example would be the reconstruction of an *"in situ* chemostat experiment", where the objective is to identify the structure of the relationship between substrate and phytoplankton growth. In this particular example the skill of the analyst would lie in arranging the analysis such that extraneous interference with the "experiment" — for simplicity, extraneous hydrodynamical disturbances — could be filtered out. At first sight, this is a rather attractive view of the true purpose of system identification and time-series analysis, but it presupposes, of course, that the part of the model required to compensate for the "experimental environment" is known *a priori* with sufficient confidence to permit the full power of the analysis to be concentrated on the problem of substrate/phytoplankton interaction. Such assumptions themselves have to be evaluated. The distinction between what is "known well" and what is "speculation" (as in eqn. 4) thus becomes vanishingly small. In a holistic sense it is difficult to claim, however tempting it may be, that there is *one* "experiment" and its complementary "environment". Instead, it is only possible to state that a number of more or less significant "experiments" are proceeding in parallel.

Clearly, complexity, and not only uncertainty, is a universal and inescapable feature of the modeling of water quality–ecological systems. From the discussion of Figure 9 in Section 3.2 it is apparent that the best way of taking complexity into account is to start from a simple model and progressively increase model complexity when the diagnostic evidence of analysis precludes acceptance of any simpler model structure. Of course, we are well aware that uncertainty and complexity would soon impose constraints on the depth of such an analysis; however, the alternative of starting with a complex model and identifying those components of the structure that are essentially redundant (i.e., surplus content) is an approach seemingly fraught with many more difficulties, the kind of difficulties raised, for instance, in Halfon's (1979b) analysis of a model for Lake Ontario (see Section 3.1.2). One of the key problems is that ambiguities arise in determining whether the *a posteriori* evidence supports rejection of an inadequate model structure. In the face of these ambiguities, and acknowledging the additional difficulties of interpreting large amounts of evidence, the analyst should respond by making particularly prudent choices for the postulated model structures. If the model is a vehicle for asking questions about the nature of "reality", then it is advisable to make those questions as few – at least initially – and as unambiguous as possible.

3.4 A Concluding Comment

Many recent exercises in water quality–ecological modeling have been conducted without serious consideration of the deeper significance of calibration. This should not be considered a mere backwater to the mainstream developments in water quality modeling. It only becomes so if one chooses to attach great confidence to *a priori* theory, thereby renouncing, in effect, much of the questioning that should accompany calibration. This choice, albeit often made subconsciously, is inherent in the present dominant approach to modeling where heavy reliance is placed upon extrapolations from laboratory or "equivalent" field systems. One might choose, in complete contrast, to put aside *a priori* theory altogether, and in view of the manifest difficulties in determining the governing mechanisms of behavior for even a well-documented case study (see for example, van Straten et al., 1979; van Straten and Somlyódy, 1980), perhaps this could be justified. Any resulting model, which in this extreme case of exclusive dependence on the *a posteriori* evidence of the sample field observations would be a true black-box model, would certainly attract a common criticism: namely, that the model should not be used for making extrapolations outside the range of conditions for which it was developed. Unfortunately, such extrapolation is precisely what is required of most models. Equally unfortunately, exactly the same criticism can be leveled at the use of models based upon extrapolations from laboratory or "equivalent" field systems. But this does not imply a stalemate in a conflict between the dominant approach and the not-so-dominant approach discussed in this section. Rather the two approaches are complementary and one way out of the seeming impasse is suggested by the ability of the not-so-dominant approach to evaluate those extrapolations characteristic of the dominant approach. The existence of genuinely complementary approaches, in the context of calibration, will be confirmed in the next section in the context of prediction. There is, however, one fundamental asymmetry in the relationship between the two approaches: namely, that any attempt at simplifying the problem of model structure identification by (conceptually) subdividing the system into smaller components (the analysis of single "experiments") is an exercise of dubious value.

4 PREDICTION AFTER IDENTIFICATION

The preceding section has wrestled at length with the extremely difficult problem of acquiring *understanding* of a complex system's behavior, irrespective of any intended application for the associated model. The stimulus for presenting what we have called a "not-so-dominant" approach originates with the concerns expressed in Section 2 about other more conventional approaches to model development and calibration. Yet in spite of such concerns it is the complementary character of the approaches, rather than their points of conflict, that is the emerging theme of Section 3. Undoubtedly, we should expect much more debate about the advantages and disadvantages of one approach or another. To a large extent, however, debates about how to acquire understanding of complex systems are best conducted at a general philosophical level (see for example, Battista, 1977). Of more immediate and specific relevance is the debate about how a model is likely to perform when applied to the problem of *prediction* of future behavior patterns *under substantially changed conditions*. This section, therefore, addresses that issue. We shall first discuss qualitatively the question of accounting for uncertainty in the relationship between the identification (calibration) of a model and its application to prediction of the future (in Section 4.1). This is clearly an important logical connection in the underlying argument of the paper as a whole. Section 4.2 gives a brief review of the methods available for analyzing the propagation of forecasting errors. We have already noted that this is a particularly active field of current research, although our interest in the methods themselves is here somewhat secondary. Rather, as in Section 4.2.1, the object is to illustrate certain important aspects of the relationship between identification and prediction. Hence, in Section 4.3, we shall conclude with a dilemma that captures some limiting features of both the dominant and the not-so-dominant approaches to water quality–ecological modeling.

4.1 Accounting for Uncertainty

Let us suppose that in an ideal study the problem of model structure identification has been solved and that it merely remains for calibration to be completed by estimation of the model parameter values. After a successful calibration exercise it would be expected that the degree of uncertainty in any given parameter estimate would be less than the uncertainty associated with the prior estimate of that parameter value before calibration. The amount by which the uncertainty in the parameter estimate is reduced should be roughly consistent with the degree of relevance that the parameter — and its associated sector of the model's behavior patterns — has to the observed system behavior. The reduction in the uncertainty of the parameter estimates will also be approximately inversely related both to the number of field observations and to the levels of uncertainty and error associated with those observations. But the *a posteriori* estimates of the parameters will still be subject to uncertainty: their estimation errors are, as it were, a kind of "fingerprint" of the calibration procedure; and the effects of these errors will propagate forward with predictions about the future.

To be more specific, let us examine the possible changes in the parameter estimation error variance–covariance matrix (as a measure of the uncertainty in the parameter estimates) that might occur during the process of recursively estimating the parameter values. The covariance matrix of the *a priori* estimation errors will be denoted by

$$\mathbf{P}^p(t_0|t_0) = \mathbf{E}\left\{[\boldsymbol{\alpha}(t_0) - \hat{\boldsymbol{\alpha}}(t_0|t_0)][\boldsymbol{\alpha}(t_0) - \hat{\boldsymbol{\alpha}}(t_0|t_0)]^T\right\} \tag{12}$$

where $\mathbf{E}\{\cdot\}$ is the expectation operation and t_0 is the time at the beginning of the period for which experimental data are available. Under the assumption that the calibration exercise is successful in yielding improved estimates of the parameters with a lower error variance, we could expect that

$$p_{ii}^p(t_N|t_N) < p_{ii}^p(t_0|t_0) \tag{13}$$

where the subscript *ii* indicates the *i*th diagonal element of the matrix $\mathbf{P}^p$ and t_N is the time at the end of the period covered by the experimental data. In other words $p_{ii}^p(t_N|t_N)$ is the *a posteriori* error variance for parameter α_i. But just how "successful" the calibration exercise is requires an important qualification, for which purpose two nominal illustrative trajectories for p_{ii}^p are given in Figure 21. For the trajectory of p_{11}^p a significant reduction in the uncertainty of the parameter estimate $\hat{\alpha}_1$ is achieved, and the rate at which this uncertainty is reduced is especially rapid during the period Δt. We might suggest that over this period Δt such an accelerated rate of decrease in error variance is due to the existence of a substantial amount of information in the data that refers to the system behavior associated with parameter α_1. The trajectory of p_{22}^p, however, displays a negligible decrease in the uncertainty of the related parameter estimate, $\hat{\alpha}_2$. Assuming the opposite of the argument used for the p_{11}^p trajectory, it might be concluded that there is virtually no information in the data that confirms the type of behavior simulated by α_2 and its associated sector of the model.

At this point it is appropriate to revive the concepts associated with the set diagram of Figure 9. In Figure 22, therefore, the set A again denotes the sample behavior observed in the historical field data and M characterizes the set of behavior patterns simulated by the model. It is not difficult to imagine that actual (A) and simulated (M) behavior do not correspond exactly so that there is only a partial overlap between A and M (although strictly speaking this suggests that the problem of model structure identification has not

Uncertainty, system identification, and water quality prediction

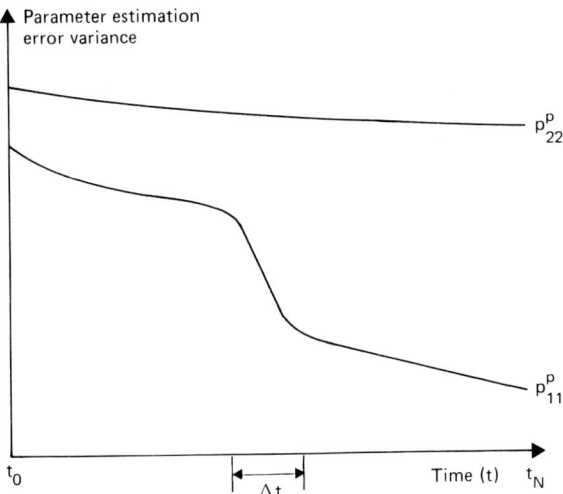

FIGURE 21 Two examples of changes in parameter estimation error variances during calibration (calibration is assumed to refer to the period of observations from t_0 to t_N).

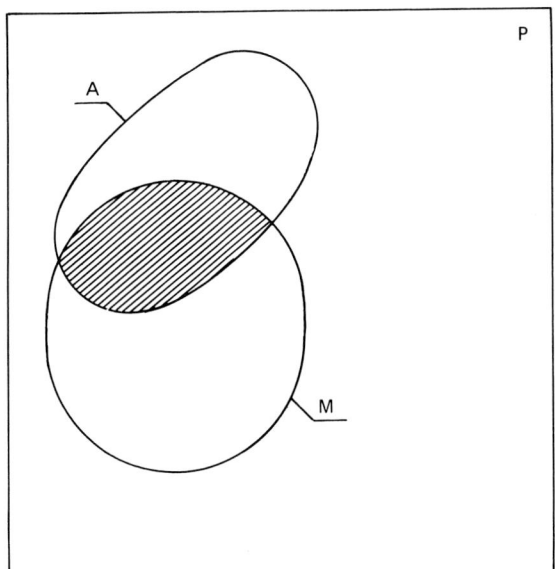

FIGURE 22 Calibration: A is the historically observed pattern of behavior; M is the set of behavior patterns simulated by the model.

been satisfactorily resolved). Transferring the argument from Figure 21 to Figure 22, let us say that parameter α_1 is associated with a part of the behavior covered by the shaded area of Figure 22, while α_2 is related to that part of M that does not intersect with A.

When the model is calibrated against the field data one would expect the uncertainty of parameter estimates associated with the intersection of A and M to decrease significantly. But for parameters associated with the nonintersecting remainder of M estimation error variances should not decrease because there is no information in the historically observed data with which to evaluate such behavior. That is to say, parameters such as α_2 are not identifiable for the given model structure from the given data; this is simply a reiteration of the same point that appeared earlier in the discussions of Section 3.1.1 (sensitivity analysis and a dispersion model for the Rhine River) and Section 3.3.1 (identification of a Streeter–Phelps-type model for the Bedford-Ouse River).

The variances of the *a posteriori* parameter estimation errors indicate, among other things, the relative degrees of uncertainty in the various sectors of the model relationships. From the summarizing picture of Figure 23 it is clear that they are a key factor connecting identification with prediction, at least in terms of accounting for uncertainty. What is the most likely influence of the *a posteriori* parameter estimation errors on the error bounds of forecasts about the future? Yet again, a Venn diagram is a useful starting point. Figure 24 shows a possible situation in which, for example, the future behavior of the system lies within the set of patterns represented by F. The sets P, A, and M have the same interpretations as previously, although the definition of M may be further qualified by stating that it represents simulated behavior in both the past and the future. Let us

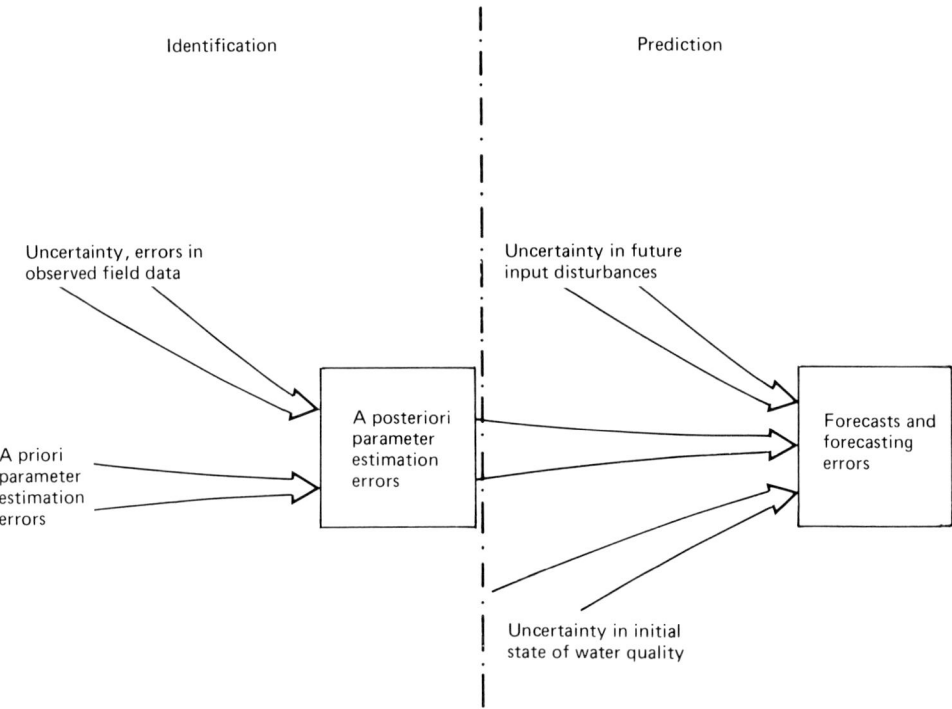

FIGURE 23 Sources of uncertainty and the connection between identification and prediction.

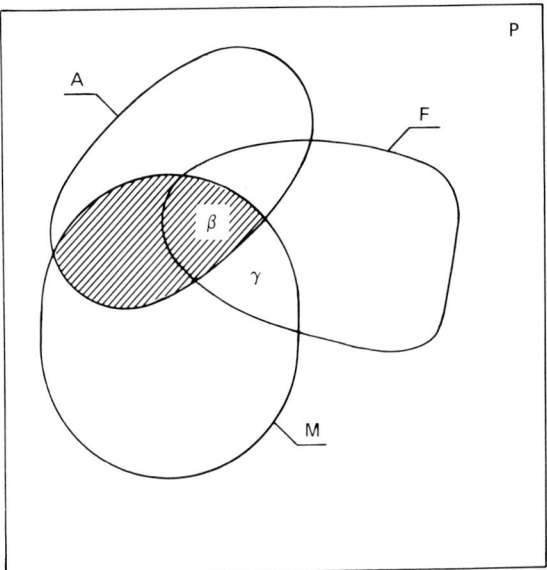

FIGURE 24   Prediction with the calibrated model: A and M are as defined for Figure 22; F represents the set of possible future behavior patterns of the actual system; β represents behavior patterns associated with a well-identified part of the model; γ represents behavior patterns associated with a poorly-identified part of the model (i.e., surplus content).

consider, in particular, what happens when at initial time t_f (with respect to the forecasting period) the model simulates behavior that is characteristic of the set β (M ∩ A ∩ F) and then at time $t_f + \tau$ it simulates behavior characteristic of the set marked γ (M ∩ F) in Figure 24. In other words, a well-calibrated sector of the model (i.e., a pattern of behavior observed in the past) is initially dominant in the simulated behavior, although subsequently a poorly identified sector of the model becomes dominant in the simulated behavior. With a nonlinear model such a transition could be easily brought about, for example, by a slightly modified combination of commonplace input disturbances that force the state of the model into a quite different region of the state space (as already noted in the discussion of Section 3.1.1). Figure 25 illustrates the associated, hypothetical trajectory of one of the state variable forecasts, $\hat{x}$, and its error bounds, which here are simply denoted by $\hat{x} \pm \sigma$, where σ is the standard deviation of the forecasting error. As the state variable trajectory crosses the "boundary" between "past" and "future" behavior patterns the error bounds on the forecast expand rapidly because the response of the model is becoming especially sensitive to relatively uncertain parameter estimates and their respective sectors of the model. Of course, it might also be that the future forcing functions are unlikely events, in which case the sudden loss of confidence in the model forecasts arises both from the uncertainty of these functions and from the parameter estimation errors.

To summarize, let us note that a most important feature, from the forecaster's point of view, is that when forecast-error bounds are computed it is possible to deduce where the model is making predictions for which there is very little historical empirical

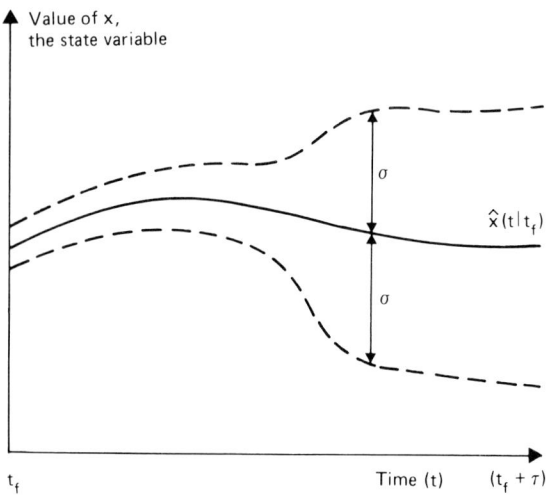

FIGURE 25 Hypothetical state variable trajectory $\hat{x}(t|t_f)$ and prediction error propagation (σ is the standard deviation of the prediction errors) for the "scenario" of Figure 24.

justification. Likewise, when calibrating large models against (probably inadequate) field data it will not be at all obvious which sectors of the model are properly calibrated, if the *a posteriori* parameter estimation errors are not calculated.

4.2 The Propagation of Forecasting Errors

Only two quantitative methods have been used for computing the propagation of forecasting errors, namely Monte Carlo simulation and first-order (possibly higher-order) error analysis, although at first sight the number of recent publications on this topic would perhaps suggest otherwise. However, a point of convergence is discernible, both in terms of the desire to compare the performance of the two methods (Scavia et al., 1981; Gardner et al., 1980a) and in the equivalent ways of stating the equations of a first-order error analysis. The application of Monte Carlo simulation in various studies is well represented, for example, by the papers of O'Neill and Gardner (1979), Gardner et al. (1980b), Whitehead and Young (1979), Hornberger (1980), and Fedra et al. (1981), and we shall not comment further on it here.

A first-order error analysis, or statistical sensitivity analysis, appears to have been first applied to models of water quality–ecological systems by Argentesi and Olivi (1976), following similar applications in other fields (see for example, Burns, 1975; Atherton et al., 1975). Reckhow (1979) and van Straten (1983) state their equations for error covariance propagation in the same form as Argentesi and Olivi (1976), in the sense that the matrix of sensitivity coefficients $[\partial x_i/\partial \alpha_j]$ appears explicitly. McLaughlin (1983) derives the error covariance propagation equations for a fairly general class of distributed-parameter models and states them in a partitioned or disaggregated form (i.e., distinguishing between various sources of error). Inspection of McLaughlin's equations provides an obvious correspondence with the algorithms of Beck et al. (1979), which are derived from recursive filtering theory and also stated in partitioned form. Canale et al. (1980), in their study of sampling strategies and error propagation with respect to modeling water quality variations in the Great Lakes, make use of the same filtering algorithms as Beck et al. (1979). To a less obvious extent, the work of Scavia et al. (1981) also draws upon a background of filtering theory.

A detailed discussion of the conditions under which the statistical analysis of Argentesi and Olivi (1976), for example, is equivalent to the covariance equations of the extended Kalman filter quoted by Beck et al. (1979) is given elsewhere (Beck, 1982). For our present purposes, it suffices to state these recursive partitioned equations (in the form used by Beck et al., 1979) as:

$$\mathbf{P}^s(t_{j+1}|t_f) = \mathbf{\Phi}_{11}\mathbf{P}^s(t_j|t_f)\mathbf{\Phi}_{11}^T$$

{Uncertainty in the state variable predictions} {Uncertainty propagated from the current state of water quality}

$$+ (\mathbf{\Phi}_{11}\mathbf{P}^c(t_j|t_f)\mathbf{\Phi}_{12}^T + \mathbf{\Phi}_{12}[\mathbf{P}^c(t_j|t_f)]^T\mathbf{\Phi}_{11}^T)$$

{Uncertainty derived from correlated state-parameter errors}

$$+ \mathbf{\Phi}_{12}\mathbf{P}^p(t_j|t_f)\mathbf{\Phi}_{12}^T$$

{Uncertainty propagated from the *a posteriori* parameter estimation errors}

$$+ \mathbf{\Gamma}^s \mathbf{S}(t_j)[\mathbf{\Gamma}^s]^T$$

{Uncertainty contributed by future input disturbance estimation errors}

$$+ \mathbf{Q}^s$$

{Uncertainty arising from other factors e.g., residual errors of model calibration}

(14a)

where

$$\mathbf{P}^c(t_{j+1}|t_f) = \mathbf{\Phi}_{11}\mathbf{P}^c(t_j|t_f) + \mathbf{\Phi}_{12}\mathbf{P}^p(t_j|t_f) \quad (14b)$$

$$\mathbf{P}^p(t_{j+1}|t_f) = \mathbf{P}^p(t_j|t_f) \quad (14c)$$

In eqn. (14) $\mathbf{P}^s$, $\mathbf{P}^c$, $\mathbf{P}^p$, and $\mathbf{S}$ are, respectively, the covariance matrices for the state prediction errors, correlated state-parameter prediction errors, parameter errors (which in this case are assumed to be propagated with constant covariance according to eqn. 14c), and the errors in the estimated future input disturbances (u). The matrices $\mathbf{\Phi}_{11}$,

Φ_{12}, and Γ^s are dependent upon the state predictions, $\hat{x}(t_j|t_f)$, the parameters $\hat{\alpha}(t_j|t_f)$, and the estimated inputs $\hat{u}(t_j)$, and must therefore be evaluated at each time step t_j. The elements of these matrices are related to the partial derivatives $[\partial f_i/\partial x_j]$, $[\partial f_i/\partial \alpha_j]$, and $[\partial f_i/\partial u_j]$ of the function $\mathbf{f}\{\cdot\}$ in eqn. (1a), which are notably not the same derivatives as those of the sensitivity coefficients. The solution for $\hat{x}(t_j|t_f)$ can be obtained from eqn. (1a) with the initial conditions $\hat{x}(t_f|t_f)$ for the period of prediction and with $\xi(t) = 0$ for all $t \geq t_f$.

Before proceeding to an application of these covariance propagation equations to illustrate the qualitative discussion of Section 4.1, it is helpful to draw together some of the threads of the foregoing arguments. First, at the very beginning of the discussion of calibration problems in Section 3.1.1, the notion of a sensitivity analysis was introduced informally. To this original idea a statistical component has now been added, in the sense that contributions to the uncertainty of a state variable prediction (x_i) are the products of a sensitivity coefficient ($\partial x_i/\partial \alpha_j$, for example) and an error covariance (p_{jj}^p) associated with a particular source of uncertainty (see also Argentesi and Olivi, 1976). Once again, there is a close connection between sensitivity analysis and the design of experimental and monitoring programs (see also Canale et al., 1980).

Second, in Section 3.2.1 eqn. (5) represents the essential element of a recursive parameter estimation algorithm. Equation (14) can be derived from the same type of algorithm (see Beck et al., 1979), indeed from the same algorithm as that used to provide the results for the Cam and Bedford-Ouse case studies of model structure identification in Sections 3.2.2 and 3.3.1. Such a natural, quantitative link between the problems of calibration and prediction permits in principle a more formal exploration of the questions raised in a qualitative manner in Section 4.1 and this undoubtedly has significant implications for the dilemma to follow in Section 4.3. The connection crystallizes around the obvious choice of setting the *a priori* error covariances for the prediction period equal to the *a posteriori* error covariances of the calibration period, for example,

$$\mathbf{P}^p(t_f|t_f) = \mathbf{P}^p(t_N|t_N) \tag{15}$$

which is clearly suggested by Figure 23. Fedra et al. (1981) provide a corresponding interpretation of the calibration–prediction connection under the somewhat different conditions of sparse data situations (recall the categorization given in the introduction of Section 3).

Lastly, given our bias toward examining the relationship between calibration and prediction, rather than examining the accuracy of algorithms for prediction error propagation, it is reasonable to ask whether the approach behind the statement of eqn. (14) yields any additional insight. In three directions the answer appears to be yes. First, let us recall that the model of the system's dynamics also includes a representation, eqn. (1b), of the output observation y whose significance for prediction error propagation has thus far been overlooked. There is at least one example in which it is sensible to assume a value for all future observations $y(t_j)$. Suppose we have a *closed* system with three interacting state variables (nutrient x_1, phytoplankton x_2, and zooplankton x_3); then if the initial total concentration of an element (phosphorus, for instance, or nitrogen) distributed among these three states is measured as

$$y_1(t_f) = x_1(t_f) + x_2(t_f) + x_3(t_f) + \eta(t_f) \tag{16}$$

it is reasonable to assume a value for $\hat{y}_1(t_j|t_f)$ for all future t_j. Since eqn. (14) is part of a recursive state-parameter estimation algorithm, the computation of prediction error propagation may be conducted *as if* it were actually a calibration exercise. Second, and as a more general complement of this argument, assumptions about the covariance of future measurement errors $\eta(t_j)$ can be made independently of making any similar assumptions about future values of y. Prediction error propagation can then be studied as a function of possible future monitoring programs as discussed by Canale et al. (1980). Third, once a covariance matrix for $\eta(t_j)$ is assumed, it is also possible to compute the gain matrix $G(t_j|t_f)$ appearing in eqns. (5) and (10), a matrix whose role in the estimation of parameters has been discussed briefly in Section 3.3.2. One might accordingly speculate that the behavior of $G(t_j|t_f)$ will be associated with measures of the effectiveness of future measurement strategies with respect to parameter estimation, although this can also be considered more explicitly in other ways (Canale et al., 1980). We may note in passing, however, that the original idea of a sensitivity analysis, subsequently expanded to accommodate a statistical component, can thus be generalized further to include not only analysis based solely upon assumptions about the model and its uncertainties but also analysis incorporating assumptions about measurement strategies.

4.2.1 An Illustrative Example

When discussing problems of uncertainty in complex models, the choice of a simple example, however desirable for reasons of clarity, is nevertheless restrictive. Moreover, a great deal of time could be spent in constructing the perfect example ("perfect" in the sense that it illustrates all the points to be made). However, since the essence of this paper is the analysis of real, and not hypothetical systems, we shall avoid the possible sterility and unreality of the "perfect" case and content ourselves with the following modest example.

Suppose we have a three state-variable model (again, nutrient x_1, phytoplankton x_2, and zooplankton x_3) representing a lake system with inflow and outflow (see Figure 26). Let us assume, without going into detail, that over the period of calibration there was no significant observed zooplankton activity and that consequently, as indicated by the line of demarcation in Figure 26, the part of the model associated with zooplankton activity is relatively uncertain. In quantitative terms, this assumption might imply for the prediction period the following error covariance assignments:

(a) For the initial state estimation errors, $\mathbf{P}^s(t_f|t_f)$, variances equivalent to coefficients of variation of, say, 5% for the nutrient and phytoplankton states and 100% for the zooplankton state.
(b) For all the model parameter estimates associated with zooplankton activity (phytoplankton grazing, excretion, and zooplankton mortality), error variances, $\mathbf{P}^p$, equivalent to coefficients of variation of 100% and for the remaining parameters coefficients of variation of 10%.
(c) For the three input disturbances (influent discharge, influent nutrient, and phytoplankton concentrations), constant error variances, $\mathbf{S}$, equivalent to

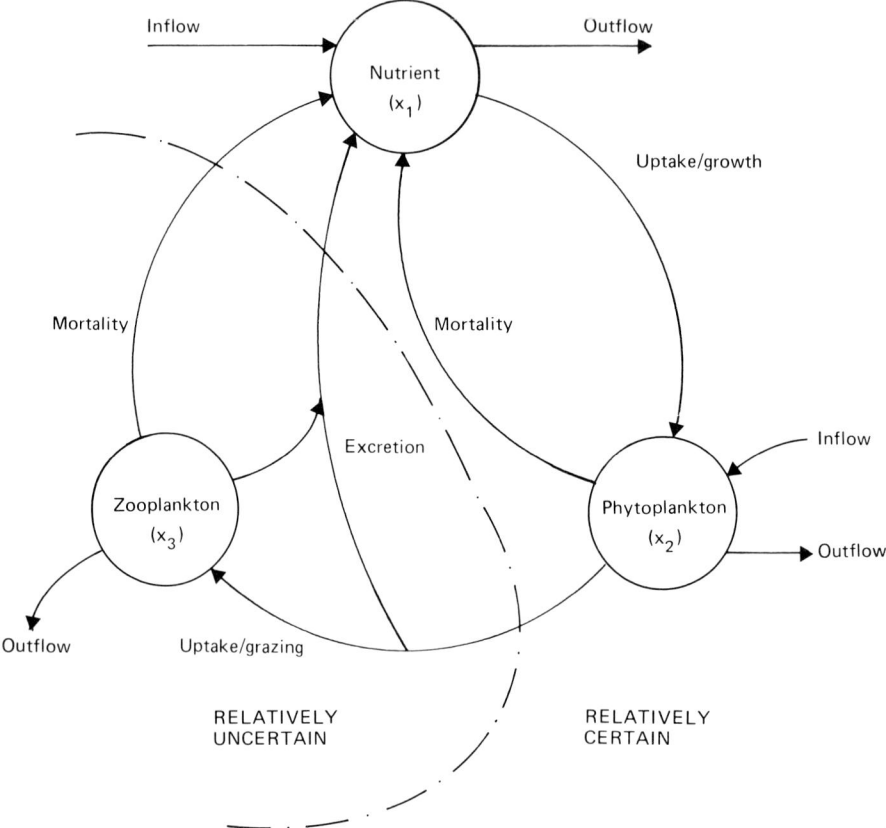

FIGURE 26 Illustrative example system for the analysis of prediction error propagation; the zooplankton dynamics are assumed to be (relatively) highly uncertain (i.e., surplus content).

coefficients of variation of 4%, 16%, and 10%, respectively, for the initial input values $u(t_f)$.

This last assignment is tantamount to the assumption that the disturbances expected for the future will be essentially similar to those observed in the past. The input variations actually used for this example are shown in Figure 27; they are intended to reflect the nature of the "smooth", unexceptional changes characteristic of the earlier discussion of Figure 4 in Section 3.1.1. There is a slow fall in the influent nutrient concentration, a temporary rise in the "seed" phytoplankton population of the inflow, and the influent discharge exhibits a response to a precipitation event. Prediction under substantially changed conditions will thus amount to assessing the effects of significant zooplankton activity in the future.

There are, in fact, a host of items emerging from the discussion of Section 3 to which the present example could be addressed. We shall, however, concentrate on illustrating essentially two groups of problems:

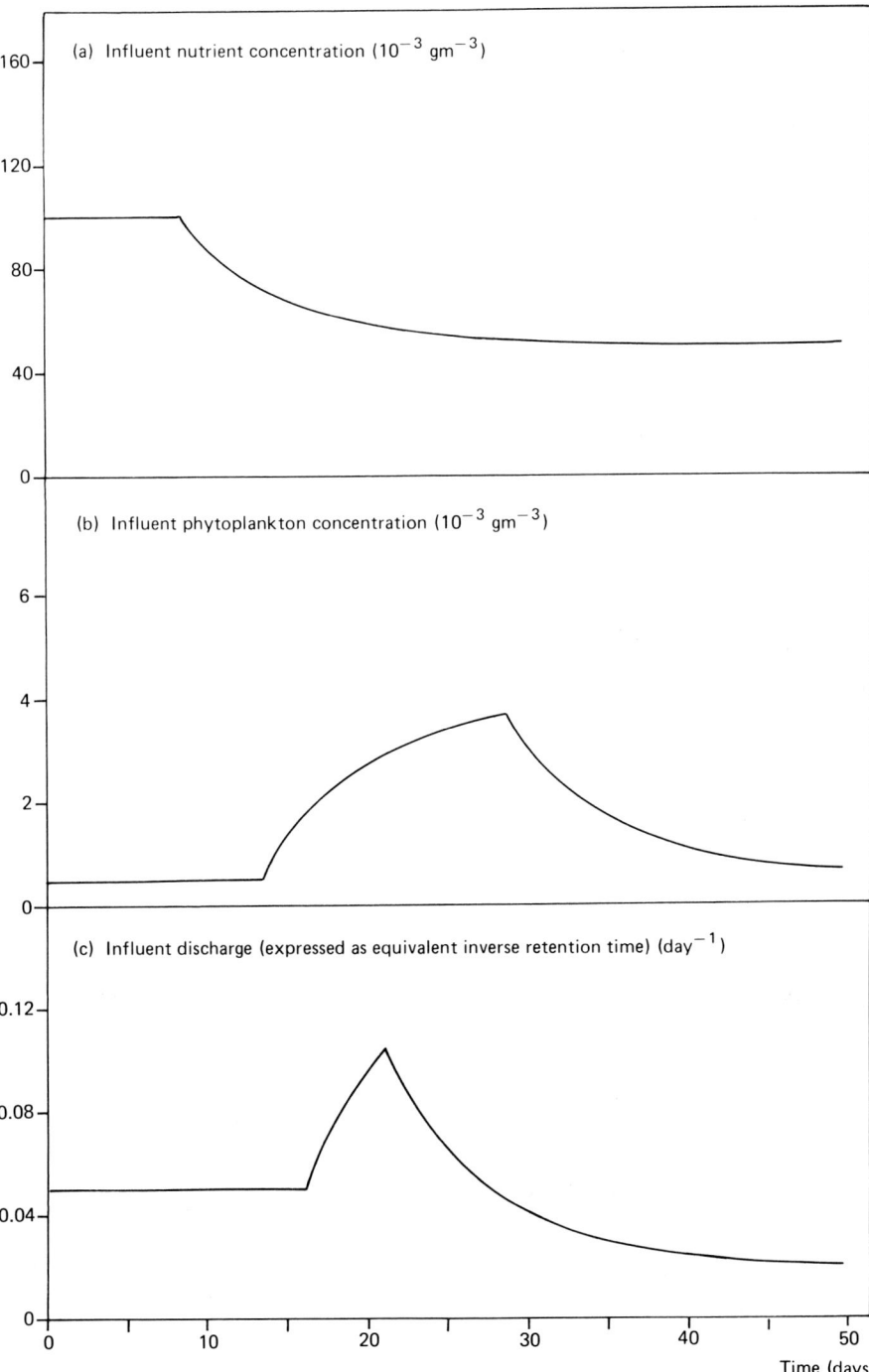

FIGURE 27 Input variable variations ($\hat{u}$) for the illustrative example.

Problem 1. If the activities of zooplankton have not been significantly observable in the past, then their inclusion in the model is equivalent to an "act of faith". In other words, as with the discussion of the Streeter–Phelps assumptions for the example of the Bedford-Ouse study in Section 3.1.1, there was no information in the historical data with which to estimate the parameters associated with zooplankton activity and such behavior, if it occurred, was dominated by other modes of behavior. The model thus contains highly uncertain *surplus content* (a term introduced in Section 3.1.2).

Problem 2. As a minor variation on the theme of Problem 1, we may point out that the model, since it contains surplus content, is probably over-parametrized and will have suffered from problems of *identifiability*. In order to illustrate this point it is necessary to be somewhat more specific. Let us introduce the following expression from the representation of the zooplankton dynamics in the model

$$\hat{\gamma} = [\hat{\alpha}_6 \hat{\alpha}_4 \hat{x}_2(t_j|t_f) - \hat{\alpha}_5] \tag{17}$$

where α_4 is the growth-rate constant for zooplankton [in day^{-1} (10^{-3} gm^{-3} phytoplankton)$^{-1}$], α_5 is the zooplankton death-rate constant (day^{-1}), and α_6 is the fraction of the phytoplankton component absorbed into the zooplankton cells. The term γ is therefore a lumped (growth–death)-rate parameter and typically, while reasonably accurate estimates for γ might be obtainable during calibration (given suitable data), it is highly unlikely that a *uniquely* "best" combination of estimates for $\alpha_4, \alpha_5, \alpha_6$ is identifiable. (This is really the same as the problem in the examples of Bierman et al. (1980) and Halfon (1979b), which were discussed in Section 3.1.2.) Many combinations of values for α_4, α_5, and α_6 give the same value of γ and, from the point of view of the observed zooplankton dynamics, the model reponse is probably more sensitive to the value of γ, and not α_4, α_5, or α_6. A characteristic result of such problems of identifiability is that covariances among the estimation errors of different parameters are very significant (see for example, Di Toro and van Straten, 1979; van Straten, 1983; and Young et al., 1980).

It is the implications of these two problems, which are intrinsic to the process of model calibration, that are important for the illustrative example of prediction error propagation.

Figure 28 shows the state variable trajectories and the propagation of prediction errors when the pattern of future input disturbances from Figure 27 is assumed. Two cases are considered: a base case (scenario 1) defined by the covariance assignments introduced above; and a case (scenario 2), which is otherwise identical with the base case, except for the complete removal of zooplankton activity from the model. It is clear that uncertainty deriving from the estimates of parameters associated with zooplankton activity (in scenario 1) dominates the propagation of errors, providing a rather dramatic illustration of the previous discussion of Figure 25. Confidence in the phytoplankton prediction is entirely eroded as soon as zooplankton grazing becomes significant although some measure of reasonable "predictability" for the base case is restored toward the end of the prediction period when zooplankton activity subsides. The second set of predictions in Figure 28 is a strikingly confident statement about the future and one could tentatively conclude that toward the end of the prediction period the two scenarios for

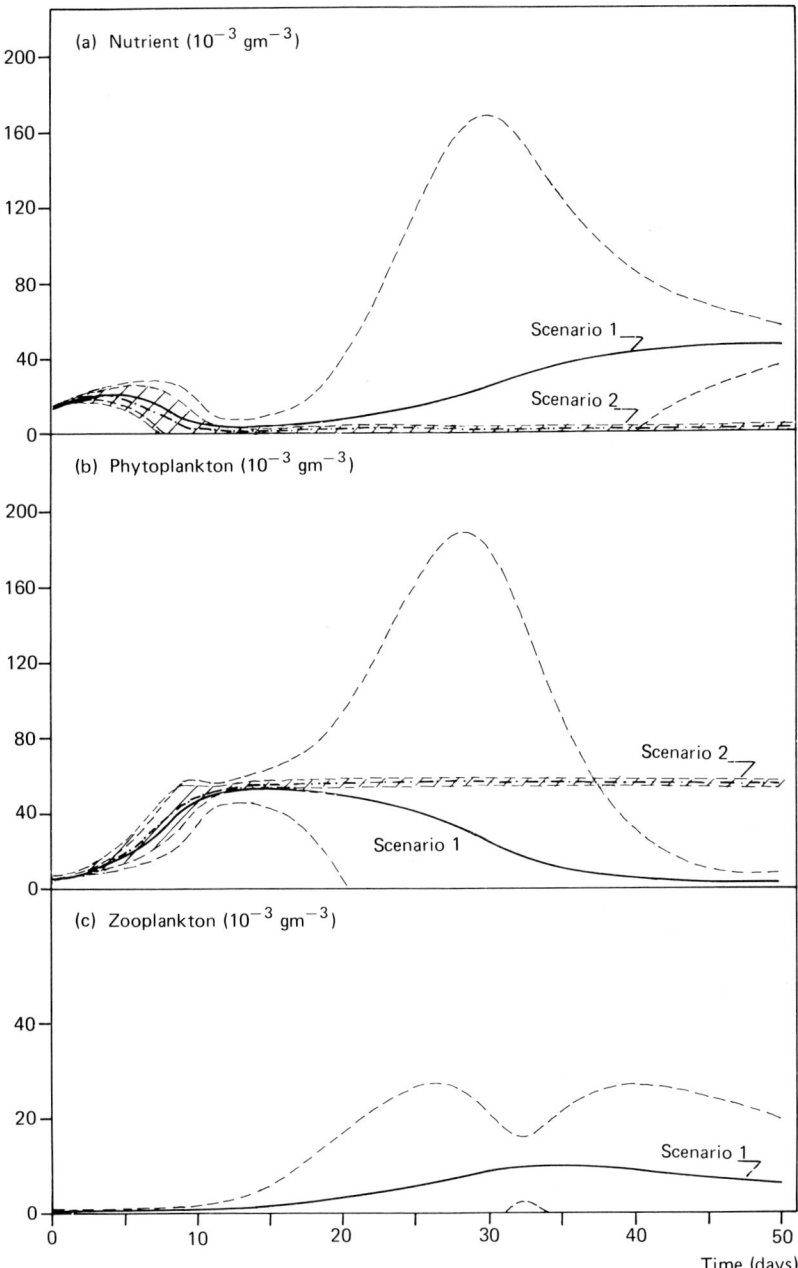

FIGURE 28 Scenarios 1 and 2 for the illustrative analysis of prediction error propagation: nominal reference trajectories are given together with the bounds representing (±) the standard deviations of the prediction errors. The shaded areas represent these bounds for scenario 2; all units for the three state variables are expressed in terms of the nutrient element contained in each compartment.

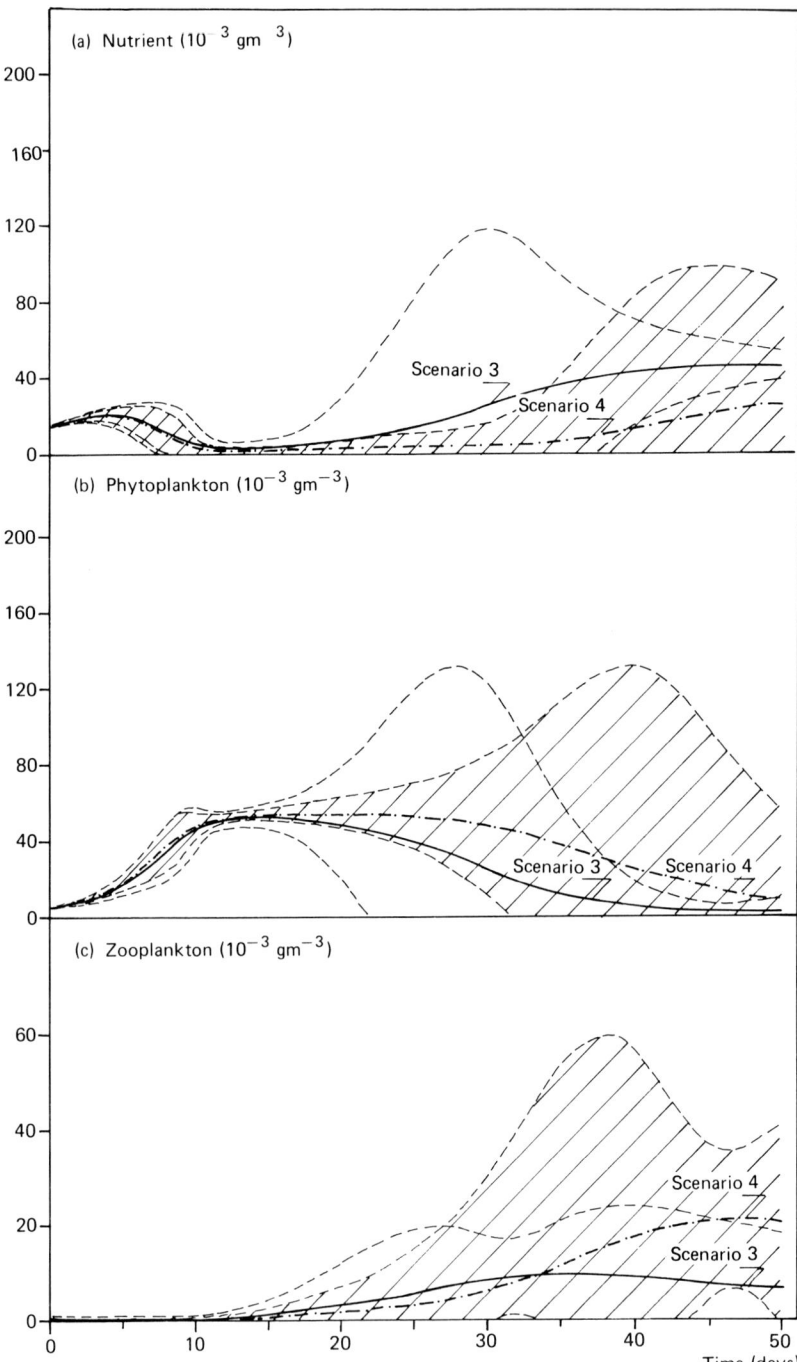

FIGURE 29 Scenarios 3 and 4 for the illustrative analysis of prediction error propagation: nominal reference trajectories are given together with the bounds representing (±) the standard deviations of the prediction errors. The shaded areas represent these bounds for scenario 4; all units for the three state variables are expressed in terms of the nutrient element contained in each compartment.

TABLE 1 Parameter estimates and parameter estimation error covariance assumptions for Problem 2 (see also Figure 29).

Parameter	Estimate		Correlation coefficient of estimation errors
	Scenario 3	Scenario 4	
α_4	0.05	0.02	$\rho(\tilde{\alpha}_4, \tilde{\alpha}_5) = 0.4$
α_5	0.03	0.025	$\rho(\tilde{\alpha}_4, \tilde{\alpha}_6) = -0.5$
α_6	0.1	0.2	$\rho(\tilde{\alpha}_5, \tilde{\alpha}_6) = 0.33$
$x_2(t_f|t_f)$	5.0	5.0	–
γ	–0.005	–0.005	–

prediction are distinctly different, an important point to which we shall return in the following section.

Figure 29 deals with Problem 2 as defined above. For these two sets of predictions (scenarios 3 and 4) the parameter estimates assumed for eqn. (17) and their associated error covariances are given in Table 1. Scenario 3 is identical with scenario 1, the base case, except for the assumption of nonzero covariances among the parameter estimation errors. Scenario 4 has the same estimation error structure as scenario 3 but different parameter estimates for α_4, α_5, and α_6 (which, nevertheless, give the same value for γ) and estimation error variances adjusted to maintain coefficients of variation of 100% for these parameter estimates. Note that a comparison of scenarios 1 and 3 in Figures 28 and 29 demonstrates the reduction in prediction error magnitudes due to the assumption of correlated parameter estimation errors. As expected, in this illustration the nominal trajectories (means) of the predicted state variables are distinctly different. Had these been confident, or even deterministic predictions, it might have been concluded that the ambiguities of model calibration result in unavoidably ambiguous statements about the future. As it is, however, the uncertainty in both scenarios is sufficient that such a conclusion is not justified, a point which otherwise may not have been apparent if the prediction errors had not been computed.

4.3 A Dilemma

This simple example attempts to show two things: that the results of a calibration exercise, which depend partly upon the nature of the model and partly upon the nature of the data, can have a decisive influence on the propagation of prediction errors; and that the association of prediction errors with a prediction can influence one's judgment about the significance of differences among alternative statements about future behavior patterns.

Referring to Figure 30, let us assume that the set of behavior patterns M_1 belongs to a model characteristic of the class of large simulation models — the type of model that simulates a much greater variety of behavior patterns than has actually been observed in the historical field data, A (i.e., a large part of M_1 does not intersect with the set A). For such a model the many parameters not associated with those modes of behavior in the set A (i.e., that part of M_1 lying outside A) would have, as we have already discussed, relatively large *a posteriori* estimation errors. The complement, or opposite, of the large

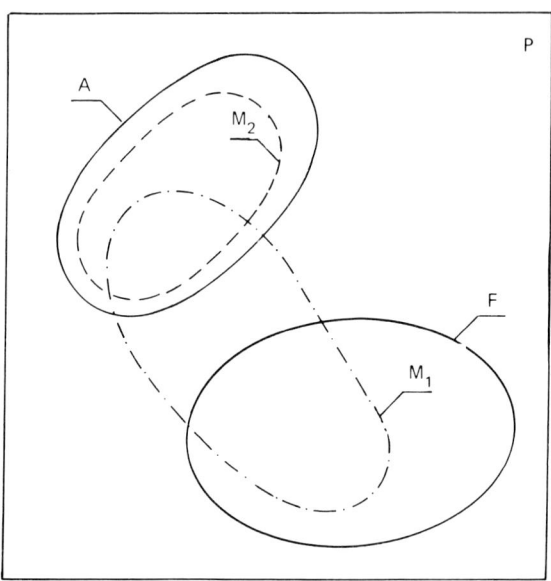

FIGURE 30 Uncertainty and prediction after identification: A is the set of historically observed patterns of behavior; M_1 is the set of behavior patterns simulated by the typical "large simulation model"; M_2 is the set of behavior patterns simulated by the small fully-identified model; F is a set of possible future behavior patterns of the actual system.

simulation model is the more compact kind of model that would typically result from the "not-so-dominant" approach to model building discussed in Section 3. Optimistically, this latter fully identified model might be represented by the set M_2 in Figure 30. Its *a posteriori* parameter estimates ought to be much less uncertain than many of those of M_1; and since this model contains no surplus content, the set M_2 is contained completely in the set A.

How might these two models perform when applied to the problem of prediction? The most interesting and challenging case to consider is that in which future input disturbances of the lake or river, such as different meteorological conditions and modified effluent discharges, force the variations of water quality into patterns of behavior (say F in Figure 30) quite different from the historically observed patterns. If our arguments from Sections 4.1 and 4.2 are sound, then for model M_1 it would be expected that predictions of behavior characteristic of F would be strongly dependent upon highly uncertain sectors of that model (since $M_1 \cap A$ is disjoint from $M_1 \cap F$). These predictions would accordingly be highly uncertain. In contrast, would a small model that captures only the dominant modes of past behavior (as does the model M_2 in Figure 30) tend not to predict different future conditions? After all, its parameter values have been well identified and would thus be associated with relatively small estimation errors. Hence, given the kind of argument presented earlier, we might be mistakenly confident about its predictions. There is, for example, no intersection between M_2 and F in Figure 30, which suggests that F is outside the scope of behavior patterns simulated by M_2.

We have in fact a dilemma; indeed, it was already foreshadowed in the illustrative predictions of Figure 28. With a large model (M_1) it may well be possible to predict the "correct" future, but one would have little or no confidence in that prediction. In contrast, with a small model (M_2) it may be that a quite "incorrect" future is predicted, and, worse still, one might place considerable confidence in that prediction.

Admittedly this dilemma has perhaps been stated in an exaggerated and overly simplistic fashion. And once again our argument has fallen into the trap of using the words "large" and "small" to qualify the models developed by their respective approaches. However, since this discussion relates back to calibration and the problems of surplus content in a model, such words are not as misleading as they might have been earlier. The intention of simplification was to sharply define the problem, not to obscure the inevitable grey areas between the black-and-white statements about the problem. For example, one might consider the rather provocative extension of Figure 30 represented by Figure 31. In this case the large simulation model (M_1) has a set of behavior patterns that stretches outside the frame of all possible (true) behavior patterns, P, of the system. That it to say, M_1 contains *spurious content* that has no parallel with reality.

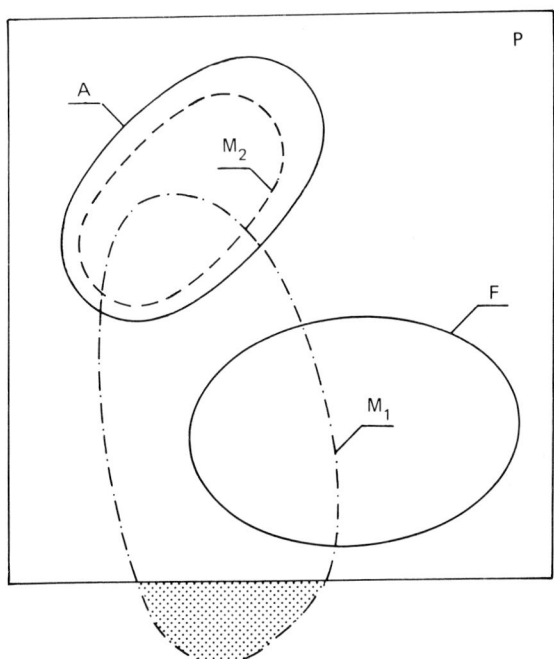

FIGURE 31 Uncertainty, prediction after identification, and spurious content: all sets are as previously defined for Figure 30.

The analysis of prediction error propagation can be viewed as an *a posteriori* sensitivity analysis. It provides a check on the *relative* levels of confidence associated with the assumptions made in developing, calibrating, and applying a model; and it should reveal when, and to what extent, the model's predictions rely upon these assumptions and upon each component of uncertainty. It ought to be possible to distinguish among the effects of uncertainty propagated from surplus content in the model, the effects of unresolved ambiguities of calibration, and the effects of uncertainty associated with the extrapolation of knowledge about the behavior of laboratory systems to knowledge about the behavior of the field system.

The major point of the dilemma is that it captures some limiting features of *both* the approaches to modeling discussed in this paper.

5   CONCLUSIONS

There has always been uncertainty. Our present concern with it is partly because it has come to be recognized (or "perceived") as one of *the* problems in water quality–ecological modeling. Again, as in the opening paragraphs of this paper, reference can be made to Orlob's review of water quality modeling, where "model reliability" is listed first among the "areas for improvement" (Orlob, 1983b). Two aspects of the problem of uncertainty have been dealt with here: uncertainty in the structure of the mathematical relationships hypothesized for a particular model; and uncertainty associated with the predictions obtained from models. These are considered to be closely related aspects that link together the subjects of model calibration (or system identification) and prediction. We have also classified approaches to water quality–ecological modeling into two types, a "dominant" and a "not-so-dominant" approach. To address the concerns expressed about the dominant approach in Section 2, we have discussed at length a complementary alternative, which we call the not-so-dominant approach (Section 3). Our arguments for linking identification and prediction have then been used to conclude with a dilemma, which in turn allows us to reiterate some the complementary features of these two approaches (Section 4).

Our main concern with the dominant approach to water quality modeling is that it tends to ignore the deeper significance of model calibration. Calibration should not be an unimportant afterthought to a model development exercise. It may only be considered in this way if one chooses, as is the tendency with the dominant approach, to attach great confidence to *a priori* theory, thereby renouncing much of the questioning that should accompany calibration. Somlyódy (1981) provides insight into why certain disciplinary conventions may encourage the view that calibration is not really important. Model complexity and (apparent) completeness cannot be equated with accuracy. Given the current limitations of the data available for calibration, larger models, with many parameter values to be estimated, are likely to lead to hidden ambiguities in the model predictions. Indeed, it is only in the context of applying a model for prediction that one can begin to challenge the belief that a large model must be correct — for how (so runs the rhetorical question) can it be incorrect if every detail of conceivable relevance has been included? This concern is not restricted to water quality–ecological modeling, but is encountered in other adjacent fields (see, for example, Kossen, 1979).

The "questioning" process of model calibration, to which such importance is attached, is what we have called model structure identification. The model is a vehicle for asking questions about the nature of "reality" and the process of model structure identification is ultimately concerned with acquiring understanding about a system's behavior. For this reason we have examined the equivalence between procedures for solving the problem of model structure identification and current theories of the scientific method (for example, Popper, 1959). The procedure of Section 3 has two basic features: (a) the use of recursive estimation algorithms; and (b) "experimentation" with the model structure to satisfy alternately the two objectives of falsifying confident hypotheses and creatively speculating about uncertain hypotheses. The notion of "experimentation", and the idea that the controlled conditions of an experiment can be reconstructed during the analysis of field data, have considerable appeal, particularly because of the suggested association with laboratory science (from which, incidentally, many of the extrapolations inherent in the dominant approach are ultimately drawn). But in a holistic sense it is difficult to claim, however tempting it may be, that there is *one* "experiment" and its complementary "environment". Rather, it is only possible to state that a number of more or less significant "experiments" are proceeding in parallel. The primary value of the approach discussed in Section 3 is its ability to provide a means for direct evaluation of those extrapolations, from laboratory or "equivalent" field systems, characteristic of the dominant approach. The not-so-dominant approach is certainly not, however, without its own difficulties. Indeed, there are serious difficulties in absorbing and interpreting the results from the analysis associated with model structure identification. However, these difficulties are somewhat independent of the particular approach taken; they arise fundamentally from the combination of complexity with uncertainty.

The real debate about approaches to modeling is, at least for the purposes of this paper, more appropriately considered in terms of the question of predicting future behavior patterns under substantially changed conditions. The arguments employed in Section 4, however, cannot be divorced from the preceding discussion of calibration and model structure identification. One of the most important issues in examining the propagation of prediction errors is the effect of highly uncertain surplus content in a model. In other words, this is an effect resulting from problems of identifiability where prior estimation of unique values for the model parameters has not been possible; the model contains too many parameters in terms of the data available for calibration. The results of a calibration exercise can have a decisive influence over the propagation of prediction errors. Equally the association of prediction errors with a prediction can determine one's judgment about the significance of differences among alternative statements about future behavior patterns.

We have closed this paper with a dilemma, the same dilemma as presented in the much shorter companion paper (Beck, 1981). Its usefulness, for the time being, is the sharp focus into which it brings the problems from which both the approaches to modeling suffer. No doubt, it also expresses some of the limits to modeling.

REFERENCES

Adachi, N. and Ikeda, S. (1978). Stability Analysis of Eutrophication Models. RM-78-53. International Institute for Applied Systems Analysis, Laxenburg, Austria.

Akaike, H. (1974). A new look at statistical model identification. IEEE Transactions on Automatic Control, 19: 716–722.

Andrews, J.F. (1977). Dynamics and control of wastewater treatment plants. In J.H. Sherrard (Editor), Fundamental Research Needs for Water and Wastewater Treatment Systems. Proceedings of the NSF/AEEP Conference, Arlington, Virginia, December, pp. 83–92.

Andrews, J.F. and Stenstrom, M.K. (1978). Dynamic models and control strategies for wastewater treatment plants — an overview. In Y. Sawaragi and H. Akashi (Editors), Environmental Systems Planning, Design, and Control, Volume 2. Pergamon, Oxford, pp. 443–452.

Argentesi, F. and Olivi, L. (1976). Statistical sensitivity analyses of a simulation model for the biomass–nutrient dynamics in aquatic ecosystems. In the Proceedings of the Summer Computer Simulation Conference, 4th. Simulation Councils, La Jolla, California, pp. 389–393.

Åström, K.J. and Eykhoff, P. (1971). System identification — a survey. Automatica, 7:123–162.

Atherton, R.W., Schainker, R.B., and Ducot, E.R. (1975). On the statistical sensitivity analysis of models for chemical kinetics. AIChE Journal, 21(3):441–448.

Battista, J.R. (1977). The holistic paradigm and general system theory. General Systems, XXII:65–71.

Beck, M.B. (1978a). A Comparative Case Study of Dynamic Models for DO–BOD–Algae Interaction in a Freshwater River. RR-78-19. International Institute for Applied Systems Analysis, Laxenburg, Austria.

Beck, M.B. (1978b). Random signal analysis in an environmental sciences problem. Applied Mathematical Modeling, 2(1):23–29.

Beck, M.B. (1979a). Model structure identification from experimental data. In E. Halfon (Editor), Theoretical Systems Ecology. Academic Press, New York, pp. 259–289.

Beck, M.B. (1979b). System Identification, Estimation and Forecasting of Water Quality — Part 1: Theory. WP-79-31. International Institute for Applied Systems Analysis, Laxenburg, Austria.

Beck, M.B. (1979c). On-Line Estimation of Nitrification Dynamics. PP-79-3. International Institute for Applied Systems Analysis, Laxenburg, Austria.

Beck, M.B. (1980). Applications of system identification and parameter estimation in water quality modelling. In the Proceedings of the Oxford Symposium on Hydrological Forecasting, IAHS–AIHS Publication No. 129, pp. 123–131.

Beck, M.B. (1981). Hard or soft environmental systems? Ecological Modelling, 11:233–251.

Beck, M.B. (1982). The Propagation of Errors and Uncertainty in Forecasting Water Quality, Part II: Some Simple Examples and Case Studies. Working Paper. International Institute for Applied Systems Analysis, Laxenburg, Austria. (Forthcoming.)

Beck, M.B. and Young, P.C. (1975). A dynamic model for DO–BOD relationships in a non-tidal stream. Water Research, 9:769–776.

Beck, M.B. and Young, P.C. (1976). Systematic identification of DO–BOD model structure. Journal of the Environmental Engineering Division, American Society of Civil Engineers, 102 (EE5): 902–927.

Beck, M.B., Latten, A., and Tong, R.M. (1978). Modelling and Operational Control of the Activated Sludge Process of Wastewater Treatment. PP-78-10. International Institute for Applied Systems Analysis, Laxenburg, Austria.

Beck, M.B., Halfon, E., and van Straten, G. (1979). The Propagation of Errors and Uncertainty in Forecasting Water Quality, Part I: Method. WP-79-100. International Institute for Applied Systems Analysis, Laxenburg, Austria.

Bedford-Ouse Study (1979). Final Report. Anglian Water Authority, Huntingdon, UK.

Bierman, V.J., Dolan, D.M., Stoermer, E.F., Gannon, J.E., and Smith, V.E. (1980). The Development and Calibration of a Spatially Simplified Multi-Class Phytoplankton Model for Saginaw Bay, Lake Huron. Contribution No. 33. Great Lakes Environmental Planning Study, US Environmental Protection Agency, Grosse Ile, Michigan.

Box, G.E.P. and Jenkins, G.M. (1970). Time-Series Analysis, Forecasting and Control. Holden-Day, San Francisco, California.

Burns, J.R. (1975). Error analysis of nonlinear simulations: applications to world dynamics. IEEE Transactions on Systems, Man, and Cybernetics, 5(3):331–340.

Canale, R.P., DePalma, L.M., and Powers, W.F. (1980). Sampling Strategies for Water Quality in the Great Lakes. Report No. EPA-600/3-80-55. US Environmental Protection Agency, Environmental Research Laboratory, Duluth, Minnesota.

Chan, C.W., Harris, C.J., and Wellstead, P.E. (1974). An order-testing criterion for mixed autoregressive moving average processes. International Journal of Control, 20:817–834.

Cobelli, C., Lepschy, A., and Romanin-Jacur, G. (1979). Structural identifiability of linear compartmental models of ecosystems. In E. Halfon (Editor), Theoretical Systems Ecology. Academic Press, New York, pp. 237–258.

Di Toro, D.M. and van Straten, G. (1979). Uncertainty in the Parameters and Predictions of Phytoplankton Models. WP-79-27. International Institute for Applied Systems Analysis, Laxenburg, Austria.

Dobbins, W.E. (1964). BOD and oxygen relationships in streams. Journal of the Sanitary Engineering Division, American Society of Civil Engineers, 90:53–78.

Eberhardt, L.L. (1977). Applied systems ecology: models, data, and statistical methods. In G.S. Innis (Editor), New Directions in the Analysis of Ecological Systems. Simulation Council Proceedings Series, 5(1):43–55.

Eykhoff, P. (1974). System Identification — Parameter and State Estimation. Wiley, Chichester.

Fedra, K., van Straten, G., and Beck, M.B. (1981). Uncertainty and arbitrariness in ecosystems modeling: a lake modeling example. Ecological Modelling, 13:87–110.

Fleming, G. (1979). Total river basin assessment of sediment–erosion–transport–deposition processes by mathematical model. In Pilot Zones for Water Quality Management. Proceedings of a Seminar, Institute for Water Management, Budapest, pp. 262–280.

Gardner, R.H., O'Neill, R.V., Mankin, J.B., and Carney, J.H. (1980a). A Comparison of Sensitivity Analysis and Error Analysis Based on a Stream Ecosystem Model. Technical Report. Environmental Sciences Division, Oak Ridge National Laboratory, Oak Ridge, Tennessee.

Gardner, R.H., O'Neill, R.V., Mankin, J.B., and Kumar, D. (1980b). Comparative error analysis of six predator–prey models. Ecology, 61(2):323–332.

Goh, B.S. (1979). Robust stability concepts for ecosystems models. In E. Halfon (Editor), Theoretical Systems Ecology. Academic Press, New York, pp. 467–487.

Gustavsson, I. (1975). Survey of applications of identification in chemical and physical processes. Automatica, 11:3–24.

Halfon, E. (1979a). The effects of data variability in the development and validation of ecosystems models. In B.P. Zeigler, M.S. Elzas, G.J. Klir, and T.I. Ören (Editors), Methodology in Systems Modelling and Simulation. North-Holland, Amsterdam, pp. 335–343.

Halfon, E. (1979b). On the parameter structure of a large-scale ecological model. In G.P. Patil and M.L. Rosenzweig (Editors), Contemporary Quantitative Ecology and Related Econometrics. International Cooperative Publishing House, Fairland, Maryland, pp. 279–293.

Holling, C.S. (Editor) (1978). Adaptive Environmental Assessment and Management. Wiley, Chichester.

Hornberger, G.M. (1980). Uncertainty in dissolved oxygen prediction due to variability in algal photosynthesis. Water Research, 14:355–361.

Hornberger, G.M. and Spear, R.C. (1980). Eutrophication in Peel Inlet — I. Problem-defining behavior and a mathematical model for the phosphorus scenario. Water Research, 14:29–42.

Hornberger, G.M. and Spear, R.C. (1981). An approach to the preliminary analysis of environmental systems. Journal of Environmental Management, 12(1):7–18.

International Association of Hydrological Sciences (1978). Modelling the Water Quality of the Hydrological Cycle. Proceedings of the Baden Symposium. IAHS–AIHS Publication No. 125.

International Association of Hydrological Sciences (1980). Hydrological Forecasting. Proceedings of the Oxford Symposium. IAHS–AIHS Publication No. 129.

Isermann, R. (Editor) (1980). Identification and System Parameter Estimation. Proceedings of the IFAC Symposium, 5th. Pergamon, Oxford.

Jakeman, A.J. and Young, P.C. (1980). Towards optimal modeling of translocation data from tracer studies. Proceedings of the Biennial Conference of the Simulation Society of Australia, 4th, pp. 248–253.

Jazwinski, A.H. (1970). Stochastic Processes and Filtering Theory. Academic Press, New York.

Jørgensen, S.E. and Harleman, D.R.F. (1978). Hydrophysical and Ecological Modeling of Deep Lakes and Reservoirs. CP-78-7. International Institute for Applied Systems Analysis, Laxenburg, Austria.

Kahlig, P. (1979). One-dimensional transient model for short-term prediction of downstream pollution in rivers. Water Research, 13:1311–1316.

Kempf, J. and van Straten, G. (1980). Applications of Catastrophe Theory to Water Quality Modelling. CP-80-12. International Institute for Applied Systems Analysis, Laxenburg, Austria.

Kossen, N.W.F. (1979). Mathematical modelling of fermentation processes: scope and limitations. In A.T. Bull, D.C. Ellwood, and C. Ratledge (Editors), Microbial Technology: Current State, Future Prospects. Cambridge University Press, Cambridge, pp. 327–357.

Lack, T.J. and Lund, J.W.G. (1974). Observations and experiments on the phytoplankton of Blelham Tarn, English Lake District. Freshwater Biology, 4:399–415.

Leonov, A. (1980). Mathematical Modeling of Phosphorus Transformation in the Lake Balaton Ecosystem. WP-80-149. International Institute for Applied Systems Analysis, Laxenburg, Austria.

Lewandowska, A. (1981). Structural Properties and Frequency Response Analysis of Simplified Water Quality Models: the Case of Time-Invariant Coefficients. WP-81-116. International Institute for Applied Systems Analysis, Laxenburg, Austria.

Lewandowski, A. (1981). Issues in Model Validation. WP-81-32. International Institute for Applied Systems Analysis, Laxenburg, Austria.

Ljung, L. (1979). Asymptotic behavior of the extended Kalman filter as a parameter estimator for linear systems. IEEE Transactions on Automatic Control, 24:36–50.

Maciejowski, J.M. (1979). Model discrimination using an algorithmic information criterion. Automatica, 15:579–593.

Maciejowski, J.M. (1980). A Least-Genericity Principle for Model Selection. Report No. 91. Control Theory Centre, University of Warwick, Coventry.

Magee, B. (1973). Popper. Fontana, London.

Mankin, J.B., O'Neill, R.V., Shugart, H.H., and Rust, B.W. (1977). The importance of validation in ecosystem analysis. In G.S. Innis (Editor), New Directions in the Analysis of Ecological Systems. Simulation Council Proceedings Series, 5(1):63–72.

Marsili-Libelli, S. (1980). The role of microprocessors in water quality management: problems and prospects. In M.B. Beck (Editor), Real-Time Water Quality Management. CP-80-38. International Institute for Applied Systems Analysis, Laxenburg, Austria, pp. 162–183.

McLaughlin, D.B. (1983). Statistical analysis of uncertainty propagation and model accuracy. In M.B. Beck and G. van Straten (Editors), Uncertainty and Forecasting of Water Quality. This volume, pp. 305–319.

Olsson, G. (1980). Estimation and identification problems in wastewater treatment. In E.F. Wood (Editor), Real-Time Forecasting/Control of Water Resource Systems. Pergamon, Oxford, pp. 93–108.

O'Neill, R.V. and Gardner, R.H. (1979). Sources of uncertainty in ecological models. In B.P. Zeigler, M.S. Elzas, G.J. Klir, and T.I. Ören (Editors), Methodology in Systems Modelling and Simulation. North-Holland, Amsterdam, pp. 447–463.

Orlob, G.T. (1983a). Introduction. In G.T. Orlob (Editor), Mathematical Modeling of Water Quality: Streams, Lakes, and Reservoirs. Wiley, Chichester, pp. 1–10.

Orlob, G.T. (1983b). Future directions. In G.T. Orlob (Editor), Mathematical Modeling of Water Quality: Streams, Lakes, and Reservoirs. Wiley, Chichester, pp. 503–509.

Park, R.A., Scavia, D., and Clesceri, N.L. (1975). CLEANER: the Lake George model. In C.S. Russell (Editor), Ecological Modelling in a Resource Management Framework. Working Paper QE-1. Resources for the Future, Washington, D.C., pp. 49–81.

Patten, B.C. (Editor) (1975). Systems Analysis and Simulation in Ecology, Volume III. Academic Press, New York, p. 622.

Patten, B.C. (1976). Systems Analysis and Simulation in Ecology, Volume IV. Academic Press, New York, p. 608.

Popper, K.R. (1959). The Logic of Scientific Discovery. Hutchinson, London.

Reckhow, K.H. (1979). The use of a simple model and uncertainty analysis in lake management. Water Resources Bulletin, 15:601–611.

Russell, C.S. (Editor) (1975). Ecological Modeling in a Resource Management Framework. Working Paper QE-1. Resources for the Future, Washington, D.C.

Scavia, D. and Robertson, A. (Editors) (1979). Perspectives on Lake Ecosystem Modeling. Ann Arbor Science Publishers, Ann Arbor, Michigan.

Scavia, D., Powers, W.F., Canale, R.P., and Moody, J.L. (1981). Comparison of first-order error analysis and Monte Carlo simulation in time-dependent lake eutrophication models. Water Resources Research, 17(4):1051–1069.

Shastry, J.S., Fan, L.T., and Erickson, L.E. (1973). Non-linear parameter estimation in water quality modeling. Journal of the Environmental Engineering Division, American Society of Civil Engineers, 99(EE3):315–331.

Siljak, D.D. (1979). Structure and stability of model ecosystems. In E. Halfon (Editor), Theoretical Systems Ecology. Academic Press, New York, pp. 151–181.

Söderström, T. (1977). On model structure testing in system identification. International Journal of Control, 26:1–18.

Söderström, T., Ljung, L., and Gustavsson, I. (1974). On the Accuracy of Identification and the Design of Identification Experiments. Report 7428. Department of Automatic Control, Lund Institute of Technology, Lund, Sweden.

Somlyody, L. (1977). Dispersion measurement on the Danube. Water Research, 11:411–417.

Somlyody, L. (1981). Water Quality Modelling: A Comparison of Transport-Oriented and Biochemistry-Oriented Approaches. WP-81-117. International Institute for Applied Systems Analysis, Laxenburg, Austria.

Spear, R.C. and Hornberger, G.M. (1980). Eutrophication in Peel Inlet – II. Identification of critical uncertainties via generalized sensitivity analysis. Water Research, 14:43–49.

Streeter, H.W. and Phelps, E.B. (1925). A Study of the Pollution and Natural Purification of the Ohio River. Bulletin No. 146. US Public Health Service, Washington, D.C.

Thé, G. (1978). Parameter identification in a model for the conductivity of a river based on noisy measurements at two locations. In G.C. Vansteenkiste (Editor), Modeling, Identification and Control in Environmental Systems. North-Holland, Amsterdam, pp. 823–830.

Thomann, R.V., Winfield, R.P., and Segna, J.J. (1979). Verification Analysis of Lake Ontario and Rochester Embayment Three-Dimensional Eutrophication Models. Report EPA-600/3-79-094. US Environmental Protection Agency, Environmental Research Laboratory, Duluth, Minnesota.

Tong, R.M., Beck, M.B., and Latten, A. (1980). Fuzzy control of the activated sludge wastewater treatment process. Automatica, 16:659–701.

Ulanowicz, R.E. (1979). Prediction, chaos, and ecological perspective. In E. Halfon (Editor), Theoretical Systems Ecology. Academic Press, New York, pp. 107–117.

Unbehauen, H. and Göhring, B. (1974). Tests for determining model-order in parameter estimation. Automatica, 10:233–244.

van den Boom, A.J.M. and van den Enden, A.W.M. (1974). The determination of the orders of process and noise dynamics. Automatica, 10:245–256.

van Straten, G. (1980). Analysis of Model and Parameter Uncertainty in Simple Phytoplankton Models for Laka Balaton. WP-80-139. International Institute for Applied Systems Analysis, Laxenburg, Austria.

van Straten, G. (1983). Maximum likelihood estimation of parameters and uncertainty in phytoplankton models. In M.B. Beck and G. van Straten (Editors), Uncertainty and Forecasting of Water Quality. This volume, pp. 157–171.

van Straten, G. and Golbach, G. (1982). Frequency-Domain and Time-Domain Analysis of Dispersion in the River Rhine. (In preparation.)

van Straten, G. and Somlyódy, L. (1980). Lake Balaton Eutrophication Study: Present Status and Future Program. WP-80-187. International Institute for Applied Systems Analysis, Laxenburg, Austria.

van Straten, G., Jolankai, G., and Herodek, S. (1979). Review and Evaluation of Research on the Eutrophication of Lake Balaton – A Background Report for Modeling. CP-79-13. International Institute for Applied Systems Analysis, Laxenburg, Austria.

Vansteenkiste, G.C. (Editor) (1975). Computer Simulation of Water Resources Systems. North-Holland, Amsterdam, p. 686.

Vansteenkiste, G.C. (Editor) (1978). Modeling, Identification and Control in Environmental Systems. North-Holland, Amsterdam, p. 1028.

Wellstead, P.E. (1976). Model order testing using an auxiliary system. Proceedings of the Institution of Electrical Engineers, 123:1373–1379.

Wellstead, P.E. (1978). An instrumental product moment test for model order estimation. Automatica, 14:89–91.

White, K.E., Lee, P.J., and Belcher, A.S.B. (1980). Time-of-travel and its significance. In M.J. Stiff (Editor), River Pollution Control. Ellis Horwood, Chichester, pp. 275–288.

Whitehead, P.G. (1979). Application of recursive estimation techniques to time-variable hydrological systems. Journal of Hydrology, 40:1–16.

Whitehead, P.G. (1983). Modeling and forecasting of water quality in non-tidal rivers: the Bedford-Ouse study. In M.B. Beck and G. van Straten (Editors), Uncertainty and Forecasting of Water Quality. This volume, pp. 321–337.

Whitehead, P.G. and Young, P.C. (1979). Water quality in river systems: Monte Carlo analysis. Water Resources Research, 15(2):451–459.

Whitehead, P.G., Young, P.C., and Hornberger, G.M. (1979). A systems model of flow and water quality in the Bedford-Ouse River, Part I: streamflow modelling. Water Research, 13:1155–1169.

Woodside, C.M. (1971). Estimation of the order of linear systems. Automatica, 7:727–733.

Young, P.C. (1974). A recursive approach to time-series analysis. Journal of the Institute of Mathematics and its Applications, 10:209–224.

Young, P.C. (1978). General theory of modeling for badly defined systems. In G.C. Vansteenkiste (Editor), Modeling, Identification and Control in Environmental Systems. North-Holland, Amsterdam, pp. 103–135.

Young, P.C. (1979). Self-adaptive Kalman filter. Electronics Letters, 15(2):358–360.

Young, P.C. (1983). The validity and credibility of models for badly defined systems. In M.B. Beck and G. van Straten (Editors), Uncertainty and Forecasting of Water Quality. This volume, pp. 69–98.

Young, P.C., Jakeman, A.J., and McMurtrie, R.E. (1980). An instrumental variable method for model order identification. Automatica, 16:281–294.

THE VALIDITY AND CREDIBILITY OF MODELS FOR BADLY DEFINED SYSTEMS

Peter Young*
Centre for Resource and Environmental Studies, Australian National University, Canberra (Australia)

1 INTRODUCTION

If the dictionary definition were the sole criterion, a model would be considered valid if it was found to be well grounded, sound, cogent, logical, and incontestable. Similarly, a model would be deemed credible if it was deserving of or entitled to belief, or if it was plausible, tenable, or reasonable. All of these characteristics are, of course, desirable in a mathematical model of a physical system; but when used as the basis for the definition of model adequacy, they are clearly too subjective to provide useful and rigorous criteria for model evaluation.

In this paper, we will consider validity and credibility as desirable properties of a model which should follow from close adherence to a systematic and comprehensive model-building procedure. This systematic procedure is evolved naturally when the model-building problem is considered within the hypothetico-deductive interpretation of the scientific method (see, for example, Popper, 1959), and it forms the basis of a "method theory" for modeling dynamic systems which would appear to have wide applicability (Young, 1977). This method theory is concerned not only with mathematical analysis techniques but also with the successful integration of mathematical analysis and data collection in whatever form may be most appropriate, whether it be active field and laboratory experimentation or passive monitoring exercises.

Within this setting, model validity — or, more correctly, conditional validity — follows the satisfactory outcome of a validation phase of the analysis in which attempts to falsify the model as a theory of system behavior are found to be unsuccessful. In contrast, model credibility is a property which depends upon success in all phases of the model-building procedure from model formulation through model structure identification to parameter estimation and validation. In this sense credibility remains a somewhat subjective concept: I believe, however, that whether a model is credible or not will always

---

* Present address: Department of Environmental Science, University of Lancaster, Lancaster LA1 4YQ, UK.

depend, to some extent, on the background of the adjudicator. For example, if the scientific establishment is firmly committed to a particular type of model for a physical system, then it may at first be difficult for the systems analyst to gain acceptance and credibility for a less conventional representation, even if the model-building procedure has been rigorous and is seen to conform with the basic tenets of the scientific method.

The modeling procedure described in this paper can be applied to any dynamic system, whether man-made or naturally occurring, but it is designed specifically for systems which can, in some sense, be considered badly defined. This poor definition usually arises for two major reasons. First, the size and complexity of many natural systems, such as those encountered in environmental and economic research, are such that the mechanisms governing the change in the observed system variables and their interrelationships are rarely fully understood *a priori*. There can, in other words, be a basic ambiguity; a situation in which a number of possible explanations for the observed behavior seem feasible but where little *a priori* evidence exists as to which of these explanations is the most plausible. The monodisciplinary expert may well be able to list the various mechanisms that could be operative in that part of the system with which he is acquainted, but he will be unlikely to proffer advice on which of these mechanisms are likely to dominate the behavior of the entire system under study.

Such limitations in *a priori* knowledge are not, of course, restricted to complex natural systems: research into many physical and biological processes often starts from a position in which little prior knowledge is available. But in the case of complex natural systems the position is exacerbated by a second problem — the difficulty, if not impossibility, of performing planned experiments. This difficulty, which is perhaps the major difference between research in the environmental and social sciences and research in the more conventional physical and biological sciences, is compounded by the associated problems entailed in even collecting adequate quantities of *in situ* data during the "normal operation" of such systems. In addition, normal operational data, even when available, are likely to be scarce and subject to some degree of uncertainty. Thus, the analyst with prior experience in the physical sciences might expect to improve this scant *a priori* knowledge of the system by careful analysis of reasonable-quality, planned experimental data to yield a much improved *a posteriori* situation. In practice, however, he is forced into the position of attempting to explain the troublesome ambiguities by reference to very restricted observational sets using conventional, and not necessarily appropriate, analytical techniques.

Faced with the dilemma of the badly defined process, systems analysts have attempted various different approaches to the problem. These approaches are often dictated almost completely by the specific background experience of the research worker, and, for this reason, there seems to be no unified approach, merely a collection of *ad hoc* procedures with various degrees of sophistication and complexity. To the outside observer at least, the only thing the procedures seem to have in common is their use of highly esoteric mathematics and an inevitable dependence upon the electronic computer. Certainly it is the complicated, although sometimes rather naive, computer-aided exercises in model-building and systems analysis that have attracted most attention in recent years and that seem most to typify the "systems approach". Furthermore it is these same exercises that have contributed most to the recent, widespread criticism of applied systems analysis and environmental model-building (see, for example, Hoos, 1972; Brewer, 1973; Ackerman et al., 1974; Philip, 1975; Berlinski, 1976).

There seem to be two main reasons for the present rather unsatisfactory state of affairs. First, the applied mathematicians who have taken a major part in the development of systems methodology appear sometimes to be attracted more by the elegance of the mathematical tools rather than the need to solve adequately the real problems at hand. Second, it is all too easy to use a computer, particularly for the "simulation modeling" of complex dynamic system behavior, and to forget the true nature of the problem in such exercises: the complex simulation model and the subsequent analysis are defined within the confines of the available mathematical methods and computer programs, while the many statistical problems of model-building often receive only cursory attention. It is almost as if the model has become more important than the problem at hand, even that the model in some sense *is* the system.

This problem is exacerbated by a tendency for most simulation-modeling methodology to be based on a rather restricted "reductionist" philosophy. Here the system is repetitively subdivided into elemental components that are assumed to have physical significance to the modeler and can be analyzed as relatively separate entities. Having evaluated the "physical" parameters associated with each of the elemental models, usually by experimentation either *in situ* or in a laboratory, the modeler then reassembles the model components in a manner which he and his advisers perceive to be appropriate, with the numerical values for the parameters inserted in accordance with this preconceived, but usually untested, perception of the system and its behavioral mechanisms.

Such a reductionist approach is rarely, however, accompanied by sufficient evaluation of the resulting model as a complete entity. "Holistic" validation (see, for example, Rigler, 1976) is normally restricted to exercises in deterministic "model fitting" in which overall "calibration" of the model is achieved using manual or automatic methods of parameter "tuning" or "optimization"; an approach that is sometimes enhanced by deterministic sensitivity analysis* in which the sensitivity of the model outputs to variations in the parameters is examined using various analytic procedures (see, for example, Miller et al., 1976).

Although such analysis is perfectly respectable, it must be used very carefully; the dangers inherent in its application are manifold, but they are not, unfortunately, always acknowledged by its proponents. It is well known that a large and complex simulation model, of the kind that abounds in current ecological and environmental system analysis, has enormous explanatory potential and can usually be fitted easily to the meager time-series data often used as the basis for such analysis. Yet even deterministic sensitivity analysis will reveal the limitation of the resulting model; many of the "estimated" parameters are found to be ill defined and only a comparatively small subset is important in explaining the observed system behavior.

Of course, over-parameterization is quite often acknowledged, albeit implicitly, by the reductionist simulation model-builder. Realizing the excessive degrees of freedom

---

* Stochastic sensitivity analysis, in which sensitivities are calculated in relation to stochastic variations in the parameter, usually by resort to Monte Carlo analysis, is preferable (see Section 2.1) but is not currently very popular, probably because it demands more comprehensive data analysis.

available for fitting the model to the data, he will often fix the values of certain "better known" parameters and then seek to fit the model by optimizing the chosen cost function (usually the sum of the squares of the difference between the model outputs and the observations) in relation to the remaining parameters only, and these are normally few. In this manner, the analyst ensures that the cost function–parameter hypersurface is dominated by a clearly defined optimum (a minimum in the least-squares case), so that estimation of the parameters which define the optimum becomes more straightforward.

But what is the value of this optimization exercise in relation to the specification of the overall model? Clearly a lower-dimensional parameter space has been located which allows for the estimation of a unique set of parameter values. However, this has been obtained only at the cost of constraining the other model parameters to fixed values that are assumed to be known perfectly and are defined in relation to the analyst's prior knowledge of the system. As a result, the model has a degree of "surplus content" not estimated from the available data, but based on a somewhat *ad hoc* evaluation of all available prior knowledge of the system and colored by the analyst's preconceived notions of its behavioral mechanisms.

On the surface, this conventional simulation-modeling approach seems quite sensible; for example, the statistician with a Bayesian turn of mind might welcome its tendency to make use of all *a priori* information available about the system in order to derive the *a posteriori* model structure and parameters. On the other hand, he would probably be concerned that the chosen procedures could so easily be misused: whereas the constrained parameter optimization represents a quantitative and relatively objective approach, it is submerged rather arbitrarily within a more qualitative and subjective framework based on a mixture of academic judgment and intuition. Such a statistician would enquire, therefore, whether it is not possible to modify this framework so that the analyst cannot, unwittingly, put too much confidence in *a priori* perceptions of the system and so generate overconfidence in the resulting model.

Consideration of the modeling problem from this kind of Bayesian statistical standpoint is the stimulus behind the present paper. The need to choose a model that is efficiently parameterized and compatible with the identifiability of the system (in relation to the available data) is a major requirement of the model-building procedure discussed here: it is clearly foolhardy to attempt the statistical estimation of parameters in a model if the model has excess content (in the form of surplus structure and/or parameters) which cannot be validated against the observed data. However, it is possible to blend the *a priori* information on the system and the subsequent analysis of the time-series data into an objective model-building exercise aimed specifically at either obviating these difficulties or, at least, identifying where the limitations of the resulting model may reside. In this manner, the main impediments to the use of the model, either as a predictive device or for control and management system design, will often become more apparent and the possibility of its misuse in such applications will be minimized.

Given the Bayesian stimulus behind the proposed model-building procedure, it is appropriate that the main statistical tool used in the analysis suggested here can be considered as the physical embodiment of Bayesian estimation, namely the "recursive" or "probabilistic iterative" estimation algorithm (see for example, Young, 1976b, 1981).

Recursive estimation in its simplest recursive, least-squares form was first developed at the beginning of the nineteenth century by the famous mathematician K.F. Gauss and described in his collected works (1821–1826), which appeared under the title *Theoria Combinationis Erroribus Minimum Obnoxiae* (Bertrand, 1855; also see, for example, Sprott, 1977; Young, 1981). In the *Theoria Combinationis,* Gauss shows how it is possible "to find the changes which the most likely values of the unknowns [the parameter estimates] undergo when a new equation [observation] is adjoined and to determine the weights [standard errors] of these new determinations" (with our comments in square brackets). In other words, and to utilize more contemporary terminology, he developed a statistical algorithm for sequentially or recursively updating the least-squares estimates on receipt of additional data.

The recursive methods of time-series analysis used in the present model-building procedure are logical successors to the algorithms of Gauss and also owe much to later work on recursive estimation by Plackett (1950) and Kalman (1960). In their latest form (Young, 1976a; Jakeman and Young, 1979; Young and Jakeman, 1979, 1980) they represent a general recursive method of time-series analysis for time-series models of the "errors-in-variables" type (see for example, Kendall and Stuart, 1961). As such they provide a robust and generally applicable procedure for identifying and estimating parametric change in stochastic models of dynamic systems.

In subsequent sections we will see how these recursive methods of time-series analysis can be of major importance in both the identification of an efficiently parameterized model structure and the estimation of the parameters which characterize this structure. In this sense, they provide the methodological cornerstone of the proposed systematic approach for modeling badly defined systems.

2   THE PHASES OF MODEL-BUILDING

If it is to be formulated in accordance with the scientific method, a model-building procedure must start with an analytical phase aimed at generating working hypotheses about the nature of the system under study. Normally, as indicated in Figure 1, such hypotheses will themselves be in the form of mathematical models which (a) attempt to embody all prior information and knowledge about the system, (b) make use of all appropriate historical data, and (c) are related closely to the objectives of the model-building exercise.

2.1   Model Formulation and the Generation of Working Hypotheses

In the early stages of an investigation into a badly defined system, time-series data are likely to be scarce. In this situation, the only way to progress is to utilize some form of simulation model in the hypothesis-generating role. "Simulation model" here implies one whose structure and parameters are explicitly related to the physical, chemical, biological, or socioeconomic processes that are assumed, *a priori*, to characterize the system.

If the selected simulation model is relatively simple and parsimonious in its parameterization, then its use as a hypothesis-generating device is fairly straightforward, as will be seen in later sections. Indeed, such a simulation model could form a basis for

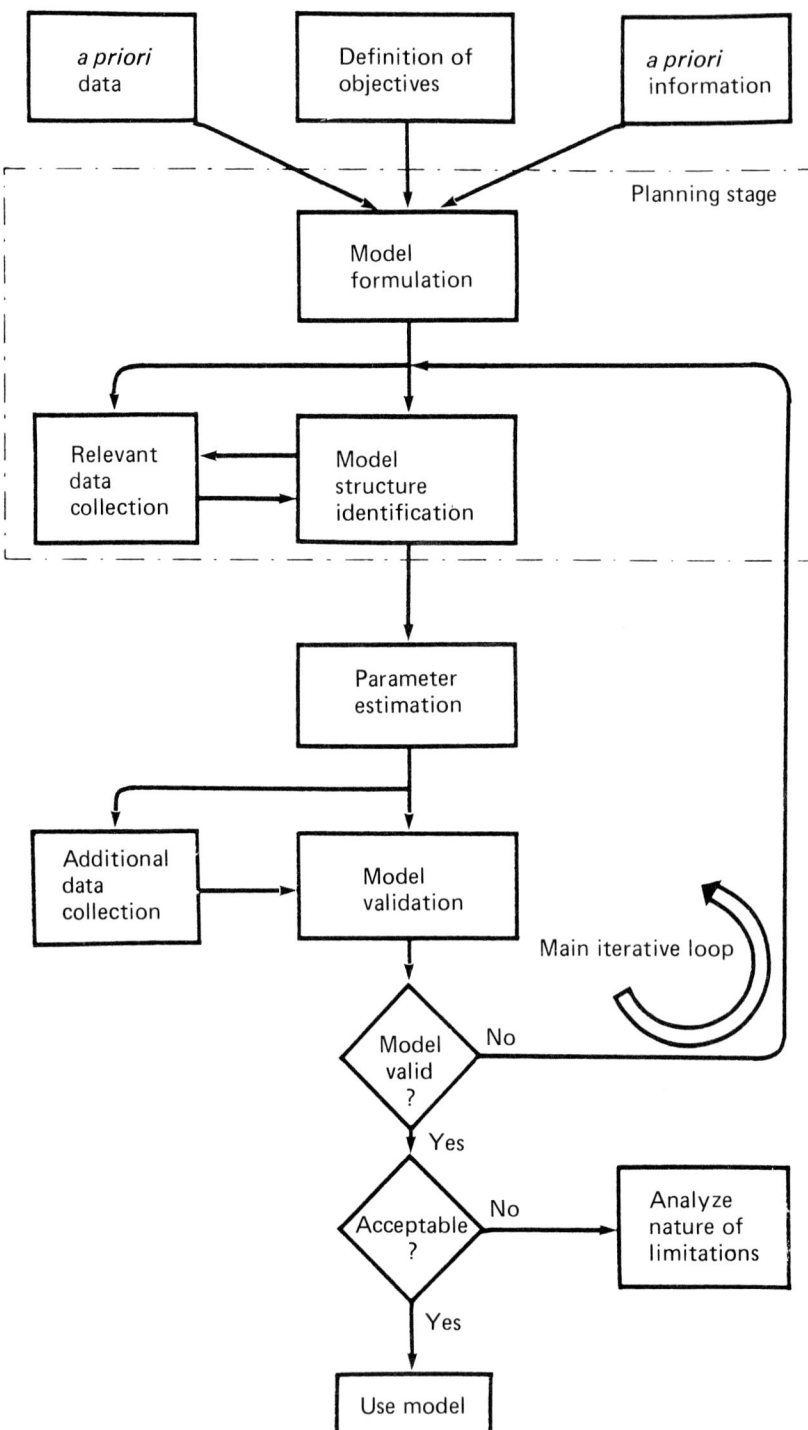

FIGURE 1 The model-building procedure.

the subsequent exercises in model-structure identification and parameter estimation shown in Figure 1. But if the model is based on normal reductionist principles then it may, as pointed out in the previous section, be rather complex, with many parameters, state variables, and nonlinear relationships.

I maintain that such complex models can only be useful in the initial analysis of badly defined systems if they are considered within a probabilistic context. That is, given the model and the inherent uncertainties in structure and parameter values, the only meaningful and safe analyses must focus on the probabilities of various behavioral patterns. Most importantly, they must focus on the probable structures and parametric relationships that appear consistent with the dominant modes of behavior associated with the "problem" under consideration (Young, 1977).

If this notion is formalized, it implies that the simulation model should not be evaluated as a single entity; in other words, not as a fixed structure characterized by a set of constant parameters defined in terms of point-estimates. Rather the parameters should be considered as inherently uncertain and, therefore, definable only in terms of statistical probability distributions. This Bayesian interpretation leads naturally to the study of a whole ensemble of models defined by the various selected structures and their associated parametric probability distributions.

In order to pursue this idea, consider a general class of systems which can be represented by the following, nonlinear, state-space differential-equation representation of the system in continuous time:

$$dx/dt = \dot{x} = f_i(x, \alpha, u^c, u^d, \xi, t) \tag{1}$$

Here,

t = time;
$x = [x_1, x_2, \ldots, x_n]^T$ is an n-vector of state variables which describe the system behavior in the "state space";
$\alpha = [\alpha_1, \alpha_2, \ldots, \alpha_q]$ is a vector of (possibly time-variable) parameters or coefficients which characterize the system in the state space;
$u^c = [u_1, u_2, \ldots, u_m]^T$ is an m-dimensional "control input" vector whose elements are capable of manipulation in some manner;
$u^d = [d_1, d_2, \ldots, d_l]^T$ is an l-vector of deterministic* but uncontrollable disturbances which affect the system; and
$\xi = [\xi_1, \xi_2, \ldots, \xi_n]^T$ is an n-vector of stochastic disturbances whose statistical properties may or may not be known, depending on the level of *a priori* information available about the system.

It will be noted that eqn. (1) is an ordinary differential equation or "lumped-parameter" representation, which can be contrasted with the partial differential equation or

---

* In the present context "deterministic" is used loosely to mean measurable in some manner; such disturbances are equivalent to the "exogenous" variables of socioeconomic systems (see, for example, Johnston, 1963).

"distributed-parameter" alternative which is more popular in some areas of environmental systems analysis. This reflects my view that the lumped-parameter description is of greater practical utility in systems analysis aimed at solving control or management problems; it also serves to emphasize that it is the modeling of badly defined systems for such control applications that is the principal concern of this paper.

The vector function f_i in eqn. (1) is nominally nonlinear and nonstationary — assumptions which reflect the idea that badly defined systems will, in general, exhibit nonlinear and possibly changing behavioral patterns. The continuous-time formulation is chosen because many physical relationships, such as mass or energy conservation laws, are stated more naturally in continuous time; thus it is likely that *a priori* assumptions about the nature of physical environment problems, for example, will fit more easily within the continuous-time framework. This is not a limiting assumption, however, since it is straightforward to consider the representation of system (1) in discrete time (see Young, 1977).

In relation to eqn. (1), the probabilistic approach requires the evaluation of model behavior for different vector functions f_i with the associated parameter vector $\hat{\alpha}$ represented in terms of the probability distribution which is chosen to encompass the complete range of "possible" values for the coefficients that compose the vector. In addition, because the disturbance vector ξ allows for random disturbances to the system, it is clearly necessary to allow for this input uncertainty in the evaluation of the model. It might also be desirable, depending upon the circumstances, to consider the model behavior for different deterministic inputs u^c and u^d, which are representative of the kind of inputs met in practice. For example, if u^d represented a vector of rainfall inputs to a river water-quality model, then "wet", "dry", and "average" conditions could be accommodated with different representative deterministic sequences (see, for example, Whitehead and Young, 1979).

This conceptual base of an uncertain or stochastically defined simulation model can be exploited in methodological terms by recourse to Monte Carlo simulation analysis. Put simply, such analysis consists of repeated solution of the model equations with the uncertain parameters and inputs specified by sampling at random from their assumed parent probability distributions. This analysis results in a large number of random simulations (or realizations), each providing a unique state trajectory $x(t)$. The set of trajectories is then examined statistically to investigate the properties of the whole ensemble of simulation models; that is, statistical procedures are utilized to infer certain properties of the ensemble from the finite sample of trajectories obtained from the random simulation experiments. It is, in other words, a method of bypassing the difficulties associated with the analytic solution of nonlinear, stochastic differential equations, albeit at some cost in computational terms.

The general aspects of the use of Monte Carlo methods to investigate the properties of an ensemble of simulation models are discussed by Spear (1970). Monte Carlo methods have been used previously in environmental and socioeconomic systems analysis (see, for example, Barrett et al., 1973; Young et al., 1973; Whitehead and Young, 1979) but, in these earlier approaches, the ensemble properties were considered mainly in terms of the propagation through time of the probability distribution associated with the state trajectory $x(t)$ itself. Here, an alternative procedure is proposed in which the state trajectory $x(t)$ obtained from each randomly selected solution of the model equations is examined to see if it is characterized by a behavioral pattern relevant to the problem under consideration. For example, in a socioeconomic simulation, the occurrence of high inflation simultaneously with a stagnant economy and high unemployment would define the existence, for that run, of the "stagflation" problem. If the state trajectory does appear to exhibit a problem behavioral pattern, then it is considered that the model parameter vector $\hat{\alpha}$ does give rise to "the behavior B": alternatively, if $x(t)$ does not exhibit such characteristics, then α is associated with "not the behavior $\bar{B}$". The end result is N simulation runs in which M parameter vectors led to the behavior and $N-M$ did not.

This kind of Monte Carlo analysis is described in detail by Spear and Hornberger (1978). For present purposes, it suffices to say that the aim is to ascertain which elements of vector α are important in giving rise to the problem behavior. This is achieved by evaluating the sample cumulative probability distributions associated with these elements (the model parameters) in both the "behavior set B" and the "not the behavior set $\bar{B}$". A parameter is then deemed important if there is a statistically significant difference between the two distributions and not important if this difference is statistically insignificant. The two procedures for assessing the significance of differences in this sense are, first, the application of conventional nonparametric tests such as the Kolmogorov–Smirnov two-sample test and the Mann–Whitney test (see, for example, Spear, 1970), and second, the use of principal-component methods based on eigenvalue–eigenvector analysis of the covariance matrices associated with the parameter vectors (see, for example, Kittler and Young, 1973).

Evaluation of the results of the Monte Carlo analysis in the above manner should yield a better understanding of the system in terms of those mechanisms and parameters that appear important to the problem at hand. Such additional insight can be useful in a number of ways. Most importantly, it can lead to the specification of hypotheses about the system behavior that can be tested by further planned experiments or monitoring exercises; in other words, it can help in the planning of further data collection in the study of the system. As will be seen, it can also indicate to the analyst the possible dominant modes of behavior of the system, information that is of crucial importance in subsequent stages of model-building.

Of course, since the analysis is not limited to a single *a priori* model structure, it may result in a number of different hypotheses, each of which will need to be tested, and a number of possible "dominant mode" descriptions, each of which will need to be evaluated during subsequent time-series analysis. It is seen, therefore, that the Monte Carlo methodology is indeed a very effective hypothesis-generating device which is based on a relatively objective analysis of all *a priori* information available about the system. This latter point helps to emphasize that the simulation models developed in this initial stage of model-building should not be considered in the same light as more conventional deterministic simulation models. Also, for the reasons outlined in the previous section they will rarely, in themselves, form the basis for subsequent exercises in time-series analysis. In the case of badly defined systems, I strongly advocate that mechanistic simulation models should be viewed principally (although not entirely) within the ensemble context. As such, their use in time-series terms is mainly as a vehicle for indicating dominant-mode mechanisms and descriptions — descriptions which will, in general, be much simpler than the original simulation-model description and can,

2.2 Time-Series Model Identification and Estimation

Time-series analysis is a systematic, statistically based approach to the problem of model development which provides an objective method of constructing both black-box (input–output) and mechanistic (internally descriptive) models from time-series data. Despite assertions to the contrary, time-series analysis is much more than just model "fitting", as currently practised in many areas of systems analysis and simulation modeling. This is emphasized by the fact that, in time-series analysis, the degree to which the model fits the data is not, in itself, used as an indication of model adequacy: other factors, such as the estimated uncertainty on the model parameters, are equally important and are, as we shall see, an indispensable part of the analysis (see Young et al., 1980).

The use of time-series methods as the basis for modeling badly defined dynamic systems has been described at length by Young (1977). As in the previous subsection, therefore, the procedures will not be discussed in depth. Rather, an attempt will be made to explain their role in the overall modeling process and to show how they follow quite naturally from the initial Monte Carlo based model-formulation and identification exercises.

It has been shown how the stochastic simulation-model experiments can reveal in the assumed model those parameters which appear important in relation to the "problem" behavior under consideration. In this way, these experiments can also help the analyst to appreciate better the relative importance of the various dynamic mechanisms in the model, to a point where he is able to identify those dynamic modes of behavior that seem dominant in characterizing the problem. Young (1977) suggested that it is these dominant modes of behavior that are important in the subsequent time-series analysis for, if the model is indeed representative of the system, then it is these modes that will be most "identifiable" from the observed data.

There is no proof at present that such a dominant-mode theory of dynamic behavior is generally applicable but experience with practical dynamic systems suggests that it is a reasonable conjecture; indeed it could be argued that the definition of a "problem" behavior is, in itself, an acceptance of some form of modal dominance. But whether or not the analyst subscribes to such a theory he will, in any specific case, be able to examine the model for evidence of such behavior. As will be shown in a subsequent example, evidence of this type can be obtained by quite straightforward exercises in systems analysis applied to one of the model realizations that exhibit the problem behavior. This may entail both evaluation of the model structure (for example, by linearization) and analysis of the model response $x(t)$ (for example, using time-series methods).

In effect, this analysis of the model in systems terms is aimed at testing the hypothesis of modal dominance. If the hypothesis is confirmed (as I feel it will be most of the time) then the analyst will have obtained some ideas about possible simple forms of the model which can be used as the basis for further time-series analysis on data from the system itself. If there appears to be no evidence to support the hypothesis (which

I feel is unlikely in general) then the analyst will be no worse off and he should at least have a better appreciation of the dynamics of the simulation model*.

The coordinated systems-analysis–modeling–data-collection strategy whose virtues are extolled here should mean that, concurrent with the simulation modeling and systems analysis, exercises in relevant data collection will have been planned and initiated. When these data on the system become available they will allow the analyst to progress one step further in his model-building: namely to the identification of a suitably identifiable, time-series representation of the system.

A simple dominant-mode characterization of the simulation model provides an ideal starting point for time-series analysis, the first stage of which is aimed at identifying a dynamic model of the dominant modes associated with the system itself. In other words, having tested the hypothesis that the simulation model can be represented simply in dominant-mode terms, I suggest that the analyst should now proceed to test the hypothesis that such representations are appropriate to the real system. The result of this analysis is the identification of a time-series model structure which may be linear or nonlinear in dynamic terms, depending upon the nature of the system. It will, however, normally be characterized by a small set of unknown parameters which need to be estimated during the subsequent parameter-estimation phase of the analysis.

The methodology of time-series model structure identification suggested by Young (1977) is based on the use of recursive estimation procedures. The model structure is then considered "well identified" if it simultaneously satisfies the following requirements:

(1) The recursive estimates of assumed time-invariant parameters are themselves indicative of time-invariance *and* the estimates of assumed variable parameters have direct physical interpretation.
(2) The covariance matrix of estimation errors associated with the estimated parameters indicates that there are no problems of over-parameterization.
(3) The model structure satisfies certain statistical identification criteria based on test statistics associated with its ability both to explain the time-series data and, at the same time, to possess well-defined parameter estimates — these identification criteria and their application are discussed fully by Young et al. (1980).
(4) The estimated stochastic disturbances $\hat{\xi}$ are purely stochastic in form and have no systematic components attributable to some physical aspects of the system behavior.
(5) The residual-error sequence or "innovations" process associated with the model possesses "white noise" properties and is statistically independent of the deterministic inputs u^c and u^d.

---

* I would go further and suggest that this kind of systems analysis applied to the simulation model is a *sine qua non* for success in any simulation-modeling exercise applied to a badly defined system: it would certainly help to avoid some of the more naive exercises in simulation modeling that currently abound in the literature. An excellent example of its value is the analysis of the Forrester world model by the "Globale Dynamica" Group at the University of Eindhoven (see, for example, Rademaker, 1973; Thissen, 1978).

Put simply, this identification analysis is aimed at producing a model structure which has a satisfactory physical interpretation, and is identifiable from the available time-series data in the sense that it can be characterized by a unique set of well-defined parameter estimates. Implicit in these specifications is the requirement for a parametrically efficient model representation and the avoidance, therefore, of over-parameterization.

The evaluation of the recursive estimates is a most important aspect of the above identification analysis. Significant variation in the estimated parameters can arise for three main reasons:

(a) The dynamic behavior of the system is changing over the observation interval (i.e., the system is nonstationary).
(b) There are nonlinearities associated with the system behavior but not present in the mathematical model.
(c) The model is over-parameterized so that the parameters are poorly defined — as a result, the recursive estimates tend to "wander" along the indeterminate valley-like surfaces which characterize the criterion function–parameter hyper-surface in this over-parameterized situation (in this situation the model always fits the data well, with high coefficients of determination, but its parameters are characterized by high levels of uncertainty).

The identification procedure in stages (1)–(3) above is designed to eliminate the occurrence of (c) by removing the possibility of over-parameterization. Any remaining recursively estimated parameter variations are then examined to discover whether they are statistically significant and, if so, whether they arise because of (a) or (b), or a mixture of both. The model structure is then modified accordingly until the existence of a set of well-defined and, if possible, constant-parameter estimates is established.

Estimation of the parameters that characterize the model structure finally identified is a fairly straightforward exercise and numerous estimation procedures can be utilized (see, for example, Young, 1981). As pointed out in Section 1, however, the recursive instrumental variable (IV) techniques provide what appears to be currently the most flexible approach (see, for example, Young, 1974, 1976a; Jakeman and Young, 1979; Young and Jakeman, 1979, 1980) since they are statistically sophisticated yet robust and simple in application terms. In addition, as we have seen, their recursive formulation makes them useful in the previous identification phase of the analysis (Young, 1977). Also, recursive smoothing versions of the IV algorithms are now available (Kaldor, 1978; Young and Kaldor, 1978) which can enhance still further the time-varying estimation potential of the IV method.

Whatever estimation procedure is used, however, the result will be a set of estimates of the parameters characterizing the identified dominant-mode model structure, together with some indication of the uncertainty associated with these estimates (usually in terms of an estimation error covariance matrix). If the identification and estimation analysis has been successful then this set of estimates should represent either a low-dimensional or, in the best circumstances, a minimum-dimensional representation of the time-series data. In other words, the analysis will represent a data-reduction exercise in which the useful information in the data has been compressed into a few important and well-defined constant-parameter estimates. Moreover, these estimates will normally have direct physical significance because of their interpretation within the physically meaningful dynamic model structure. This could help substantially in establishing the overall credibility of the model and should aid in its use as a tool in control and management system design.

2.3 Times-Series Model Validation

The final and continuing stage in model-building is validation; here the model's forecasting ability is evaluated on data other than those used in the identification and estimation studies. If the model continues to forecast well over this test data interval, it is assumed that it is conditionally acceptable in the sense that, as far as it is possible to test it, the model seems satisfactory.

Validation is a continuing procedure since the model will need to be reassessed in the light of future developments and additional data. If major changes in the system take place, for instance, it is likely that the model will need to be modified because it will not necessarily mirror the changed dynamic behavior in the new situation. Nevertheless, the continuing process of model assessment based upon a supply of new data should indicate whether the model has become questionable in any sense and will, in these circumstances, indicate the need for further model identification, estimation, and validation. In such a situation, the recursive nature of the estimation algorithms will greatly facilitate the process of model reassessment, a process which could entail simply updating the model parameters, but which might require changes in the basic model structure.

The inherently stochastic nature of the model discussed in the previous section also helps considerably in the continuing process of model assessment because it allows for the application of statistical tests regarding the model's suitability. Such tests can help to remove some of the more subjective judgments which are often encountered with conventional model-building procedures.

Of course, the only real validation of a model is that it satisfies the purposes for which it was intended, in other words, that it "works" in practice. It is hoped that by going through the systematic procedures suggested here the analyst will maximize the probability that the model will be acceptable in this sense. But this can never be guaranteed in the case of badly defined systems; the analyst must, unfortunately, "wait and see".

2.4 Model-Building and the Scientific Method

Before discussing practical examples, it is worthwhile stressing the relationship between the model-building procedure discussed in previous sections and the scientific method (see, for example, Popper, 1959). Model formulation is simply the formulation of hypotheses about the nature of the system; model structure identification and parameter estimation represent initial steps in the deductive procedure that is used to test these hypotheses against data; model validation is the final step in that deductive procedure in which the analyst attempts to "falsify" the model (or theory) of behavior and accepts the conditional validity of the model if such attempts fail.

It may seem trite and obvious to draw this analogy but it seems necessary. Often simulation modeling of badly defined systems, as at present practised, does not necessarily conform to the principles of the scientific method; indeed I regard this limitation of many simulation-modeling exercises as the main reason for their failure to be fully successful in practical terms.

Too often, reductionist philosophy is misused in simulation modeling: the model structure is assumed to be known *a priori* and all subsequent analysis accepts this assumption, usually without serious question. Thus, the hypothesis that the model structure is correct is not tested adequately, the checks and balances of the scientific method tend to be bypassed, and surplus content within the assumed model structure is clearly a possibility. By pursuing the approach discussed here, however, serious consideration of the model structure in statistical terms becomes an important aspect of the analysis and the likelihood that the model structure is incorrect is minimized. At the very least, it should be possible for the analyst using this line of approach to identify and be aware of any surplus content and make allowance for this in any subsequent use of the model.

Note that I am not advocating here the elimination of speculative simulation modeling. On the contrary, there is no doubt that such modeling can be an extremely useful tool in applied systems analysis. Rather I am warning of the dangers inherent in the blind use and acceptance of such models, and emphasizing the need for greater care both in the development of simulation models for badly defined systems and the interpretation of the results obtained from these results.

3 APPLICATION OF THE MODELING PROCEDURE

The efficacy of a particular analytical approach to a complex problem can only be properly evaluated by applying it to problems that are meaningful in practice. In this section, a number of practical examples are discussed which illustrate how the model-building procedure described in previous sections has worked in practice. In these examples, all phases of the procedure have been tested individually but a complete exercise involving all phases applied to a single problem has not yet been undertaken. This emphasizes the point that not all problems will demand application of the entire procedure: depending on the level of *a priori* knowledge, information, and data, it may well prove possible to achieve the stated study objectives by more limited modeling activities involving only certain phases of the overall procedure. For example, if the system is relatively well defined in relation to the study objectives, it should be possible to dispense with the probabilistic simulation-modeling phase and put greater emphasis on data collection and time-series analysis.

3.1 The Use of Probabilistic Simulation Modeling for the Generation of Working Hypotheses on Macroalgal Growth in an Estuarine System

In this example, the system in question is the Peel–Harvey Estuary in Western Australia and the "problem" behavior is the excessive growth of the green alga *Cladophora*

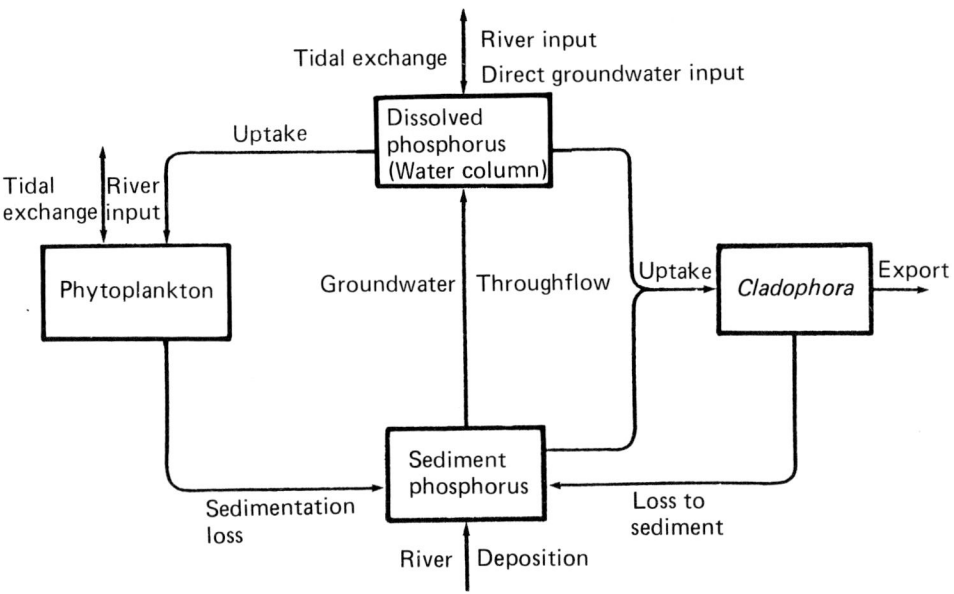

FIGURE 2 Phosphorus budget model for *Cladophora*: schematic block diagram.

that has occurred in recent years, creating certain objectionable conditions in the Peel Inlet. Between 1976 and 1980, a study team composed of a number of experts from disciplines ranging from hydrology through soil science and biology to applied systems analysis conducted scientific investigations under the sponsorship of the Western Australian Environmental Protection Authority's Estuarine and Marine Advisory Committee. The initial stages of systems analysis discussed here are, of course, only a small aspect of this larger study (Humphries et al., 1980).

The initial exercises in simulation modeling for this problem were aimed at characterizing the system under a "phosphorus scenario"; in other words, the simulation model was formulated on the basis of a phosphorus budget, under the hypothesis that phosphorus is the major nutrient of importance to the algal problem. It is these initial exercises in simulation modeling that will be considered here. It should be made clear, however, that the phosphorus scenario is not the only one that has been considered in the study: other research has been concerned with the evaluation of an alternative "nitrogen scenario". As has been emphasized, the simulation-modeling studies are "hypothesis-generating" exercises and should not, therefore, be overly restrictive in any sense.

The phosphorus budget model consists of four compartments: *Cladophora* x_1, phytoplankton x_2, soluble phosphorus in the water column x_3, and sediment phosphorus x_4. Two other equations describe the water and sediment volume balances. A schematic diagram of the whole system is shown in Figure 2. The equations for each compartment are described in detail by Spear and Hornberger (1978) and only the *Cladophora* equation, which exemplifies the model, will be considered here. The equation is nominally nonlinear and takes the form

$$\mathrm{d}x_1/\mathrm{d}t = \gamma_1 T I_b (X_c/(K_1 + X_c)) g(x_1) - a_{11} x_1 \qquad (2)$$

where

- x_1 = *Cladophora* biomass in terms of phosphorus content (μg);
- γ_1 = temperature–light–growth coefficient (cm$^2\,^\circ$C^{-1} cal^{-1} day^{-1});
- T = water temperature ($^\circ$C);
- I_b = total light at the bottom for the day (cal cm^{-2}),
  = $Ie^{-K_T Z}$, where I is actual surface-light intensity, K_T is the extinction coefficient, and Z is depth;
- X_c = available phosphorus concentration (μg l^{-1}),
  = $\alpha X_3 + (1-\alpha) X_4$, where α is a number between zero and one and X_3 and X_4 are phosphorus concentrations in water and sediment, respectively;
- K_1 = half-saturation (Michaelis) constant for phosphorus-uptake (μg l^{-1});
- $g(x_1)$ = biomass (phosphorus) available for active photosynthesis (μg) – this term is equal to x_1 for low values of biomass but asymptotically approaches a constant value x_m; and
- a_{11} = rate constant for biomass decrease from all causes, i.e., death, respiration, grazing, and export to beaches.

The various coefficients in eqn. (2) and the other model equations were either derived from the literature, inferred from measurements on the Peel Inlet system, or estimated by experts familiar with the system. The probability distributions associated with the parameters were chosen in accordance with the uncertainty in their specification: in general, they were chosen as rectangular distributions with limits selected to reflect reasonable upper and lower bounds on the parameter values. Environmental functions required to solve the equations from the specified initial conditions (e.g., temperature, irradiance, river discharge, tidal exchange) were mostly specified from existing data on the system collected during 1976, but sometimes were estimated by time-series analysis (e.g., in the case of tidal exchange).

The "problem" behavior was defined from prior knowledge of the system and was specified in terms of the state variables: in simple terms, it involved the simultaneous occurrence of high *Cladophora* biomass, low water-column nutrients, and low phytoplankton biomass. Figure 3 shows two of the variables, *Cladophora* biomass and total sediment phosphorus, as generated in three typical realizations from the Monte Carlo analysis. Also shown are plots of the irradiance I and the hydrograph of river inflow over the same period.

The Monte Carlo analysis entailed 626 random simulations of the model and out of these 281 exhibited the problem behavior. Statistical analysis of these results suggested not only that the important parameters were those connected with *Cladophora* growth (as might be expected) but also that those specifically connected with the *Cladophora* sediment-phosphorus interaction and self-limitation (due to self-shading) were particularly important. In addition, the analysis pointed to the special part played by the river input in maintaining the sediment phosphorus supply. In other words, the analysis generates the hypothesis that it is this pathway for phosphorus in the system that is dominant in causing the problem behavior. It also indicates that it is this hypothesis

Validity and credibility of models for badly defined systems 85

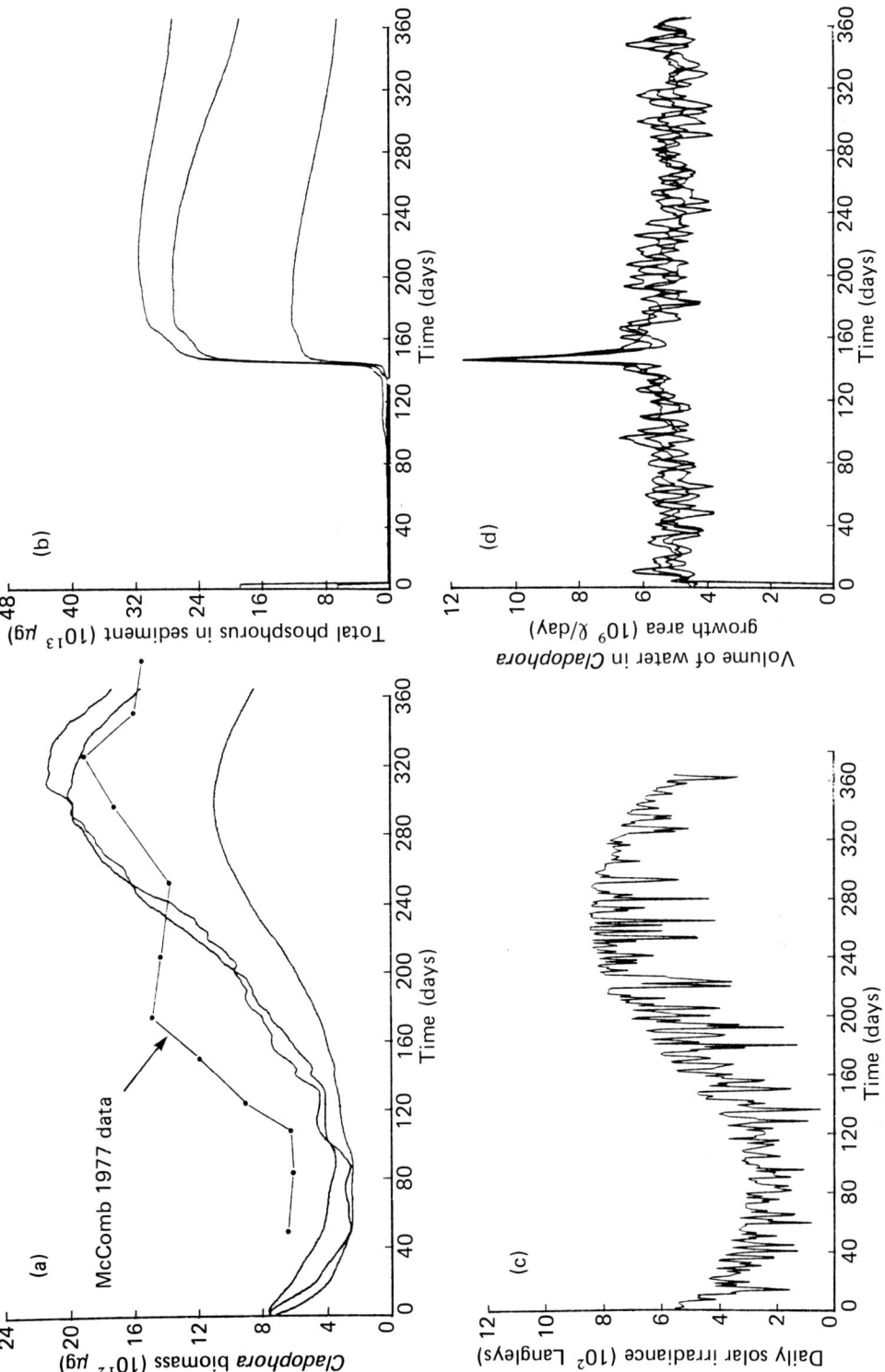

FIGURE 3 (a) Three random simulations (realizations) for *Cladophora* biomass compared with measured data. (b) Three random simulations (realizations) for total phosphorus in the sediment. (c) Variation in solar irradiance used in speculative simulation modeling. (d) Three random simulations (realizations) for water volume in *Cladophora* growth area.

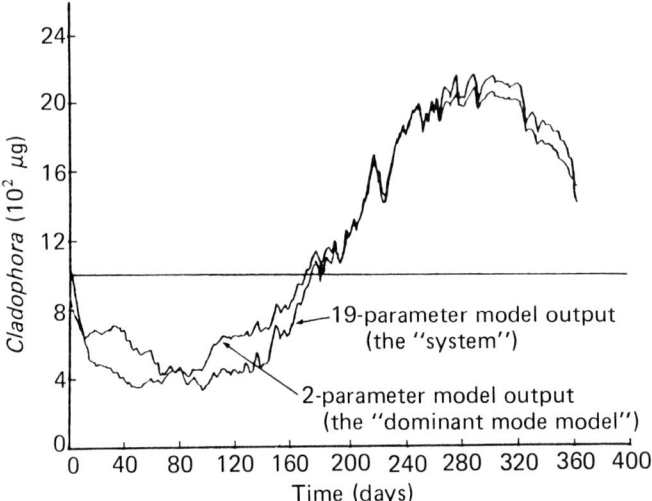

FIGURE 4 Comparison of simple two-parameter linearized time-series model output with 19-parameter nonlinear model output.

amongst all others connected with the phosphorus budget that seems to demand most attention and should be tested in subsequent field studies on the system.

An analysis of the model in systems terms also yields some interesting and useful results. For example, as might be expected from visual appraisal of graphs (a) and (c) of Figure 3, linearization of eqn. (2) shows that, for much of the time, the system behaves approximately as a first-order, linear system with irradiance I as the input and *Cladophora* biomass x_1 as the output. This is confirmed further by simple constant-parameter, time-series analysis of the model data. Figure 4, for example, illustrates the results obtained from such an analysis over the period shown in Figure 3; it is worth noting that the steady-state gain and time constant of this model are consistent with the linearization analysis. It is interesting to note also that recursive estimation of a time-variable parameter model yielded a perfect fit to this model data with the estimated coefficients exactly equal to the values obtained by the linearization analysis.

These dominant-mode results suggest that *Cladophora* growth in the model is controlled for much of the time by light limitation. Thus, while the sediment phosphorus is indicated as the important pathway through which the *Cladophora* receives its phosphorus nutrient inputs, phosphorus itself is unimportant in its effect on the modeled problem behavior, because it is almost always in plentiful supply.

The implications of these conclusions on time-series analysis are fairly serious. They mean that if the "phosphorus scenario" simulation model is representative of the real world, then observation of the system during "normal operation" will not necessarily supply much useful information on the dynamic relationship between phosphorus inputs and *Cladophora* growth. This in turn means that it would be difficult to identify and estimate time-series models for such interactions, which are of potential importance from the management standpoint. This has been confirmed by later analysis (Humphries et al., 1980).

3.2 The Flushing Dynamics of an Estuarine System

This example also derives from the Peel–Harvey study and is concerned with the evaluation of the flushing characteristics on the basis of monitored salinity variations at various sampling sites in the estuary.

There have been numerous attempts at modeling estuarine dynamics ranging from the very simple (e.g., Ellis et al., 1977) to the highly esoteric (e.g., Smith, 1980). In the Peel–Harvey study, an attempt was made to take an intermediate route and develop a model that was able to describe the behavior of the system in a manner appropriate to the requirements of the study, but without the fine detail normally demanded in more classical hydrodynamic analysis.

The system was decomposed into seven zones or compartments associated with the seven sampling sites (Figure 5) and each site was considered to be well mixed, in the sense that the sampled salinity was representative of the salinity in the compartment as a whole. By simple conservation-of-mass arguments, it can be shown that the equations controlling salinity in such a system will be of the form:

$$dS/dt = -S[(Q/V) + (1/V)(dV/dt)] + (Q_i S_i/V) \tag{3}$$

where

S = salinity;
Q = flow out of the compartment;
V = volume of the compartment;
Q_i = flow into the compartment; and
S_i = input salinity or forcing function to the compartment.

Since an adequate quantity of data (104 weeks) was available for analysis, as shown in Figure 6, time-series analysis was initiated directly in this case, with eqn. (3) providing a major motivation for the analysis. Equation (3) can be written in the form

$$dS/dt = -a(t)S + b(t)S_i \tag{4}$$

where $a(t)$ and $b(t)$ are nominally time-variable coefficients. Considering initially the relationship between salinity at Sites 2 and 1, a time-series model of the form (4) but with constant coefficients was first estimated* and Figure 7 compares the deterministic output of this model with the measured salinity at Site 1.

The highly periodic nature of the residuals is indicative that either additional inputs are affecting the system linearly or, as might be expected from eqn. (3), the system is nonlinear, with $a(t)$ and $b(t)$ time-variable functions of other environmental variables such as evaporation, river flow, and rainfall (Figure 6). This is confirmed by recursive estimation, which indicates that $a(t)$ is indeed a time-variable coefficient but that b can

---

* In fact, the discrete-time equivalent of this model was used for convenience of analysis but we will consider here only the continuous-time interpretation of this model.

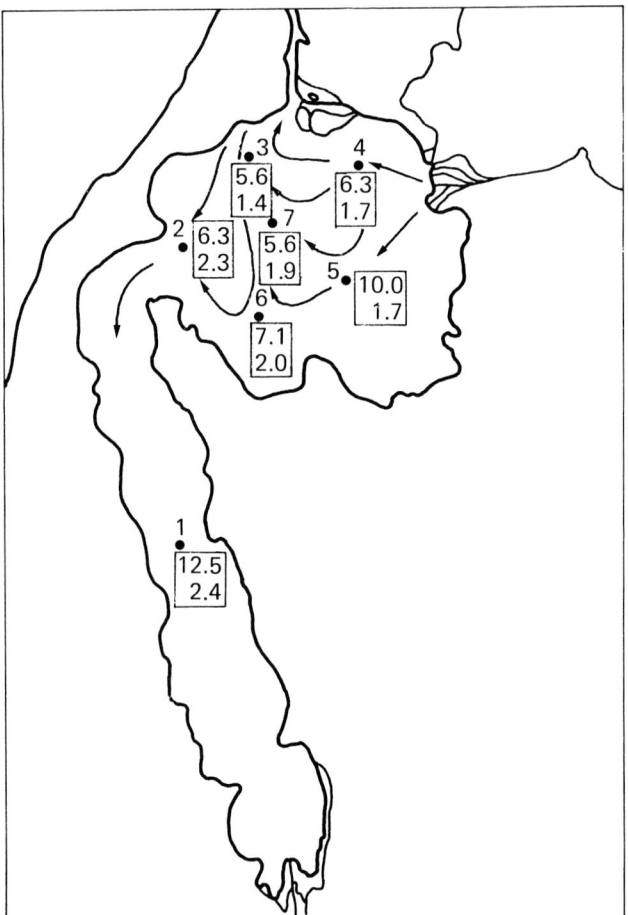

FIGURE 5 The Peel–Harvey estuarine system showing estimated circulation and flushing characteristics. The numbered points represent sampling sites; the boxed values are (top) summer maximum flushing residence time (weeks), and (bottom) winter minimum flushing residence time (weeks).

be considered constant for the purposes of the present analysis. Figure 8 compares the output of this model with the observed salinity and it can be seen that the data are explained rather well, with the residual series conforming to the requirements of the identification analysis (Section 2).

Figure 9 shows the recursive estimate $\hat{a}(t)$ of $a(t)$ and the dotted sinusoidal curve shows that the estimated variation is dominated by a periodic component with a period of one year. Considerable fluctuations about this sinusoid occur, however, particularly during weeks 0–12, 48–72, and 96–104. These periods correspond to the winter periods in Western Australia when fluvial inputs to the system are dominant (see Figure 6).

Bearing in mind eqn. (3), these results make sense: changes in volume will occur because of periodic evaporation changes, seasonal rainfall effects, and the differences in tidal height between the compartments. Put mathematically, the small perturbation equations can be written:

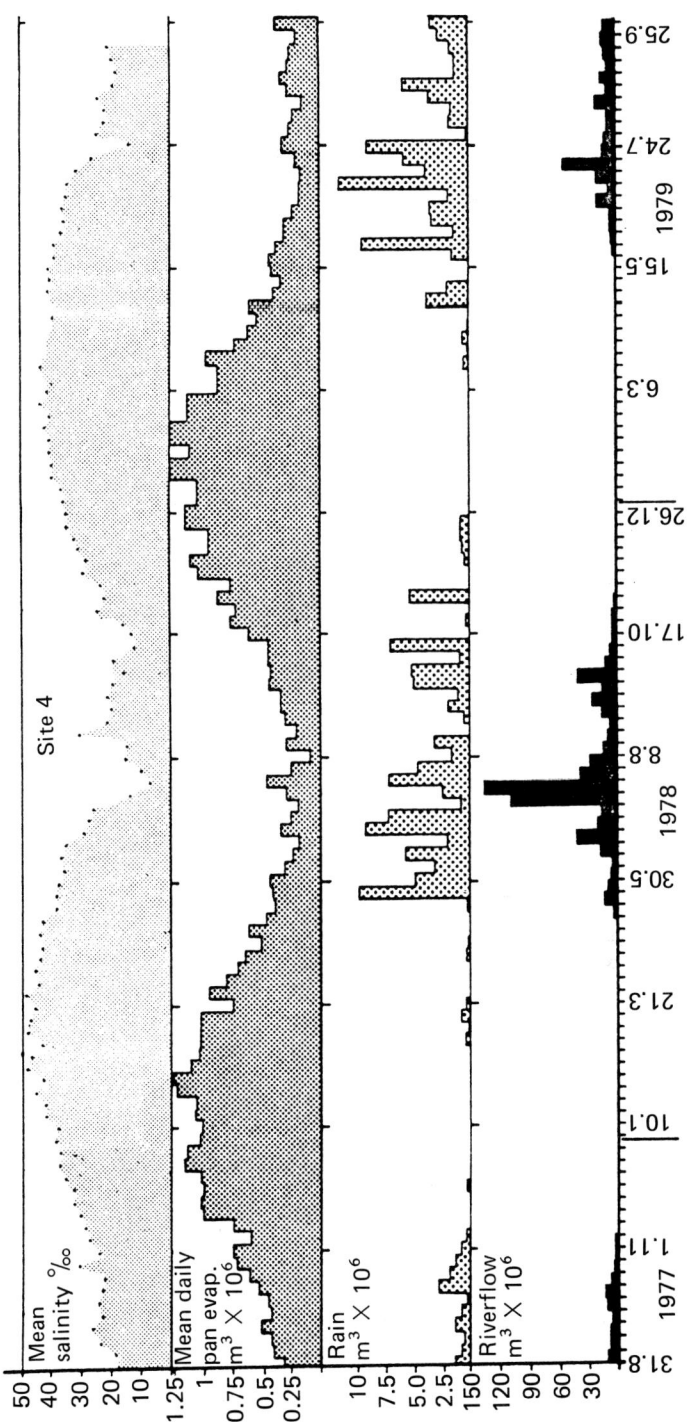

FIGURE 6 The Peel–Harvey estuarine system: variations in salinity, evaporation, rainfall and total river flow obtained from weekly monitoring over the period 1977–1979.

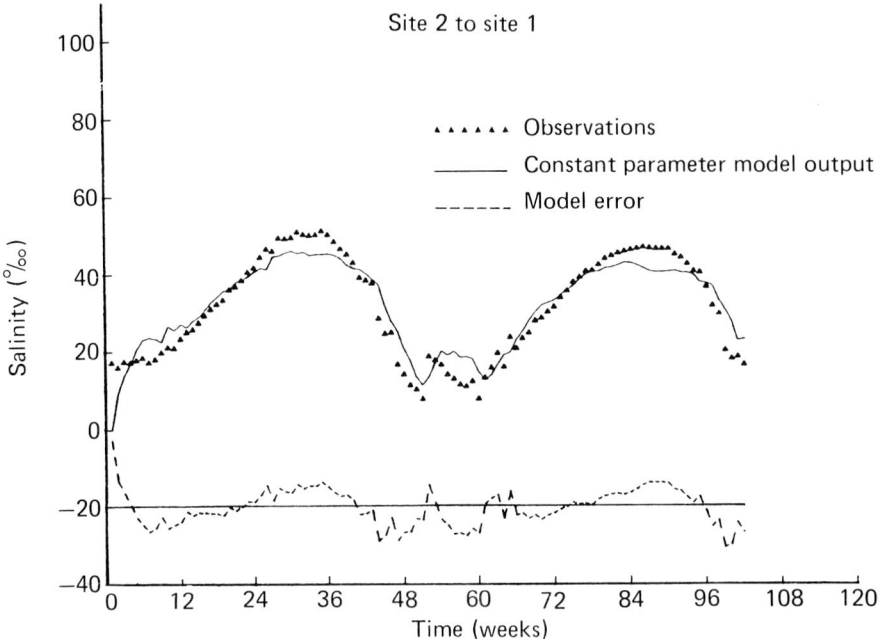

FIGURE 7 Constant-parameter time-series salinity model output compared to measured salinity.

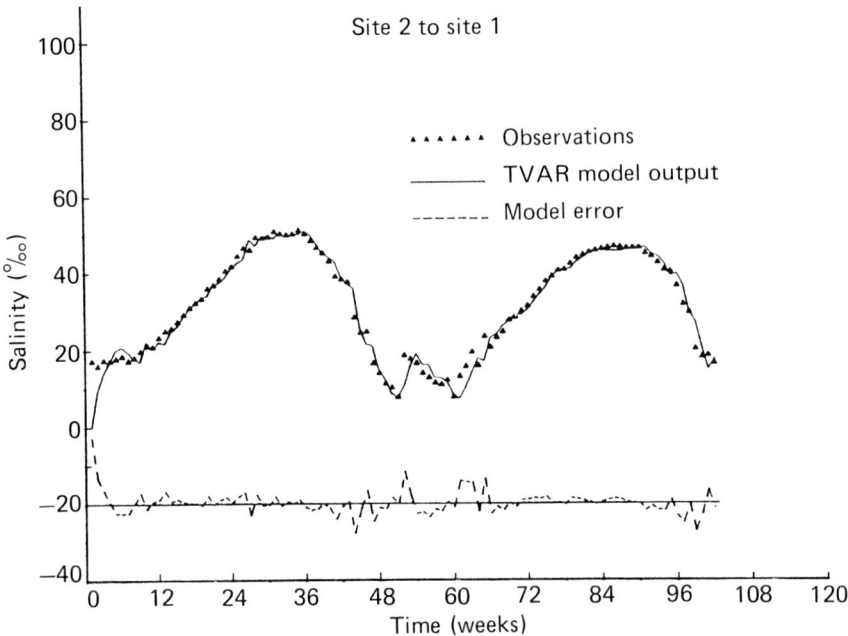

FIGURE 8 Time-variable parameter time-series salinity model output compared to measured salinity.

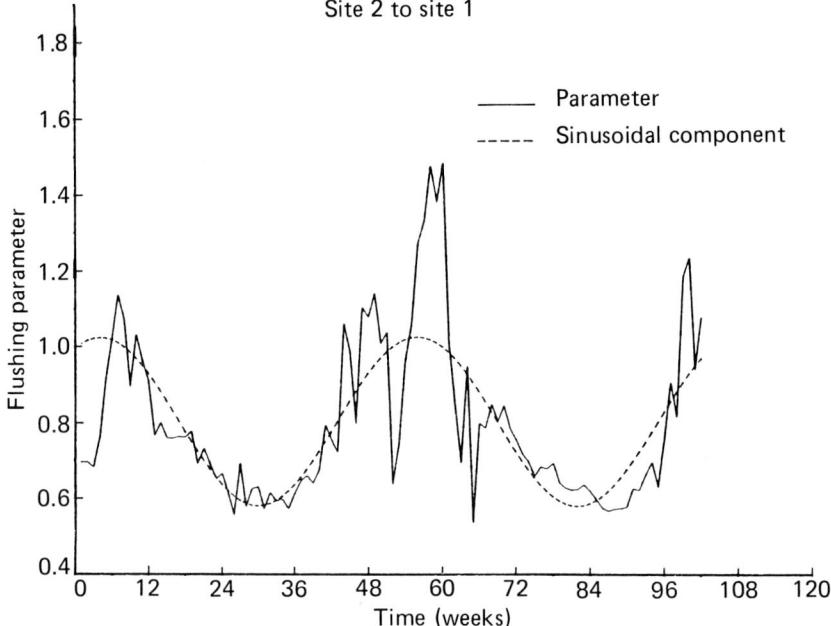

FIGURE 9 Recursive estimate of time-variable "flushing" parameter $a(t)$.

$$dV/dt = d(Ah)/dt = ((h_i - h)/R) + Q_R - Ae$$

or (5)

$$dh/dt = -h((1/AR) + (1/A)(dA/dt)) + (h_i/AR) + (Q_R/A) - e$$

where

h = depth of water in compartment;
A = surface area of compartment;
h_i = depth of water associated with the input location and measured with respect to the same height datum as h;
Q_R = river flow; and
e = effective evaporation (evaporation minus rainfall).

This identification analysis suggests that eqns. (4) and (5) provide a reasonable *a priori* model structure in this case and it would be interesting to pursue the analysis on this basis. However, in relation to the study objectives (and given the usual time restrictions on any practical study), this did not prove necessary. The estimated variation of $a(t)$, in itself, provides sufficient information both to assess the overall nature of the flushing dynamics and to help in the evaluation of nutrient budgets, as required by the study objectives. Figure 5, for example, shows the estimated maximum and minimum flushing times (obtained in performing the above analysis at each site in turn) together

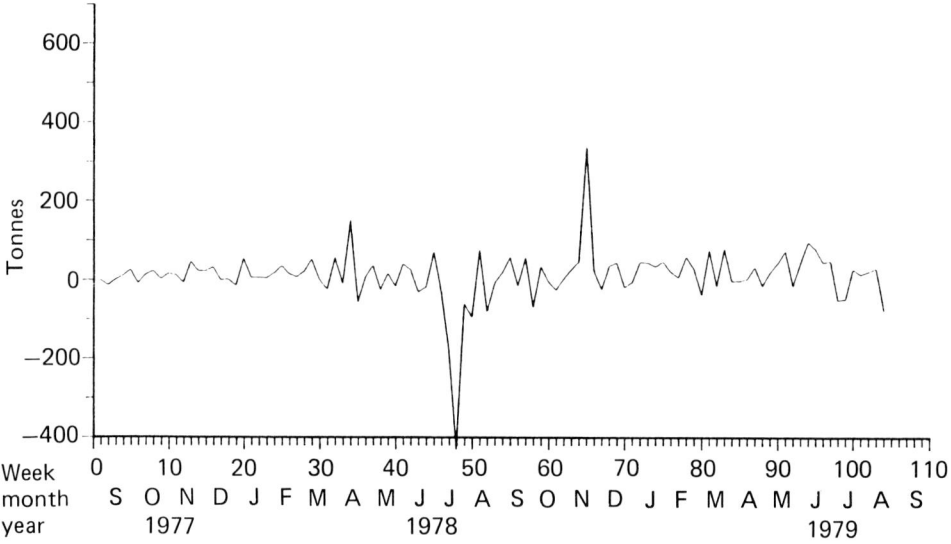

FIGURE 10 "Innovations" series computed from observed minus estimated total nitrogen load in the Peel-Harvey estuarine system.

with inferred circulation patterns: the details of this analysis are given in Humphries et al. (1980). Figure 10 is a plot of the innovations series (i.e., observed − predicted nitrogen (N) load) obtained in a subsequent nutrient budget analysis which made use of the flushing information to estimate oceanic exchange of nutrients. The fact that this series has zero-mean, serially uncorrelated characteristics is a further, independent check on the efficacy of the analysis. The two large transient deviations in the innovations series in July (negative) and November (positive) can be accounted for in terms of biological activity in the estuary. The large negative deviation is due to apparent gross sedimentation of inorganic nitrogen from the water column by a phytoplankton bloom during winter riverine enrichment of the estuarine water column; the large positive deviation occurred during a massive *Nodularia* bloom which fixed about 270 tonnes of nitrogen in the estuary.

3.3 The Transportation and Dispersion of Pollutants and Tracer Materials in Flowing Media

As a final example, we will consider a subject which is closely related to that discussed in the previous subsection but has wider implications in a scientific sense because it has relevance to various areas of research. Much has been written on dispersion in flowing media, and applications where characterization of dispersive behavior are important range from water-quality modeling to the analysis of data obtained from tracer studies in plants, animals, and man (Jakeman and Young, 1980).

Classical hydrodynamic analysis associated with this kind of problem is often related to the seminal work of Taylor (1954) on flow in pipes and involves the use of the following one-dimensional, partial-differential diffusion equation (Taylor, 1954; Fischer, 1966):

$$(\partial c/\partial t) + \langle U \rangle (\partial c/\partial x) = D(\partial^2 c/\partial x^2) \tag{6}$$

where $\langle U \rangle$ is the mean flow velocity and D the longitudinal dispersion coefficient.

Taylor's original work was based on turbulent flow in pipes and, although most textbooks subscribe to its use in the river context, reasonable arguments can be put forward to suggest that it is not strictly applicable to flow in natural streams, particularly close to the point source (see, for example, Fischer, 1966). Moreover, while it is often the accepted model in the literature, this does not mean that other mathematical representations may not provide an equally good, if not superior, description of the observed natural phenomena. For eqn. (6) is not the real world, although it may sometimes be interpreted as such; it is an approximate model of the real world and there is no guarantee that, for certain applications, it is either the best or the most useful model.

Recent work carried out at the Centre for Resource and Environmental Studies has been concerned with both an alternative approach to dispersion modeling (Beer and Young, 1980; Young, 1980) and the planning of tracer experiments to evaluate dispersion characteristics and models (Jakeman and Young, 1980). This approach, like that used in the last subsection, uses a lumped-parameter, ordinary-differential equation (ODE) compartmental model based on a combination of the plug flow and continuous stirred tank reactor (CSTR) mechanisms used so often in chemical engineering research (see also Beck and Young, 1975). However, the model is capable of a more conventional hydrodynamic interpretation (Beer and Young, 1980): in particular, it can be interpreted as the solution, at specified spatial locations down the river, of a partial differential equation (PDE) of the form:

$$(\partial c/\partial t) + U(\partial c/\partial x) = (c_i - c)/T \tag{7}$$

Here c_i is the input concentration into the reach considered and the first velocity term $U(\partial c/\partial x)$ accounts for the plug-flow characteristics*. The CSTR mechanism resides in the term on the right-hand side of eqn. (7), which can be interpreted as an "aggregated dead zone" (ADZ) effect: this arises from the aggregated effects of all those physical processes in the stream (e.g., bottom holes, turbulence, rocks, side irregularities, meanders, and pool-riffle effects) that contribute to retaining dissolved material temporarily and then releasing it, on a time-scale defined by the ADZ residence time (or time-constant) T.

Note that in eqn. (7) the dispersive effect arises completely from the ADZ terms on the right-hand side and is characterized by the residence-time parameter T; the conventional dispersion mechanism characterized by the dispersion coefficient D (eqn. (6))

---

* Here we use the symbol U rather than $\langle U \rangle$ to emphasize that the velocity coefficient does not have the same interpretation here as in eqn. (6).

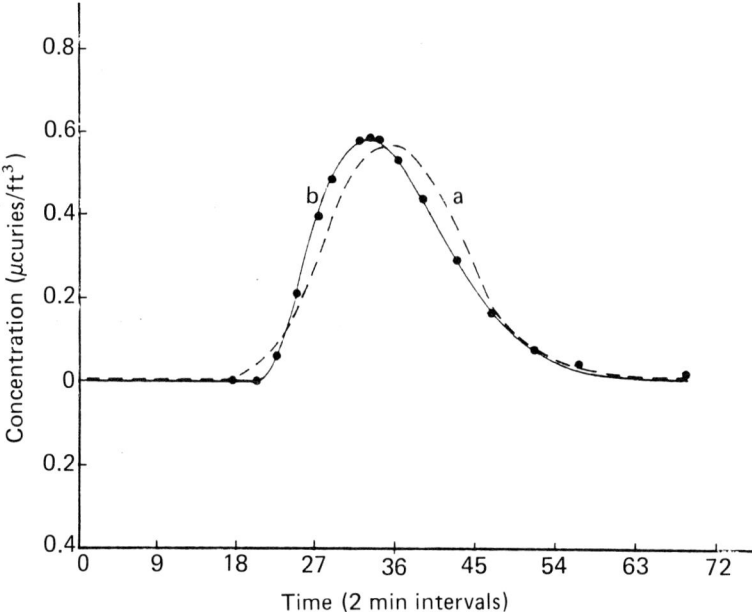

FIGURE 11 Modeling dispersion in natural streams: comparison of model results obtained from (a) Fischer's (1968) analysis based on conventional diffusion partial differential equation (shown broken), and (b) aggregated dead zone (ADZ) model obtained by time-series analysis (full line).

does not appear in the equation at all! Clearly, the argument here is not that turbulent diffusion is not taking place but rather that its effect is completely overshadowed by the much more significant dispersive effects of the aggregated dead zones.

The parameters of these ADZ models can be estimated quite easily from experimental or monitored data using the techniques of time-series analysis discussed in this paper. The resulting models consistently provide a much better explanation of the observed data than the more classical PDE representation (6) and are considerably easier to estimate and use. A typical example is shown in curve (b) of Figure 11, which compares the observed and modeled concentration of tracer material in a river system. The data in this case are those used by Fischer (1968) to test his alternative modeling procedure which is based on the more conventional PDE description (6): Fischer's results are shown as curve (a) of Figure 11 for comparison.

Despite the ability of the ADZ model to provide a better explanation of experimental data, it has not yet attracted a very favorable response from the fluid dynamics establishment, which tends to regard it as a black-box representation obtained by curve fitting. In particular, the establishment seems to consider the ADZ approach entirely devoid of the "nice" physical interpretation it associates with the classical PDE description. This criticism seems a little unfair since the ADZ description clearly does have a physical interpretation (see, for example, Buffham and Gibilaro, 1970; Beer and Young, 1980) albeit an unconventional one which does not directly involve the dispersion coefficient, D. This physical interpretation appears to allow for the model to be used in an "extrapolative" mode, so allowing for the prediction of dispersive characteristics at different river flows and conditions. In addition, it is possible to estimate an "equivalent" dispersion coefficient on the basis of the model parameters if this is required (Beer et al., 1980; Beer and Young, 1980).

I have chosen this latter example deliberately to illustrate one problem with the concept of credibility, a problem already alluded to in Section 1. By following a quite rigorous model-building procedure and satisfying all of the requirements specified in this paper, it is possible to produce dispersion models which seem highly satisfactory practical tools and can be considered valid in the strictest meaning of that word. But they still do not appear wholly credible to a large and important body in the scientific community whose different outlook on the problem makes them somewhat skeptical of the unconventional physical interpretation of the model and the methodological approach used to obtain it. Nevertheless, it might be hoped that, given time, these unconventional models will find general acceptance, not necessarily as replacements for the more classical representations but as alternatives which have merit in certain applications.

3.4 Other Examples

The model-building procedures described in this paper have been applied to numerous other examples, ranging from the analysis of fluorescence data in chemical experiments (Jakeman and Young, 1979b), through the modeling of rainfall flow characteristics in hydrology (Whitehead et al., 1979), water quality behavior in river systems (Young and Beck, 1974; Beck and Young, 1975; Whitehead and Young, 1979), and air quality (Jakeman et al., 1980; Steele, 1981), to the evaluation of economic models and data (Young et al., 1973; Young, 1977; Salmon and Young, 1978). While it has not, of course, solved all problems in these applications, it has provided a systematic approach which has helped a great deal in the overall systems analysis.

4 CONCLUSIONS

This paper has presented a comprehensive methodological approach to model-building based on a general theory of modeling for badly defined systems. Whilst it is unlikely that this approach will solve all modeling problems associated with such systems, it is felt that it will provide a satisfactory system of "checks and balances" which should at least help the model-builder and systems analyst in this most difficult of problem areas. The most important features of the proposed approach are as follows:

(1) It is consistent with the hypothetico-deductive procedures of the scientific method and can be considered within the framework of Bayesian estimation theory.
(2) It presents a fully integrated approach involving the systematic application of mathematical analysis, planned multidisciplinary monitoring, experimentation, and fieldwork, and allows for a continuous form of adaptive assessment along the lines suggested, for example, by Holling (1978).

(3) In the model-formulation phase, it makes use of a novel type of probabilistic or "speculative" simulation modeling based on Monte Carlo analysis in order to generate working hypotheses about the nature of the system; these hypotheses can then be thoroughly tested by planned monitoring or experimentation.

(4) In the later identification and estimation phases of model-building, it exploits sophisticated methods of recursive time-series analysis to detect the presence of significant model-parameter variations and hence: define any important nonlinear or nonstationary aspects of the observed system behavior which are not present in the model; define any over-parameterization or surplus content in the model; and, in this manner, derive a model which will normally be efficiently parameterized and characterized by a low-dimensional set of well-defined and, hopefully, constant-parameter estimates.

(5) It emphasizes the need for thorough validation of both model structure and parameter estimates, and stresses the need to ensure that the consequences of any surplus, unvalidated content are fully acknowledged in any subsequent application of the model.

By discussing a number of practical examples, it is hoped that the reader will better understand the procedures involved and will be encouraged to use them in practice. For it is only by practical application that the true value of any method of applied systems analysis can be assessed. Whether the application of this method will eventually lead to models that are more "valid" or "credible" is, however, another matter. Only time will tell.

ACKNOWLEDGMENTS

This paper is based in part on the paper "Modeling Badly Defined Systems: Some Further Thoughts" by P.C. Young, G. Hornberger, and R.C. Spear, presented at the SIMSIG Conference held at Canberra, Australia, in 1978. The author is grateful to all the members of his research group in the Centre for Resource and Environmental Studies whose work on various environmental projects and research on time-series analysis have provided an important stimulus to the specification of the "Method Theory" described in this paper.

REFERENCES

Ackerman, B.A., Ackerman, S.R., Sawyer, J.W., and Henderson, D.W. (1974). The Uncertain Search for Environmental Quality. The Free Press, New York.

Barrett, J.F., Coales, J.F., Ledwich, M.A., Naughton, J.J., and Young, P.C. (1973). Macro-economic modeling: a critical appraisal. Proceedings of the IFAC/IFORS Conference on Dynamic Modelling and Control of National Economies. IEE, London.

Beck, M.B. and Young, P.C. (1975). A dynamic model for DO–BOD relationships in a non-tidal stream. Water Research, 9: 769–776.

Beer, T. and Young, P.C. (1980). On the Characterization of Longitudinal Dispersion in Natural Streams. Report No. AS/R42. Centre for Resource and Environmental Studies, Australian National University, Canberra.

Beer, T., Young, P.C., and Humphries, R.B. (1980). Murrumbidgee Water Quality Study: Report on CRES Contribution to the Study. Report No. AS/R41. Centre for Resource and Environmental Studies, Australian National University, Canberra.

Berlinski, D. (1976). On Systems Analysis. MIT Press, Cambridge, Massachusetts.

Bertrand, J. (1855). Méthode des Moindres Carrés, by K.F. Gauss. Translated into French by J. Bertrand. Mallet-Bachelur, Imprimeur-Libraire de L'École Polytechnique, Paris.

Brewer, G.C. (1973). Politicians, Bureaucrats and the Consultant. Basic Books, New York.

Buffham, B.A. and Gibilaro, L.G. (1970). A unified time delay model for dispersion in flowing media. Chemical Engineering Journal, 1: 31–35.

Ellis, J., Kanamori, S., and Laird, P.G. (1977). Water pollution studies on Lake Illawarra. Australian Journal of Marine and Freshwater Research, 28: 467–477.

Fischer, H.B. (1966). A note on the one dimensional dispersion model. Air and Water Pollution, International Journal, 10: 443–452.

Fischer, H.B. (1968). Dispersion prediction in natural streams. Journal of Sanitation Engineering, ASCE, 94: 927–944.

Holling, C.S. (Editor) (1978). Adaptive Environmental Assessment and Management. Wiley, Chichester.

Hoos, I.R. (1972). Systems Analysis in Public Policy. University of California Press, Berkeley and Los Angeles.

Humphries, R.B., Young, P.C., and Beer, T. (1980). Systems Analysis of an Estuary; Report of the CRES Contribution to the Peel–Harvey Estuary Study. Bulletin No. 100, Western Australian Department of Conservation and Environment, Perth, Western Australia.

Jakeman, A.J. and Young, P.C. (1979a). Refined instrumental variable methods of recursive time series analysis. Part II: multivariable systems. International Journal of Control, 29: 621–644.

Jakeman, A.J. and Young, P.C. (1979b). Time-series methods in biological and medical data analysis. In R. Isermann (Editor), Identification and System Parameter Estimation. Pergamon Press, Oxford.

Jakeman, A.J. and Young, P.C. (1980). Towards optimal modeling of translocation data from tracer studies. Proceedings of the Biennial Conference of the Simulation Society of Australia, 4th, pp. 248–253.

Jakeman, A.J., Steele, L.P., and Young, P.C. (1980). Instrumental variable algorithms for multiple input systems described by multiple transfer functions. IEEE Transactions, Systems, Man, and Cybernetics, SMC-10: 593–602.

Johnston, J. (1963). Econometric Methods. McGraw-Hill, New York.

Kaldor, J.M. (1978). The Estimation of Parametric Change in Time-Series Models. M.A. Thesis, Australian National University, Canberra.

Kalman, R.E. (1960). A new approach to linear filtering and prediction problems. ASME Transactions, Journal of Basic Engineering, 83D: 95–108.

Kendall, M.G. and Stuart, A. (1961). The Advanced Theory of Statistics, Volume 2. Griffin, London.

Kittler, J. and Young, P.C. (1973). A new approach to feature selection based on the Karhunen–Loeve expansion. Pattern Recognition, 5: 335–352.

Miller, D.R., Butler, G., and Bramall, C. (1976). Validation of ecological system models. Journal of Environmental Management, 4:383–401.

Philip, J.R. (1975). Some remarks on science and catchment prediction. In T.G. Chapman and F.X. Dunin (Editors), Prediction in Catchment Hydrology. Australian Academy of Science, Canberra.

Plackett, R.L. (1950). On some theorems in least squares. Biometrika, 37: 149–157.

Popper, K.R. (1959). The Logic of Scientific Discovery. Hutchinson, London.

Rademaker, O. (1973). On understanding complicated models: simple methods. Presented at the American/Soviet Conference on Methodological Aspects of Social Systems Simulation, Sukhumi (USSR), October 24–26.

Rigler, F.H. (1976). Review of "Systems Analysis and Simulation in Ecology", Volume 3. B.C. Patten (Editor), Limnology and Oceanography, 21 (3): 481–483.

Salmon, M. and Young, P.C. (1978). Control methods and quantitative economic policy. In S. Holly, B. Rustem, and M. Zarrop (Editors), Optimal Control for Econometric Models: An Approach to Economic Policy Formation. MacMillan, London.

Smith, R. (1980). Buoyancy effects upon longitudinal dispersion in wide well-mixed estuaries. Philosophical Transactions of the Royal Society, Series A, 296:467–496.

Spear, R.C. (1970). Application of Kolmogorov–Renyi statistics to problems of parameter uncertainty in systems design. International Journal of Control, 11: 771–778.

Spear, R.C. and Hornberger, G.M. (1978). Eutrophication in Peel Inlet: an analysis of behaviour and sensitivity of a poorly defined system. Report No. AS/R18, Centre for Resource and Environmental Studies, Australian National University, Canberra.

Sprott, D.A. (1977). Gauss's Contributions to Statistics. Presented at the Gauss Bicentennial, Toronto. (D.A. Sprott is with the University of Waterloo, Ontario, Canada.)

Steele, L.P. (1981). Recursive Estimation in the Identification of Air Pollution Models. Ph.D. Thesis, Australian National University, Canberra.

Taylor, G.I. (1954). The dispersion of matter in turbulent flow through a pipe. Proceedings of the Royal Society, Series A, 223: 446–468.

Thissen, W. (1978). Investigations into the World 3 model: overall behaviour and policy conclusions. IEEE Transactions, Systems, Man, and Cybernetics, SMC-8: 172–182.

Whitehead, P.G. and Young, P.C. (1975). A dynamic-stochastic model for water quality in part of the Bedford Ouse river system. In G.C. Vansteenkiste (Editor), Computer Simulation of Water Resources Systems. North-Holland, Amsterdam, pp. 417–438.

Whitehead, P.G. and Young, P.C. (1979). Water quality in river systems – Monte Carlo analysis. Water Resources Research, 15: 451–459.

Whitehead, P.G., Young, P.C., and Hornberger, G.H. (1979). A systems model of streamflow and water quality in the Bedford-Ouse river, I: streamflow modelling. Water Research, 13:1155–1169.

Young, P.C. (1974). Recursive approaches to time series analysis. Bulletin of the Institute of Mathematics and its Application, 10: 209–224.

Young, P.C. (1976a). Some observations on instrumental variable methods of time series analysis. International Journal of Control, 23: 593–612.

Young, P.C. (1976b). Optimization in the presence of noise: a guided tour. In L.C.R. Dixon (Editor), Optimization in Action. Academic Press, London, pp. 517–573.

Young, P.C. (1977). A general theory of modeling for badly defined systems. Report No. AS/R9, Centre for Resource and Environmental Studies, Australian National University, Canberra. Also published in G.C. Vansteenkiste (Editor), Modeling, Identification, and Control in Environmental Systems. North-Holland, Amsterdam/American Elsevier, New York, pp. 103–135.

Young, P.C. (1980). Mining and the Natural Environment – Systems Analysis and Mathematical Modeling. In S.F. Harris (Editor), Social and Environmental Choice: The Impact of Uranium Mining in the Northern Territory. Centre for Resource and Environmental Studies, Australian National University, Canberra, pp. 64–78.

Young, P.C. (1981). Parameter estimation for continuous-time models – a survey. Automatica, 17: 23–29.

Young, P.C. (1982). An Introduction to Recursive Estimation. Lecture Notes Series, Springer, Berlin, in press.

Young, P.C. and Beck, M.B. (1974). The modeling and control of water quality in a river system. Automatica, 10:455–468.

Young, P.C. and Jakeman, A.J. (1979). Refined instrumental variable methods of recursive time series analysis, Part I: single input–single output systems. International Journal of Control, 29: 1–30.

Young, P.C. and Jakeman, A.J. (1980). Refined instrumental variable methods of recursive time series analysis, Part III: extensions. International Journal of Control, 31:741–764.

Young, P.C. and Kaldor, J.M. (1978). Recursive estimation: a methodological tool for investigating climatic change. Report No. AS/R14, Centre for Resource and Environmental Studies, Australian National University, Canberra (to be revised).

Young, P.C., Naughton, J.J., Neethling, C.G., and Shellswell, S.H. (1973). Macro-economic modeling: a case study. In P. Eykhoff (Editor), Identification and System Parameter Estimation. North-Holland, Amsterdam/American Elsevier, New York.

Young, P.C., Jakeman, A.J., and McMurtrie, R.E. (1980). An instrumental variable method for model structure identification. Automatica, 16: 281–294.

Part Two

Uncertainty and Model Identification

AN APPROACH TO THE ANALYSIS OF BEHAVIOR AND SENSITIVITY IN ENVIRONMENTAL SYSTEMS

G.M. Hornberger
Department of Environmental Sciences, University of Virginia, Charlottesville, Virginia 22903 (USA)

R.C. Spear
Department of Biomedical and Environmental Health Sciences, School of Public Health, University of California, Berkeley, California 94720 (USA)

1   INTRODUCTION

The use of simulation models for analyzing complex natural systems can be criticized in terms of the mathematical techniques normally employed (see, for example, Berlinski, 1976) and in terms of the sophistication of representations of the "realities of the natural world" (Hedgepeth, 1977). Regardless of one's philosophy concerning construction of "ecosystem" models, we argue that certain elements of such criticism must be addressed. In dealing with complex environmental problems in particular, the benefits of traditional systems analysis, if such exist, appear to be severely limited. Not only do the forcing inputs and parametric values of our models change with time and circumstance, but often so do their internal structures that yesterday appeared to best summarize the important causal relations in the system. Because it is virtually impossible to completely distinguish, let alone decouple, the system from its environment, model verification is itself a dynamic process which cannot be assumed to approach an equilibrium state. Thus, it seems unlikely that any moderately complex environmental system can be well defined in the traditional physical–chemical sense. This conclusion does not, in our view, destroy the appeal of applying systems analysis methods to environmental forecasting problems. It does, however, set a fairly short timescale over which models can be used to develop management strategies with any confidence. It seems, therefore, that the important issues pertaining to the forecasting problem relate to methods of making the best use of the diverse data available at any time to develop these short-term management strategies. This is not to say that long-term environmental planning is not necessary or profitable. We do contend, however, that long-term planning based on environmental models is of dubious value except insofar as such exercises may provoke analytical thought in a broader context.

If one accepts that the only model-based systems analysis worth doing is short-term, a practical dilemma arises immediately. The modeling of environmental systems of any complexity is usually not a short-term proposition. It will usually take several years of data collection and background work even to get a start. When a model finally emerges its authors are then loathe to regard it as anything but the revealed gospel and the notion that processes and causal relations may have changed or may be changing is resisted with vigor. Often the model becomes the system.

Over the last several years we have developed an approach to the modeling of environmental systems that has some promise as a means of circumventing this long lead-time. The approach is based on two premises: that the literature contains much information of relevance to an understanding of the problem at hand, and that it is possible to describe, at least in qualitative terms, the principal features of the behavior of the environmental system that define the problem to be managed.

Our ideas were developed while working on the analysis of a cultural eutrophication problem in the Peel–Harvey Estuary of Western Australia. The specific problem we addressed was one of research direction, that is, what were the critical uncertainties in the knowledge of the behavior of this system which required resolution before a strategy could be formulated for the management of the "nuisance" alga. The system behavior of concern was the excessive growth of the benthic alga *Cladophora*, which led to its transport, accumulation, and decay on the beaches of Peel Inlet. Research on this problem was being carried out by several groups under the overall direction of the Estuarine and Marine Advisory Committee (EMAC) of the Western Australian Environmental Protection Authority. At the time we became involved there were available somewhat over a year's data on nutrient levels, algal biomass, phytoplankton populations, etc., usually on a monthly basis. There were somewhat more extensive hydrological data and a variety of other fragmentary data as well as speculations from the various research teams. The issue was to make an assessment of these results and speculations, however preliminary, in order to guide future research.

We chose to approach this task via simulation modeling because the logic and order inherent in the model-building process so often expose causal as well as quantitative uncertainties in the system under study. Indeed, the nature and extent of the data from Peel Inlet were such that the quantitative aspects of a conventional modeling exercise would be of little benefit. However, given that we were willing to hypothesize that the factor or factors controlling *Cladophora* growth in Peel Inlet were among those common to other estuarine systems, a great deal of relevant information was available in the literature. The extent of this information, coupled with the data from the field, led us to speculate on the possibility of estimating the parameters of a model in some approximate fashion and investigating the degree to which the resulting model might mimic the qualitative behavior of the Peel system with respect to the *Cladophora* problem. More to the point in view of our overall objective, we asked if the study of such a model could lead to the generation of hypotheses or could point to critical gaps in knowledge that might not otherwise emerge until later in the life of the project.

Because the factor or factors controlling the growth of *Cladophora* in Peel Inlet remain uncertain, various models of the phenomena are possible depending on which of the competing hypotheses is chosen. The assumption of a controlling factor, and the model resulting therefrom, we term a scenario in order to emphasize its speculative nature. A comprehensive analysis must consider various scenarios. During 1978 we considered a phosphorus scenario and found it to lead to several intriguing hypotheses and to exemplify a methodology we feel has wide applicability.

Because the Peel Inlet data available for this original study did not include comprehensive time-series information on the principal variables of interest, only the qualitative behavior of the system could be defined over an annual cycle. Therefore, only qualitative contrasts were possible between model performance and that of the system. We maintain, however, that the salient qualitative aspects of the behavior could be specified and that the result of any simulation using a model consequently classified as exhibiting either "the behavior" or "not the behavior". In this study the behavior was defined by a *Cladophora* "bloom" qualitatively similar to observed conditions in Peel Inlet in the period 1976–1977.

In any simulation model of an environmental system there is substantial uncertainty surrounding the "best" values of the parameter set. At the stage of understanding of the system discussed here, the level of parametric uncertainty precludes the use of any analytical procedure which relies on point or "best" estimates. In most cases, however, it is possible to make some defensible assessment of the probable or, at least, the allowable values of the parameters. We adopted this approach in the Peel Inlet work and associated with each model parameter a statistical distribution function to represent the uncertainty in knowledge of its "best" value given the assumed model structure. The distribution function assumed for any given parameter represents our best *a priori* knowledge of its likely or allowable values based on the current literature or on the limited field data.

Taken together, the scenario and its associated parameter distributions define an ensemble of models. Using a Monte Carlo approach one can explore the degree to which the parameter space underlying this ensemble partitions under the behavioral classification. This separation under the behavioral classification forms the basis for a type of sensitivity analysis. It is intended that the results of this analysis, when interpreted in the light of the totality of current knowledge of the system, will indicate gaps in present research efforts or suggest new hypotheses and profitable avenues for the next phase of the research program.

2   SYSTEM BEHAVIOR AND THE PHOSPHORUS SCENARIO

The problem-defining behavior was principally based on *Cladophora* biomass measurements taken by Professor A.J. McComb and his associates in the Botany Department of the University of Western Australia during the period from April 1976 to April 1977 (Atkins et al., 1977). These data indicated a relatively slowly reacting system in which bloom conditions are characterized by biomass increases on the order of two to five times the minimum biomass. Further, high biomass levels exist for a relatively prolonged period. Also, during the period in which *Cladophora* biomass is high or increasing rapidly the average concentrations of phosphorus in the water column are low. Nitrate nitrogen was always less than $10 \mu g l^{-1}$ but ammonia nitrogen rose to $140 \mu g l^{-1}$ in April 1976 and declined steadily to $25 \mu g l^{-1}$ by August of that year. There were only very sparse data on phytoplankton levels in the Inlet but those that were available suggested a curious absence of phytoplankton. Few values in excess of $5 \mu g \, l^{-1}$ of chlorophyll-a had been reported.

At the time we began our work there were, among the researchers, proponents of both phosphorus and nitrogen as the principal limiting nutrient in the system. Given our modeling approach it was necessary to assume some growth factor to be limiting and we chose to pursue the phosphorus scenario. The behavior was then defined on the basis of six conditions placed on *Cladophora* biomass, phytoplankton biomass, and phosphorus concentrations in a section of Peel Inlet termed the *Cladophora* growth area. Behavior condition 5, for example, was that during the period in which *Cladophora* biomass exceeds 1.5 times its initial value the average soluble phosphorus concentration in the water column must be less than $10 \mu g \, l^{-1}$.

The choice of a model to use in the preliminary analysis of the problem of nuisance algal growth in the Peel Inlet is dictated by a number of constraints. First, the model must of necessity be mechanistic, to as great an extent as possible, because if our approach is to succeed, data available in the literature must be relied upon as a surrogate to data on the system itself. This requirement of "transferability" of information from diverse sources is the primary reason for selecting conventional model components; it is only by using such mechanistic components that information derived from vastly different problems can be utilized to construct our candidate models and to select reasonable *a priori* probability density functions for the parameters. (For example, as described below, we chose to use the Monod kinetics description of nutrient uptake because values of the half-saturation constant are routinely reported in the literature.) A second constraint is in some ways antithetical to the first: the model should be as simple as possible in recognition of the fact that available data are sparse or nonexistent. Thus, the model must be structured to provide enough detail to be capable of reproducing (in the broad sense discussed above) the behavior of the system that defines the problem but not be so excessively complex as to prohibit its use in a preliminary study. The lumped-parameter model described below was chosen in recognition of these practical considerations.

Our simplified phosphorus model consists of four compartments: *Cladophora*, phytoplankton, soluble phosphorus in the water column, and sediment. The term "sediment" as used here includes the layer of decomposed organic material that underlies much of the actively photosynthesizing *Cladophora* mat. A phytoplankton compartment is included because, as discussed above, we suspected the general absence of large populations of phytoplankton to be an important aspect of the behavior of the system. For each of the compartments a mass-balance equation for phosphorus can be developed.

The equation for dissolved phosphorus in the water column is a good example of the structure and level of detail of the entire model. It is

$$V_w(dX_3/dt) = Q_{T_i}U_T - Q_{T_o}X_3 + Q_R U_R - (1-\beta)Q_G U_G \quad \text{advection terms}$$

$$- \alpha \gamma_1 T I_b (X_3/(K_1 + X_c))g(X_1) \quad \text{\textit{Cladophora} uptake from water column}$$

$$- \gamma_2 T(ef/K_T Z)(e^{-\alpha_1} - e^{-\alpha_0})(X_3/(K_2 + X_3))X_2 V_w \quad \text{phytoplankton uptake}$$

$$+ a_{43}(X_4 - X_3) \quad \text{transfer from sediment}$$

where the state variables are:

X_1 = *Cladophora* biomass,
X_2 = phytoplankton biomass concentration,
X_3 = phosphorus concentration in the water column, and
X_4 = phosphorus concentration in the sediment.

The tidal parameters Q_{T_i} and Q_{T_o} were estimated from tide gauge data as was water volume V_w. The river flow, Q_R, was known and some idea of the magnitude of groundwater flow, Q_G, was available, as was information on the phosphorus concentrations U_T, U_R, and U_G in these inputs. The *Cladophora* and phytoplankton terms are in standard form except for the function $g(X_1)$ in the *Cladophora* growth term. This was inserted to account for the fact that only the upper layers of the *Cladophora* mat are exposed to light and, as a result, only a fraction of the mat can be actively photosynthesizing at any moment. Hence, $g(X_1)$ is a saturation-type function that contains a parameter X_m, which is an estimate of the maximum photosynthesizing biomass.

Another type of parameter used to account for uncertainty is exemplified by β. The parameter β was introduced to allow a fraction of the groundwater flow to go directly into the water column with the remaining fraction being routed through the sediment compartment. This mechanism was included in the formulation because it represented one theory of nutrient source current at the time of model development.

The primary parameters of this equation that were assigned distribution functions were then: γ_1, K_1, X_m, α, γ_2, K_2, and a_{43}. The values for a variety of the light-related parameters f, α_1, α_0, and I_b were calculated from the distributions of more basic parameters, e.g., k, the phytoplankton-shading coefficient.

There is, of course, almost endless detail involved in explaining why certain features of this model were chosen and how estimates of the various inputs and parameter distributions were developed. This documentation is available (Spear and Hornberger, 1978) but the present point is that, at the conclusion of the development process, we felt there to be surprisingly few real "holes" in our knowledge. That is, we felt it to be possible to explain, if not defend, the model and each of the various estimates to a greater or lesser extent.

3   SIMULATION RESULTS

A total of 626 simulation runs were carried out for the phosphorus scenario with the parameter distributions as given in Table 1. These required approximately 75 minutes of CPU time on a Univac 1110. The 626 runs comprised 281 in the behavior category and 345 in the nonbehavior. Figure 1 shows the time course of *Cladophora* biomass, phytoplankton biomass, soluble phosphorus in the water column, and phosphorus in the sediment for a typical behavior-producing run. Figure 2 shows the same variables for a run in which *Cladophora* biomass was insufficient and phytoplankton biomass too high to constitute a behavior. In virtually all of the runs in which the behavior did not occur, this was due to a deficiency of *Cladophora*, an excess of phytoplankton, or both occurring simultaneously. Also characteristic of the behavior-producing simulations was

TABLE 1 Limits for the rectangular probability density functions for the nineteen parameters of the phosphorus model.

Parameter	Definition	Range of values	Units
γ_1	*Cladophora* growth coefficient	$1 \times 10^{-4} - 3 \times 10^{-4}$	day^{-1} ly^{-1} °C^{-1}
K_1	*Cladophora* half-saturation constant	5–10	µg l^{-1}
X_m	*Cladophora* maximum photosynthesizing biomass	$5 \times 10^{11} - 2.5 \times 10^{12}$	µg
a_{11}	*Cladophora* loss rate	$1.25 \times 10^{-3} - 6.25 \times 10^{-3}$	day^{-1} °C^{-1}
α	*Cladophora* nutrient source parameter	0.95–1	dimensionless
γ_2	phytoplankton growth coefficient	0.05–0.15	day^{-1} °C^{-1}
I_s	phytoplankton optimal irradiance	150–500	ly
K_2	phytoplankton half-saturation constant	5–50	µg l^{-1}
a_{22}	phytoplankton loss rate	$2.5 \times 10^{-3} - 7.5 \times 10^{-3}$	day^{-1} °C^{-1}
K_W	light-attenuation coefficient (water)	0.45–1.3	m^{-1}
k	light-shading coefficient (phytoplankton)	$3 \times 10^{-3} - 9 \times 10^{-3}$	(m^{-1})(µg l^{-1})$^{-1}$
β	groundwater partitioning coefficient	0–1	dimensionless
a_{43}	sediment–water diffusion parameter	$10^6 - 5 \times 10^7$	l day^{-1}
X_4^*	maximum sediment concentration	$1 \times 10^{-3} - 3 \times 10^{-3}$	µg l^{-1}
λ_1	*Cladophora* loss to sediment	0.3–0.9	dimensionless
λ_2	phytoplankton loss to sediment	0.3–0.9	dimensionless
U_G	groundwater concentration	250–600	µg l^{-1}
ρ	tidal exchange efficiency parameter	0.5–0.9	dimensionless
s_P	river sediment–phosphorus coefficient	$5 \times 10^{-7} - 5 \times 10^{-6}$	(µg l^{-1})(l day^{-1})

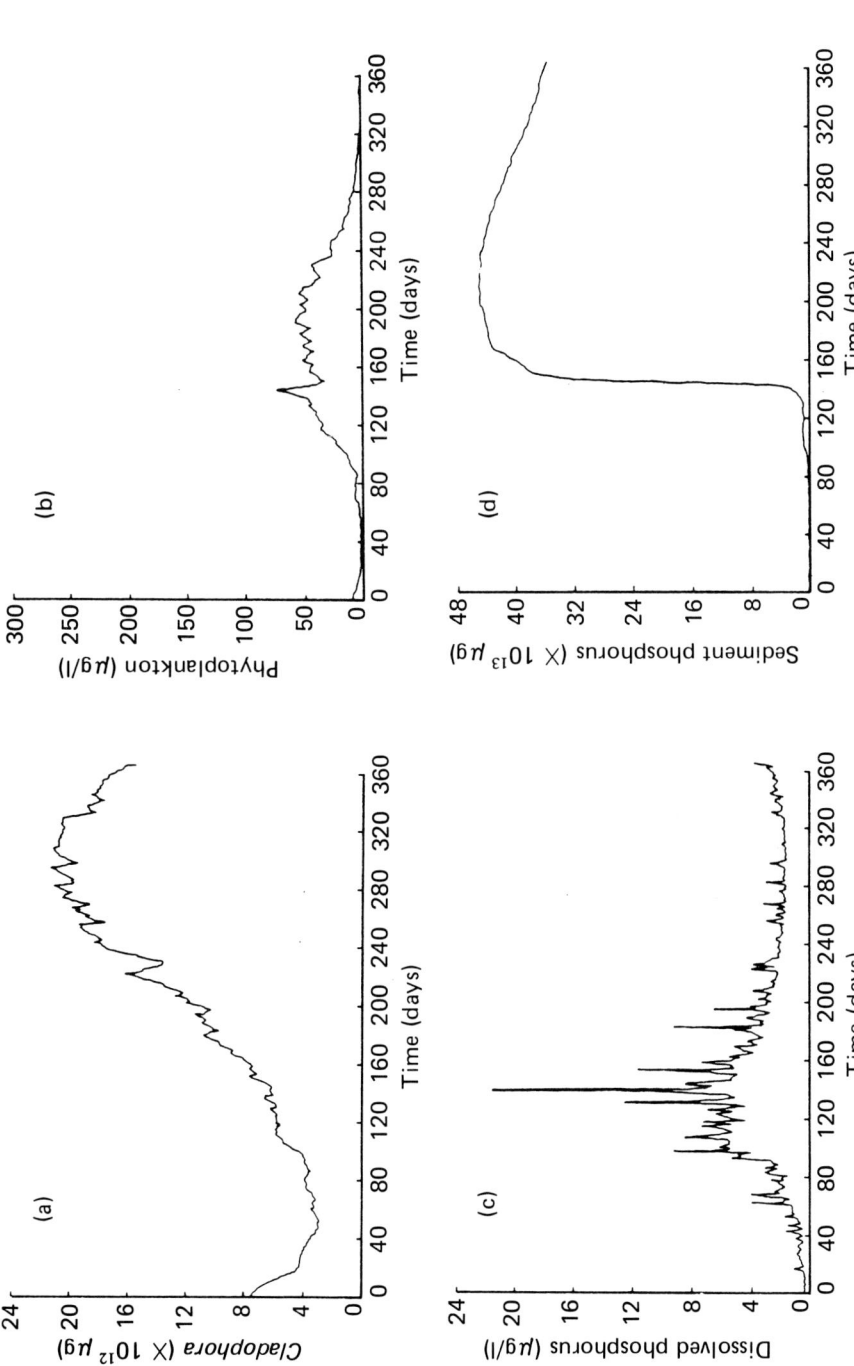

FIGURE 1 Trajectories of a simulation run that satisfied the behavior criteria. All quantities are in terms of phosphorus: (a) *Cladophora*, (b) phytoplankton, (c) dissolved phosphorus in the water column, and (d) total "available" phosphorus in the sediment.

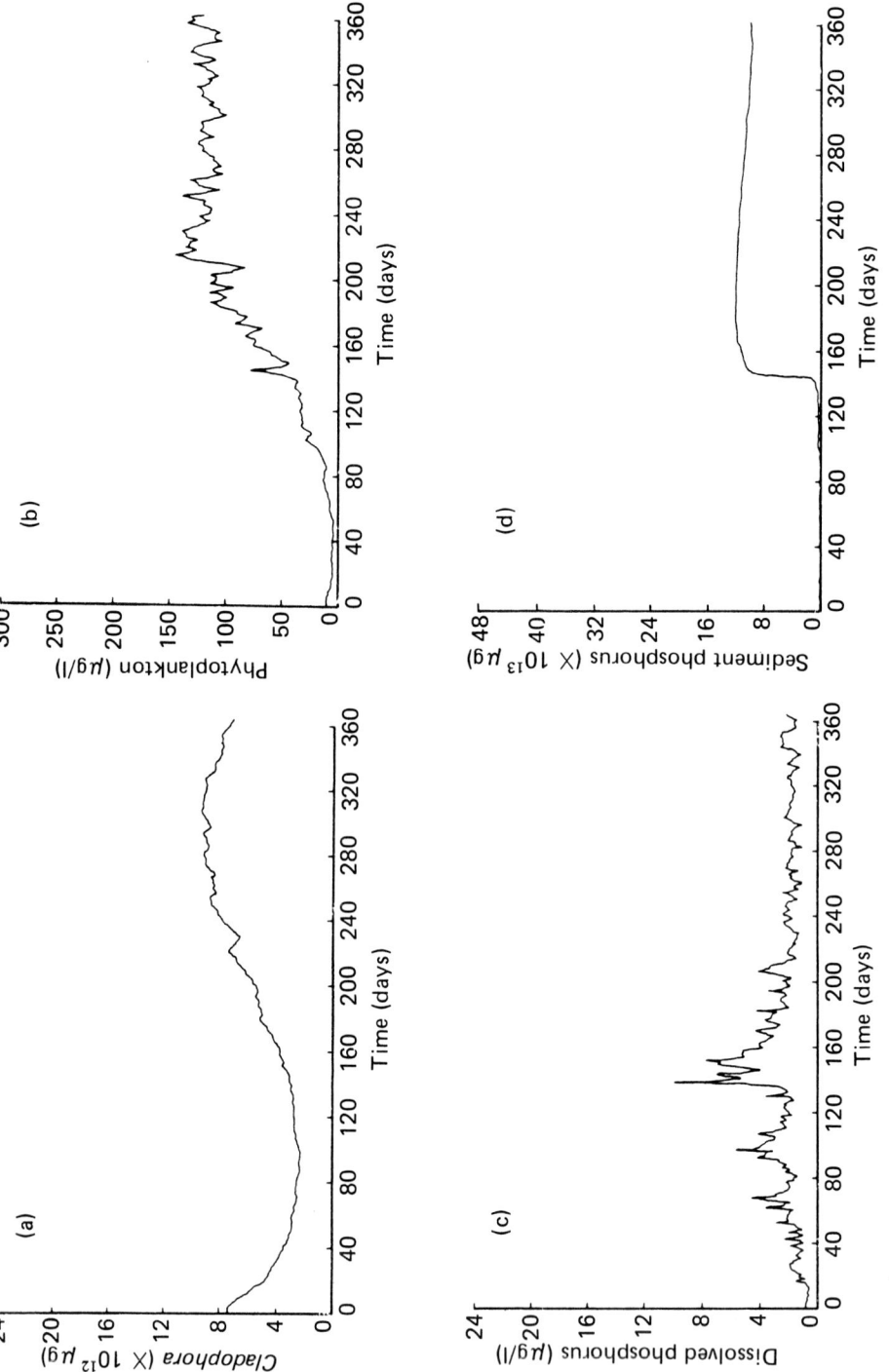

FIGURE 2 Trajectories of a simulation run that failed to satisfy the behavior criteria. All quantities as in Figure 1.

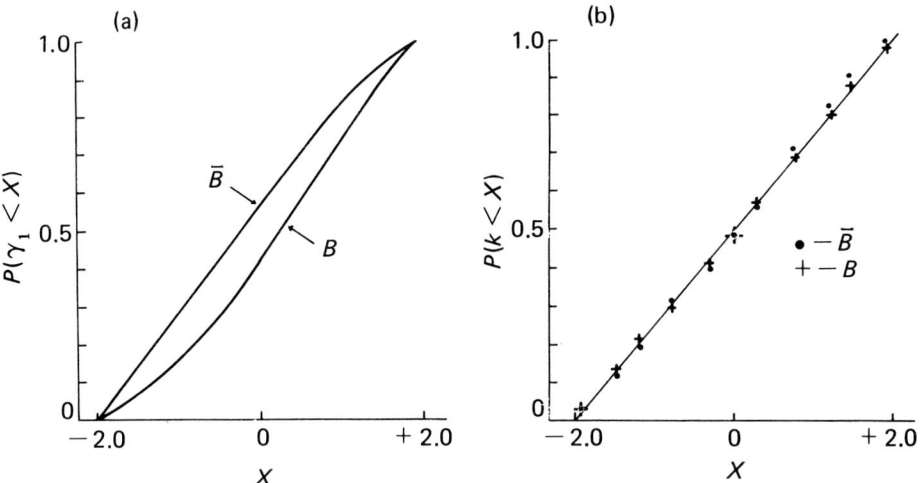

FIGURE 3 Cumulative distribution functions under the behavioral mapping for (a) γ_1, which shows a distinct separation, and (b) k, which shows no separation.

a brief phytoplankton bloom and a marked increase in the growth rate of *Cladophora* at the time of peak river flow about day 140.

Figure 3 shows values of the sample distribution functions under B and $\bar{B}$ for various values of the *Cladophora* growth coefficient γ_1 and for the phytoplankton light-shading coefficient k. Both the Kolmogorov–Smirnov statistic, $d_{m,n}$, and the Mann–Whitney statistic, U, indicated that $F(\gamma_1|B) \neq F(\gamma_1|\bar{B})$ at well above the 99% level of significance. The distributions of the light-shading coefficient k, on the other hand, differ by a maximum of 0.05, a value which corresponds to a level of significance of below 90%. We interpreted these results to indicate that, over the stipulated ranges of uncertainty, γ_1 is an important determinant of the behavior and k is not, at least in terms of the univariate tests. It must be emphasized that these results pertain only to the multidimensional region of parameter space defined by the limits of the *a priori* distributions given in Table 1.

Table 2 contains the class means and variance of each of the nineteen normalized parameters together with the values of $d_{m,n}$ and U. Also included is a classification for each parameter into one of three groups, critical, important, or unimportant. This classification corresponds to the significance levels of the Kolmogorov–Smirnov statistic of greater than 0.99, 0.90–0.99, and less than 0.90, with class 1 being of critical importance. These intervals are somewhat arbitrary since the significance level of any given value of $d_{m,n}$ is a function of sample size. On the other hand as the sample size increases, $d_{m,n}$ will converge to a constant value which is the maximum difference between the cumulative distribution functions $F(\xi_k|B)$ and $F(\xi_k|\bar{B})$, so that although the significance levels associated with the values of $d_{m,n}$ given in Table 2 will continually increase with sample size, the actual values of the statistic will be relatively stable. This is the reason that $d_{m,n}$ was used as the basis of the sensitivity classification.

As shown in Table 2, seven of the nineteen parameters are classified as unimportant. Significantly, these include the parameters related to nutrient inputs from the river

TABLE 2 Univariate results: class means, class variances, $d_{m,n}$, and the corresponding sensitivity classification.

Parameter	$\bar{\xi}_1$	$\bar{\xi}_2$	s_1^2	s_2^2	$d_{m,n}$	Sensitivity class
γ_1	0.181	−0.173	0.804	0.936	0.198	1
K_1	−0.205	0.143	0.967	0.906	0.175	1
X_m	0.223	−0.397	0.804	0.958	0.315	1
a_{11}	−0.189	0.250	1.057	0.957	0.220	1
α	−0.160	0.148	0.891	1.151	0.167	1
γ_2	−0.153	0.150	1.050	1.019	0.151	2
I_s	−0.194	−0.024	1.087	1.029	0.107	2
K_2	0.212	−0.258	0.987	1.030	0.206	1
a_{22}	0.191	−0.196	0.980	1.108	0.181	1
K_W	−0.154	0.078	1.080	1.063	0.124	2
k	−0.001	0.010	0.908	1.027	0.050	3
β	0.064	−0.002	0.938	0.894	0.062	3
a_{43}	−0.249	0.218	0.881	1.040	0.236	1
X_4^*	−0.095	1.361	0.995	0.951	0.142	2
λ_1	0.168	0.036	0.963	1.023	0.076	3
λ_2	−0.079	0.018	0.889	1.064	0.094	3
u_G	0.019	0.047	0.976	0.965	0.052	3
ρ	0.049	0.085	1.027	1.016	0.076	3
S_P	−0.026	−0.050	0.988	0.980	0.054	3

and from groundwater, S_P, u_G, and β, as well as the nutrient recycling parameters λ_1, λ_2, and ρ. These results suggest that, under the modeling assumptions, the system behavior or lack of it is not due to nutrient limitation. Indeed, the results of our analysis suggested that light-limitation may be the critical growth-limiting factor. Elsewhere, we have shown that a linear first-order dominant-mode model with light as input and *Cladophora* biomass as output can reproduce the behavior with remarkable fidelity (Young et al., 1978).

4 EVALUATION OF RESULTS IN LIGHT OF RECENT DATA

Our work on the preliminary analysis of the Peel Inlet *Cladophora* problem was completed in June, 1978 (Spear and Hornberger, 1978). The field research program under the sponsorship of EMAC has continued from that time through the present, and some of the more recent data from the study can be used to "test" a number of the assumptions that were necessary in our original study and to evaluate the results of the work. Many of the data used in this evaluation are taken from McComb et al. (1979) and the remainder were graciously provided by the Systems Analysis Group at the Centre for Resource and Environmental Studies, Australian National University. The latter group, under the direction of Peter Young, is responsible for data collation and analysis for the EMAC study.

The first stage in our *a posteriori* evaluation is to examine data that provide rough estimates for parameters about which we originally had little or no information, at least in terms of observed behavior of the Peel Inlet system itself. We undertake these comparisons not to "verify" our simulation model but rather to determine whether any new data suggest that our original phosphorus scenario is totally inappropriate. That is, if new data suggest that one of our original parameter values was underestimated or overestimated by an order of magnitude or more, the sensitivity rankings deriving from the Monte Carlo results would be suspect. On the other hand, agreement between a newly calculated value and a previously assumed value would not imply the validity of the simulation model itself, but could be interpreted as a failure to alter the speculation on priorities for further research made in the preliminary analysis. In fact agreement between a parameter derived from measurement and one found necessary to simulate the behavior would suggest, at least in terms of our analysis, that the importance of collection of further detailed data on that particular phenomenon would be diminished; we would argue that research efforts in that event should be focused on other processes singled out in the sensitivity analysis.

From the data presented by McComb et al. (1979) a value can be estimated for the following parameters that were identified in the Monte Carlo studies as important for simulating the behavior but for which little or no *a priori* information was available: X_m, the parameter that describes the maximum photosynthesizing biomass of *Cladophora* in the Peel system; γ_1, the temperature–light growth coefficient for *Cladophora*; and a combined estimate for two parameters (α and X_4^*) that describe the phosphate (PO_4) concentration in "interalgal" water.

In our preliminary analysis we hypothesized that self-shading in the *Cladophora* beds would be important and introduced the parameter X_m. The data of Pfeifer and McDiffett (1975) for a riverine species indicated that a density of $30 \text{ g dry weight m}^{-2}$ is appropriate in that situation. We speculated that higher productivity in the estuarine environment might result in a considerably larger value for X_m for Peel Inlet than that for the riverine environment and arbitrarily set the upper limit on the probability density for that parameter at $150 \text{ g dry weight m}^{-2}$. The distribution of X_m separated very clearly under the behavioral classification. The mean value of X_m in the behavior class was about $98 \text{ g dry weight m}^{-2}$. Recently McComb et al. (1979) have estimated from laboratory and field data that the compensation point for the *Cladophora* sp. in Peel Inlet is $15-20 \mu\text{E m}^{-2}\text{s}^{-1}$ and that, due to light-attenuation, this level would be reached at about 1 cm depth in the algal bed. We are not aware of data for Peel Inlet that relates the depth of the algal bed to density but Bach and Josselyn (1978) have reported that a 3-cm depth of a ball-forming *Cladophora* sp. in Bermuda corresponded to a density of $300 \text{ g dry weight m}^{-2}$. The data of McComb et al. are obviously consistent with our preliminary work in this instance.

The *Cladophora* growth coefficient, γ_1, is another parameter for which no prior information was available from the Peel system but which ranked high in our sensitivity classification. McComb et al. (1979) produced a series of curves relating oxygen productivity per gram fresh weight of *Cladophora* to the flux density of photosynthetically active radiation at four temperatures. Using irradiance values below the observed saturation values of McComb et al. and making a number of assumptions about algal composition and functioning (e.g., a photosynthetic quotient of unity, a C:N:P ratio of 18.8:2.7:1, and a fresh weight to dry weight ratio of 8:1) one can derive estimates of γ_1 at the four temperatures reported by McComb et al. (1979) of 2.9×10^{-4}, 4.3×10^{-4}, 4.0×10^{-4}, and $3.6 \times 10^{-4} \text{cm}^2 {}^\circ\text{C}^{-1} \text{cal}^{-1} \text{day}^{-1}$. Two conclusions can be drawn from these results. First, the fact that the values for the four different temperatures are reasonably close to one another argues that our simple multiplicative function for light and temperature is probably adequate. Second, the limits chosen for the rectangular distribution for γ_1 in the original study are 1×10^{-4} to $3 \times 10^{-4}\,\mathrm{cm^2\,°C^{-1}\,cal^{-1}\,day^{-1}}$ and the mean under the behavior was shifted toward the higher end of the distribution. We consider the new data to be consistent with the preliminary guesses.

Finally, McComb et al. (1979) report phosphate (PO_4) concentrations in "interalgal" water of $93\,\mu\mathrm{g\,l^{-1}}$ with a standard deviation of $25\,\mu\mathrm{g\,l^{-1}}$. In our simulations of the behavior the mean concentration available to *Cladophora*, which we interpret as equivalent to that in interalgal water, was defined primarily by two parameters, one defining maximum sediment concentration (X_4^*) and the other the proportion of sediment phosphorus "available" for growth (α). For the behavior simulations the mean values of these parameters yield a value of interalgal concentration of $53\,\mu\mathrm{g\,l^{-1}}$. The "discrepancy" between measured and assumed is again not very large, certainly not large enough to lead us to reject our original sensitivity rankings at this point.

Apart from a comparison of newly calculated parameter values with previously assumed values, a second stage in the *a posteriori* evaluation is to examine new data on the overall system and its behavior and to view this in the context of the processes that were isolated in the sensitivity analysis as deserving of further study. Figure 4(a) shows the assumed *Cladophora* growth area from our 1978 study and Figure 4(b) is a representation of data on measured percent cover of *Cladophora* in the Inlet reported by McComb et al. (1979). Our assumed area of major growth does appear to be an area of dense algal coverage. On the basis of the data used to construct Figure 4(b), McComb et al. (1979) estimated a total biomass of *Cladophora* in the system as "very approximately 20,000 tonnes dry weight" whereas the biomass for our simulated behaviors was about 7000 tonnes on a dry-weight basis.

Perhaps the most striking aspect of the system behavior in terms of *Cladophora* over the time since completion of our Monte Carlo work is the marked decline in biomass in the winter of 1978. Figure 5, after McComb et al. (1979), shows that while the biomass remained relatively constant between late 1976 and the autumn of 1978, a drastic decrease occurred in the winter of 1978. McComb et al. (1979) noted that during the period of this dramatic decline in *Cladophora* biomass "the water of the estuary became very turbid". This increase in turbidity coincides with a phytoplankton bloom that may be the result of increased river input of nutrients. As we pointed out previously, our simulation results for *Cladophora* behavior are very strongly conditioned by available light and the importance of phytoplankton in the model is that relatively small concentrations are sufficient to prevent development of massive *Cladophora* beds through the light-shading effect. It is obvious that this particular aspect of the qualitative behavior predicted by the model does seem to be an observable mode of system functioning under conditions that occur in the Inlet.

The phytoplankton blooms that were observed in Peel Inlet in the winter of 1978 also reinforce a deficiency in the model that we noted in our original report: "the model predicts that phytoplankton should be able to grow in Peel Inlet and that in doing so they should lower phosphorus concentrations below those observed". Even in the behavior-producing runs of the model, maximum phytoplankton concentrations reached $40-50\,\mu\mathrm{g\,l^{-1}}$. Such high concentrations were never observed in the Inlet during the

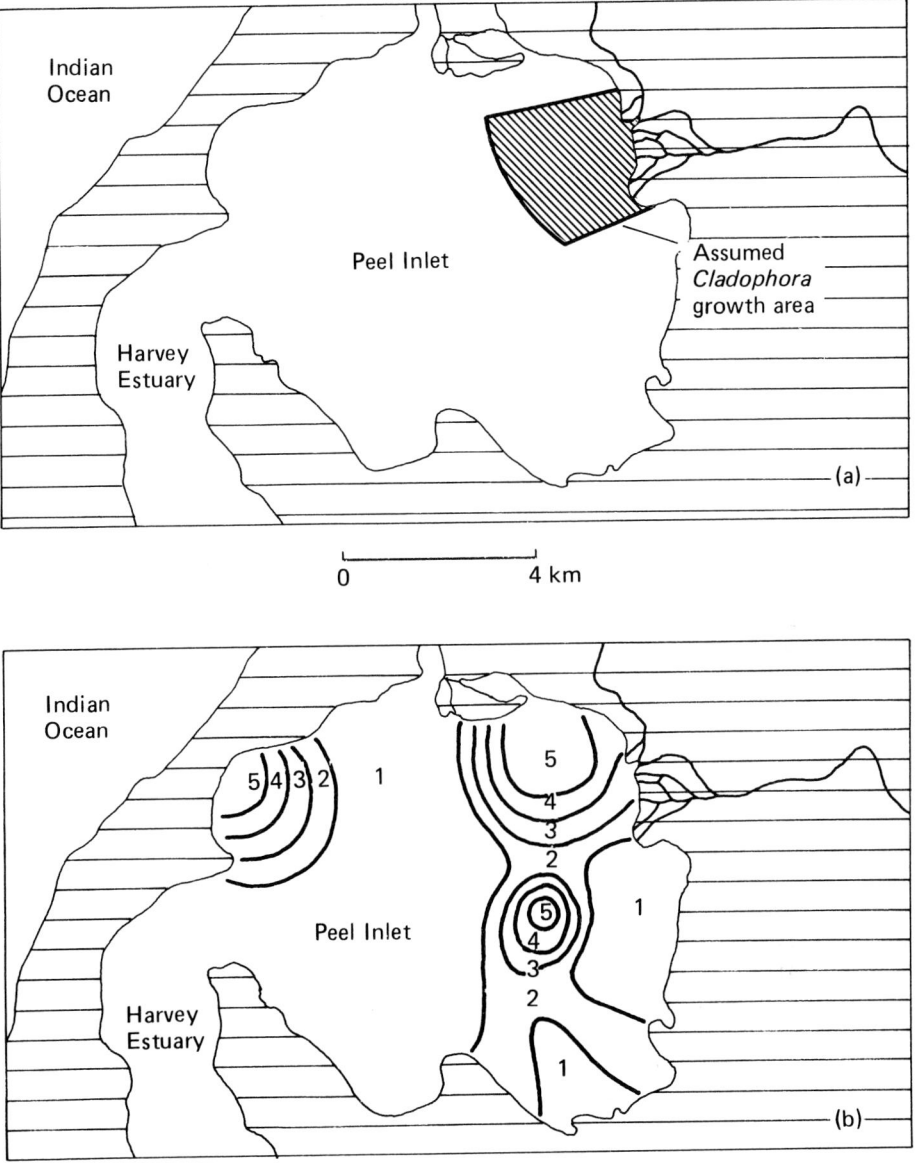

FIGURE 4 (a) Map of Peel Inlet and Harvey Estuary showing the assumed *Cladophora* growth area used in Spear and Hornberger (1978); (b) the observed distribution of *Cladophora* as reported by McComb et al. (1979).

period 1976–1977 and, even during the first bloom observed during the winter of 1978, mean levels of chlorophyll in the Inlet rose to only about $60\,\mu g\,l^{-1}$. Why the model is "wrong" is still unclear but McComb et al. (1979) argued that during the summer/autumn period inorganic nitrogen and not phosphorus limits phytoplankton growth.

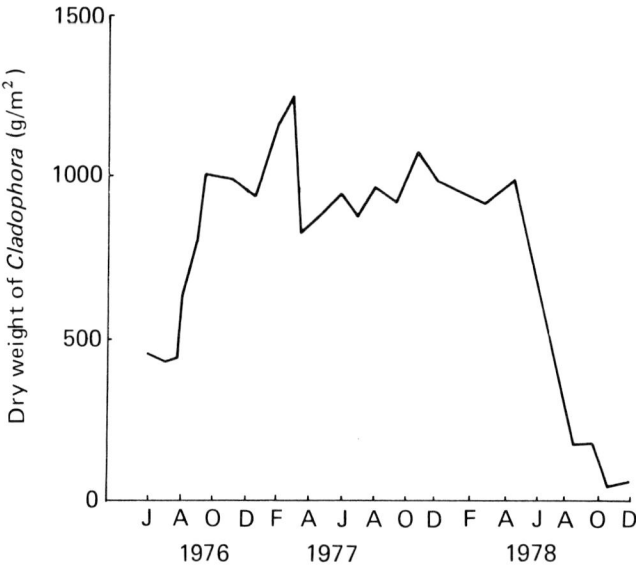

FIGURE 5   Measured *Cladophora* biomass in g dry weight m^{-2} at one site in Peel Inlet from 1976 to 1979 (after McComb et al., 1979).

Considering all aspects of the data available at present, the speculations that we derived from our Monte Carlo work are in remarkably good agreement with conclusions and/or speculations that McComb et al. (1979) derived from their recent laboratory and field measurements. Table 3 compares a number of statements from Spear and Hornberger (1978) with some from McComb et al. (1979). Two possible explanations for this agreement come immediately to mind. The first is that the phosphorus scenario that we constructed is partially correct and that the generalized sensitivity analysis did serve to isolate a number of areas of critical uncertainty. A second explanation might be that everything was very clear from the outset and that our analysis merely served as a framework for exposing obvious relationships. Regardless of the explanation accepted, it appears that the research priorities that we outlined in June, 1978 are to some extent being recognized in the ongoing work and consequently our work may have had some limited value in a practical as well as in an academic sense.

5   DISCUSSION

Although we do not think of our work on the Peel Inlet problem as a form of forecasting, it is possible to interpret it in that sense or at least to envision forecasting-like extensions to it. In principle we see no objection to the use of our approach in the forecasting context. However, because of our conviction that environmental models are, by the nature of the environment, always likely to be ill-defined, we feel it essential that the scenario concept take a prominent role in forecasting applications. In this paper, for example, we have argued that there is presently little evidence to suggest that the

TABLE 3 Statements from the original report (Spear and Hornberger, 1978) on the sensitivity analysis compared with "similar" statements from the paper by McComb et al. (1979) based on recent data.

From Spear and Hornberger (1978)	From McComb et al. (1979)
"In terms of the phosphorus model even moderate concentrations of phytoplankton are sufficient to suppress the growth of *Cladophora* by limiting the light available at the bottom of the Inlet."	"Broadly speaking, the absence of *Cladophora* from the Harvey appears to be related to the higher water turbidity there, due to phytoplankton and other suspended material."
"However, once a large quantity of phosphorus has been located in the sediment, the *Cladophora* are primarily light limited."	"There is little doubt that light must be the primary limiting factor in the estuary, even in shallow water."
". . . explanation of the behavior depends primarily on one feature of the model structure: the presumption that *Cladophora* have access to phosphorus in the sediment."	"This suggests that the alga obtains most of its nutrients not from the water column above, but from the decomposing material below."
"In the model, the second condition [a source of sediment phosphorus] requires that there be a significant input of nutrient to the sediment by the river."	"What is the long-term explanation of the accumulation of nutrients within the algal bank? . . . One possibility is that during river flow there is a deposition of particulate nitrogen and phosphorus."

phosphorus scenario is not a good explanation for the situation in Peel Inlet. However, our confidence in this result would certainly not extend to basing management decisions on phosphorus-model predictions without a thorough study of the nitrogen scenario at the very least.

A final point concerns the basic concept of water-quality forecasting. If it is assumed that the most common conception of forecasting of this sort is aimed toward managing an existing or potential problem and involves a bounded input–bounded output notion, then it is useful to keep in mind that such an approach may simply not be relevant to the short-term situation in the Peel Inlet where the single most important input appears to be light. That is, we can envision no practical way to control this input. On the other hand, other forms of managing the situation, by dredging for example, would most likely perturb the system sufficiently to destroy any confidence we may have had in the predictive abilities of our models. The moral is, perhaps, that the utility of modeling in environmental management is probably very much the same as in traditional engineering analysis: it is great when it works but a solution is much more important than the methodology used to achieve it.

REFERENCES

Atkins, R.P., Gordon, D.M., McComb, A.J., and Pate, J.S. (1977). Botanical Contributions to the Peel/Harvey Estuarine Study. Preliminary Report of Results Obtained During the First Year of the Study. Department of Botany, University of Western Australia, Perth.

Bach, S.D. and Josselyn, M.N. (1978). Mass blooms of the alga *Cladophora* in Bermuda. Marine Pollution Bulletin, 9: 34-37.

Berlinksi, D. (1976). On Systems Analysis. MIT Press, Cambridge, Massachusetts.

Hedgepeth, J.W. (1977). Models and muddles. Helgoländer wissenschaftliche Meeresuntersuchungen, 30:92-104.

McComb, A.J., Atkins, R.P., Birch, P.B., Gordon, D.M., and Lukatelich, R.J. (1979). Eutrophication in the Peel-Harvey estuarine system, Western Australia. In B. Nelson and A. Cronin (Editors), Nutrient Enrichment in Estuaries. Hanover Press, New Jersey.

Pfeifer, R.F. and McDiffett, W.F. (1975). Some factors affecting primary productivity of stream riffle communities. Archiv für Hydrobiologie, 75:306-317.

Spear, R.C. and Hornberger, G.M. (1978). Eutrophication in Peel Inlet: An Analysis of Behavior and Sensitivity of a Poorly-Defined System. Report No. AS/R18, Centre for Resource and Environmental Studies, Australian National University, Canberra, pp. 24-32.

Young, P.C., Hornberger, G.M., and Spear, R.C. (1978). Modeling badly defined systems: some further thoughts. SIMSIG Conference, Canberra, September.

DISTRIBUTION AND TRANSFORMATION OF FENITROTHION SPRAYED ON A POND: MODELING UNDER UNCERTAINTY

Efraim Halfon and R. James Maguire
National Water Research Institute, Canada Centre for Inland Waters, Burlington, Ontario L7R 4A6 (Canada)

1 INTRODUCTION

Fenitrothion* was used during the period 1969–1977 in New Brunswick, Canada, to control the spruce budworm (*Choristoneura fumiferana* [Clemens]) in the province's forests. Millions of hectares were sprayed annually with 150–300 g active ingredient per hectare. The routes and rates of its environmental transformation and disappearance are subjects of much interest (National Research Council of Canada, 1975, 1977). Maguire and Hale (1980) recently reported on the aquatic fate of fenitrothion. Surface water microlayer, subsurface water, suspended solids, and sediment samples were collected from a small pond in a spruce-fir forest in New Brunswick before and after the aerial spraying of a fenitrothion formulation for spruce budworm control; the samples were then analyzed for fenitrothion and its degradation and transformation products. Fenitrothion concentrations in the surface microlayer, subsurface water, suspended solids, and sediment fell below detectable levels two days after the spray; the only identified products were *p*-nitro-*m*-cresol in water, which persisted less than two days, and aminofenitrothion (*O,O*-dimethyl-*O*-(*p*-amino-*m*-tolyl)phosphorothionate) in sediment, which persisted less than four days. Laboratory experiments showed that chemical hydrolysis of fenitrothion and volatilization of fenitrothion from true solution were both slow processes; however, volatilization of fenitrothion from surface slicks was very fast ($t_{1/2} = 18$ min at 20°C). Thus, a large fraction of the fenitrothion that reached the pond surface appeared to volatilize rapidly, while the fraction that remained in the water disappeared or degraded within a few days, largely through photolysis and microbial reduction. The kinetics of appearance and disappearance of fenitrothion and its metabolites are the subjects of this paper.

---

* Systematic name *O,O*-dimethyl-*O*-(*p*-nitro-*m*-tolyl)phosphorothionate.

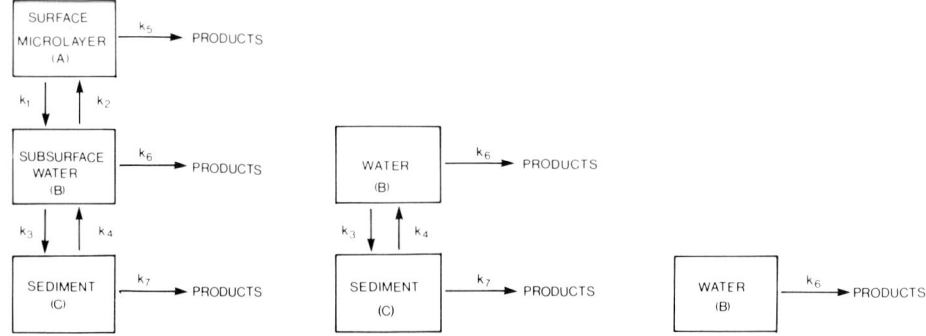

FIGURE 1 Model structures of the three models of fenitrothion in a New Brunswick pond. The relative sizes of the compartments are 28 l (A), 136,000 l (B), and 4460 l (C).

2 MODEL STRUCTURES

A three-compartment model (Figure 1) describes the behavior of fenitrothion in the surface microlayer (compartment A); subsurface water, i.e., the bulk of water in the pond, (B); and sediment (C). Although the surface microlayer contains an insignificant amount of fenitrothion compared with the other two compartments (Maguire and Hale, 1980), it is included in the model since volatization is only important from the surface microlayer. The rate constants k_1–k_4 are first-order rate constants of transfer processes between compartments, and k_5–k_7 are first-order rate constants of removal processes (assumed irreversible), physical or chemical, from each compartment. A removal rate constant may represent the sum of rate constants for a number of processes; in this model, for example, k_5 may represent volatilization and photolysis (producing p-nitro-m-cresol), k_6 may represent photolysis, and k_7 may represent reduction (producing amino-fenitrothion). The behavior of fenitrothion is computed with the following ordinary linear differential equations, where the V symbols are the effective compartment sizes (all with units of volume – Maguire and Hale, 1980) and square brackets represent concentration in $\mu g\, l^{-1}$:

$$V_A d[A]/dt = k_2 V_B[B] - (k_1 + k_5) V_A[A] \tag{1}$$

$$V_B d[B]/dt = k_1 V_A[A] + k_4 V_C[C] - (k_2 + k_3 + k_6) V_B[B] \tag{2}$$

$$V_C d[C]/dt = k_3 V_B[B] - (k_4 + k_7) V_C[C] \tag{3}$$

The effective sizes of the surface microlayer ($V_A = 28$ l) and subsurface water ($V_B = 136,000$ l) are defined as their volumes, with reference to the dimensions of the pond (Maguire and Hale, 1980). The effective size of the sediment is defined for convenience (since a large amount of interstitial water is present in the sediments) in volume units, i.e., as the volume of a 1-cm thick section of sediment over the area of the pond ($V_C = 4460$ l). Initial estimates of the rate constants k_1–k_7 were obtained with data from

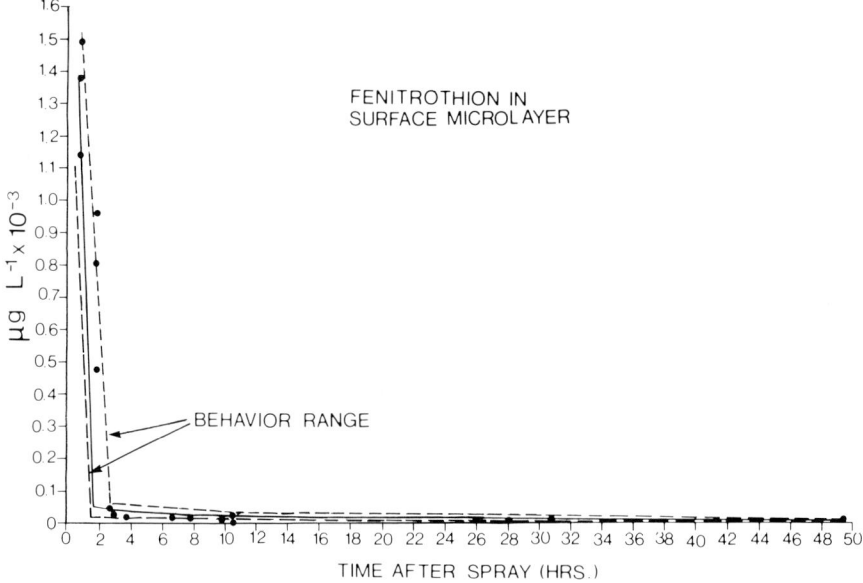

FIGURE 2 Simulation of fenitrothion in surface microlayer. Data points are from three sites in the pond. The solid line is the deterministic simulation with the "best fit" parameter values. Dotted lines define the behavior set.

laboratory and field-sampling experiments, and are shown in Table 1. Concentration–time data for fenitrothion in each compartment are presented in Figures 2–4.

Two-compartment and one-compartment models homomorphic to the three-compartment model are also presented in Figure 1. Table 1 presents relevant information on the three models. The homomorphic relation between any two models S_1 and S_2 is a triple (g, h, k) of maps $g: U_1 \to U_2$, $h: X_1 \to X_2$, $k: Y_1 \to Y_2$ such that $h\delta_1(x, u) = \delta_2(h(x), g(u))$ and $k\lambda_1(x) = \lambda_2(h(x))$ for each $x \in X_1$ and $u \in U_1$. A model is here defined as a quintuple $(U, Y, X, \delta, \lambda)$ where U is a set of admissible inputs, Y is a set of outputs, X is a set of states, $\delta: X \times U \to X$ is the state transition function, and $\lambda: X \to Y$ is the output function. Therefore, S_2 is a valid model of S_1 since S_2 is at least a homomorphic image of S_1. Note that in this particular example, the input is an impulse function which is incorporated in the model through the initial conditions. Some of the stated rules therefore do not apply. Note also that there is a homomorphic relation between the real system, i.e., the pond, and the three-compartment model: it is assumed that the model is a valid representation of the behavior of fenitrothion. The practical dynamics of the aggregation process were made in accordance with the principles stated by O'Neill and Rust (1979).

3   PARAMETER ESTIMATION

Simulations for the three-compartment model were obtained by solving eqns. (1)–(3) numerically on a computer using Hamming's modified predictor–corrector method

TABLE 1 Estimated and calculated values of initial conditions (μg l^{-1}), rate constants (h^{-1}), and compartment sizes (l) in the compartment models[a].

Definition	Estimate[b]	Three-compartment model "best fit"	Two-compartment model "optimal aggregation"	One-compartment model "optimal aggregation"
Initial concentration of fenitrothion in surface microlayer (A) 0.67 h after spray	1,350 (1,140–1,480)			
Initial concentration of fenitrothion in subsurface water (B) 0.67 h after spray	16.1 (13.9–18.3)			
Initial concentration of fenitrothion in sediment (C) at time 0.67 h	0.0			
k_1	0.5	0.864 (0.72–1.02)		
k_2	0.05	0.00571 (0.0048–0.0069)		
k_3	0.389	0.0258 (0.023–0.0282)	0.0329	
k_4	0.008	0.0717 (0.042–0.084)	0.0717	
k_5	2.05	7.385 (6.0–9.6)		
k_6	0.208	0.0704 (0.06–0.072)	0.0704	
k_7	0.0393	0.0885 (0.069–0.102)	0.0885	0.0885
Volume of A	28 ± 3 (23–33)			
Volume of B	136,000 ± 4,000 (1.2 × 10^5–1.5 × 10^5)			
Volume of C	4,460 ± 500 (3,120–5,800)			

[a] Compartment sizes have standard deviations associated with them. Numbers in parentheses are the identified ranges. These numbers have been used in all runs. Initial conditions: means and ranges of three field measurements. Compartment sizes: see text for mean values. Standard deviations and ranges arbitrary.
[b] k_1 is assumed to represent the difference between the rate constant determined for the fast clearance of fenitrothion from the surface microlayer and k_5, i.e., $k_1 = k_{\text{fast}} - k_5 = 0.5 \text{ h}^{-1}$.
k_2 is arbitrarily assumed to be $0.1 \times k_1$; no data are available on the rate of transfer of fenitrothion from bulk solution to the surface microlayer.
k_3 is assumed to be the rate constant of adsorption to the sediment; it has been assigned the value associated with the appearance of fenitrothion in the sediment, determined from the field study.

k_4 is assumed to represent the difference between the rate constant for disappearance of fenitrothion from the sediment and the rate constant for appearance of aminofenitrothion in the sediment, i.e., $k_{\text{fen. disappearance}} - k_{\text{aminofen. appearance}} = k_4 = 0.008\,\text{h}^{-1}$.

k_5 is assumed to be the rate constant for volatilization from surface slicks, and is assigned the value determined in laboratory experiments, $2.05\,\text{h}^{-1}$.

k_6 is assumed to be the rate constant for photolysis; it has been assigned the value associated with the appearance of p-nitro-m-cresol in the subsurface water, determined in the field study.

k_7 is assumed to be the rate constant for microbial reduction; it has been assigned the value associated with the appearance of aminofenitrothion in the sediment, determined in the field study.

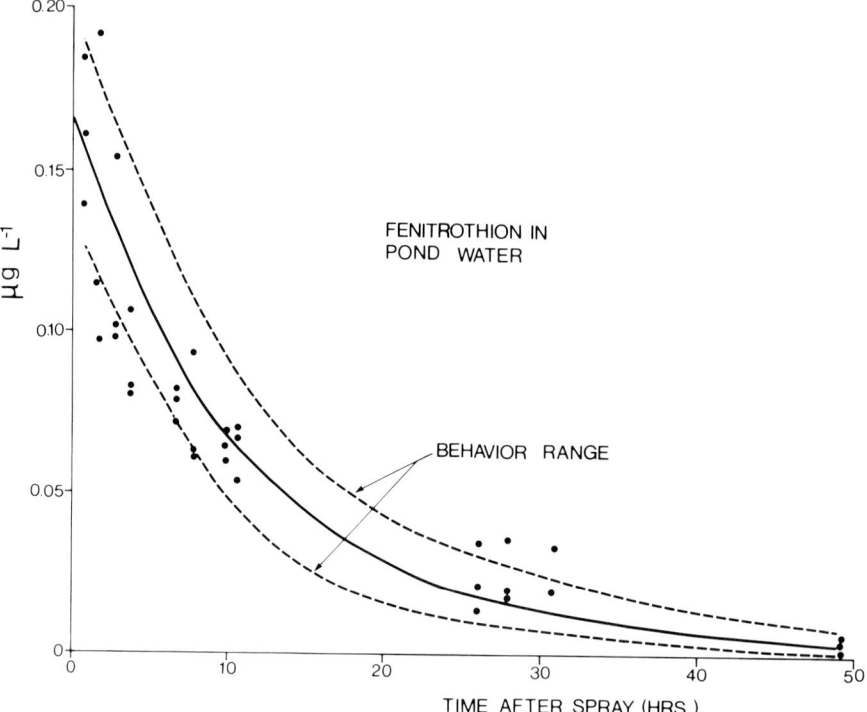

FIGURE 3 Simulation of fenitrothion in subsurface water. Other details are given in Figure 2.

(Ralston and Wilf, 1960) with IBM Scientific Subroutine HPCG. The "best fit" values of the rate constants were obtained by using a random search algorithm (Price, 1977), which minimized a weighted difference between observed and predicted concentrations. Theil's inequality function (Theil, 1970) was chosen as the objective function since it put equal weight on the goodness of fit of each compartment, regardless of size. By contrast, a linear least-squares function was unsatisfactory since it put too much weight on concentrations in the surface microlayer relative to concentrations in the other compartments. Data used in the computations are shown in Figures 2, 3, and 4. These figures show the fit of the simulations for compartment A, B, and C, respectively. Simulations for the two-compartment model were done in the same way and the results are practically identical to the relevant parts of the three-compartment model (Figures 2 and 3 are also for the two-compartment model; computed rate constants are given in Table 1). The one-compartment model produces simulations only for the water compartment (Figure 3). This simulation is slightly lower than those of the larger models (see Table 1 for coefficients).

4 AGGREGATION ERROR

The effects of the aggregation were computed by the total error, $TE = \int_0^{50 \text{ hours}} E^2(t) \, dt$. Fifty hours is the time span over which data are available. The error was

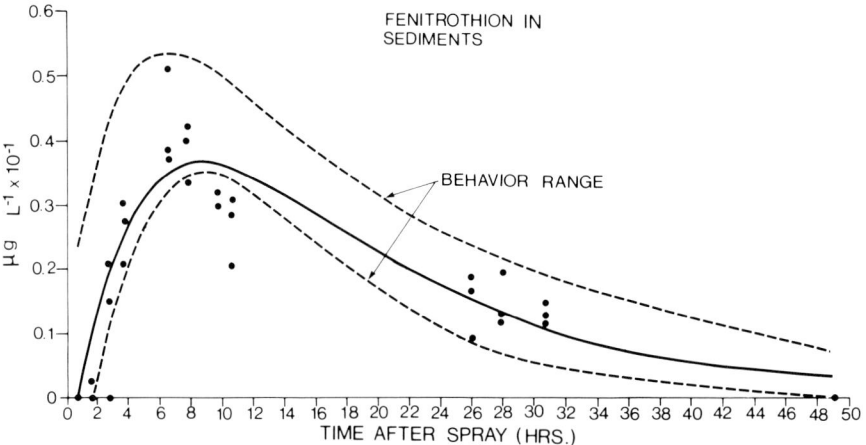

FIGURE 4 Simulation of fenitrothion in sediment. See Figure 2 for details.

computed only for the subsurface water compartment (B) since this is the only compartment common to all three models. The error E is the amount of fenitrothion in water for model j minus the amount of fenitrothion in water for model i (where $j = 2, 3$; $i = 1, 2$; $i \neq j$). The digits indicate the number of compartments in the model. The errors, expressed as percentages (i.e., $100 \times TE/$(initial amount of fenitrothion in water)) are aggregation from: three to two compartments 0.35%, two to one compartment 3.76%, and three to one compartment 5.76%. From these results we conclude that the two-compartment model is almost a perfect representation of the three-compartment model. The one-compartment model is also a valid model for the water compartment.

5 SENSITIVITY ANALYSIS THROUGH MONTE CARLO SIMULATIONS

We performed a sensitivity analysis similar to that described by Hornberger and Spear (preceding paper in this volume, pp. 101–116) to which readers are referred for the technical details. Two sets, "behavior" and "nonbehavior", were identified. The "behavior" set was defined as ranges of concentrations for the three compartments. These ranges were identified by data collected in the pond. However, since some data were too noisy (Figures 2–4), the "behavior" range was arbitrarily reduced or enlarged at points. A simulation was accepted in the behavior set if conditions for all three compartments were satisfied. Note that this approach allowed some freedom in the determination of the maximum concentration of fenitrothion in the sediments (Figure 4) since sampling may not have occurred at the time of maximum fenitrothion concentration. In Table 1 the ranges of the parameter values are presented with the "best estimate" for the three-compartment model. A uniform frequency distribution was chosen for the parameters. The initial conditions were considered to be known. The ranges of the parameters were chosen after examination of the results of the random search: minimum and maximum values of the parameters which produced a good fit through the behavior were used. Large limits would not have affected the sensitivity analysis much since values outside these limits would have belonged to the nonbehavior set anyway.

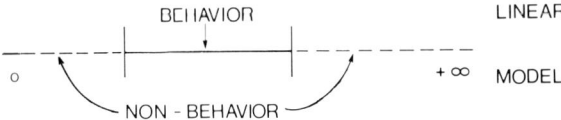

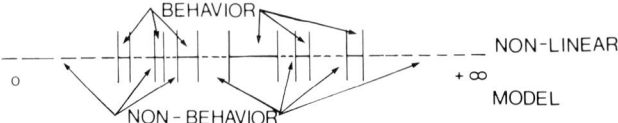

FIGURE 5 Line defining range of possible parameter values. For linear models, when only one parameter is modified at a time, the simulations remain in the behavior zone for a determined set of parameter values only (univariate case). For nonlinear models, when only one parameter is modified at a time (univariate case), the simulations may move in and out of the behavior zone.

In contrast to the procedure of Hornberger and Spear (preceding paper), we performed a multivariate sensitivity analysis. From preliminary runs we found that, for a linear model, a univariate search would be useless since simulations in the "behavior" set are obtained for a continuous but limited range for each parameter (Figure 5). The multivariate analysis (Table 2) showed that the model was most sensitive to estimates of the sediment volume. This result was expected since no direct measure of the amount of actively adsorbing sediment was possible. The model was also relatively sensitive to the volume estimate of the surface microlayer and relatively insensitive to all other rate constants. Among the "true" parameters k_1–k_7, the parameters k_1, k_5, and k_6 showed a somewhat larger influence than the others. These results were obtained from 1000 runs of the model with 121 runs falling in the "behavior" set and 879 in the "nonbehavior" set. Note that in these runs the volumes were also considered model "parameters" and that these results should be considered for values within the ranges indicated in Table 1.

6 PREDICTION UNDER UNCERTAINTY

The most important goal of the field-research program was the identification of the behavior of fenitrothion and its degradation products. The chemistry of fenitrothion and its metabolites in the pond has been described by Maguire and Hale (1980). From field data and laboratory investigations, they noted that these chemical reactions had a degree of natural variability which made predictions through a deterministic model somewhat unreliable. Following Halfon (1979) and O'Neill and Gardner (1979) we decided to use Monte Carlo simulations to assess the frequency distribution of the time needed for 99% of fenitrothion to disappear from the pond. Two sets of 1000 Monte Carlo runs were performed, one with the parameters having a triangular frequency distribution, as suggested by Tiwari and Hobbie (1976), and the other with a uniform frequency distribution, signifying our uncertain knowledge of the rates of the chemical reactions. The initial conditions had a uniform distribution, with the limits derived from the data

TABLE 2 Multivariate results of sensitivity analysis through Monte Carlo simulations: class means, class standard deviations, Kolmogorov–Smirnov maximum distance between the "behavior" and "nonbehavior" distribution functions ($d_{n,m}$), and the corresponding sensitivity classification[a] for the parameters.

Parameter	Mean behavior	Mean nonbehavior	Standard-deviation behavior	Standard-deviation nonbehavior	$d_{n,m}$	Sensitivity class[a]
k_1	0.8659	0.8599	0.0554	0.0579	0.0871	3
k_2	0.0058	0.0059	0.0005	0.0005	0.0755	4
k_3	0.0256	0.0257	0.0015	0.0015	0.0478	4
k_4	0.0727	0.0727	0.0065	0.0066	0.0737	4
k_5	7.5333	7.3705	0.5760	0.5821	0.0845	3
k_6	0.0672	0.0679	0.0022	0.0024	0.0888	3
k_7	0.0884	0.0885	0.0057	0.0059	0.0429	4
Volume of A	29.2267	27.8060	2.6160	2.8667	0.1352	2
Volume of B	133,658.9586	136,206.2664	8,226.0447	8,024.8455	0.0843	3
Volume of C	4,521.3060	4,519.3376	410.4819	817.6992	0.2497	1

[a] Sensitivity classes: 1, parameter very significant ($p > 99\%$); 2, significant ($99 \geqslant p > 90\%$); 3, relatively important ($90 \geqslant p > 50\%$); 4, unimportant ($p \leqslant 50\%$). The percentages indicate the probability that the two sets "behavior" and "nonbehavior" are different due to the given parameter.

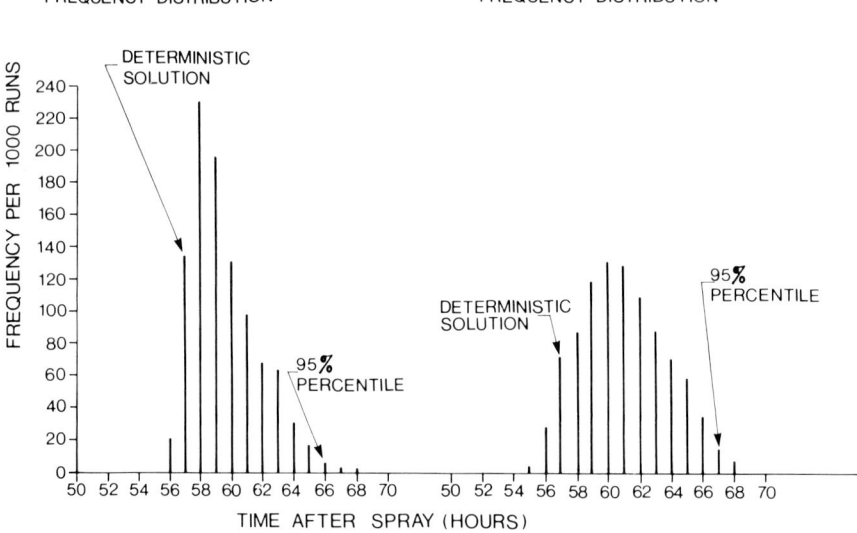

FIGURE 6 Histograms describing the frequency distribution of times when 99% of fenitrothion has disappeared. The left figure describes the case when the parameters have a triangular frequency distribution; the right, when parameters have uniform distribution. Time is in hours.

(Table 1), since we did not know exactly how much fenitrothion actually fell on the pond, and the volumes had a normal distribution (Table 1).

Results (Figure 6) showed that the two sets of runs produced equal ranges with 99% degradation occurring at not earlier than 54 and not later than 68 hours. When parameters belong to a triangular frequency distribution, we can confidently state that most fenitrothion will be eliminated by the 59th hour; in the second case (uniform distribution), by hour 64. The deterministic simulation predicts removal of 99% of fenitrothion in 56.9 hours, which is about a 20% underestimate of the worst case (68 hours).

7 DISCUSSION

Pesticides volatilize and degrade according to their chemical properties and those of the environment (e.g., pH) where they are sprayed. Transformations by biota are relatively unimportant to a mass balance of many pesticides. For this reason, most models of toxic substances are linear, as in this instance. Also, modelers commonly follow the degradation of the pesticide only in water, where it is most easily detected, and then compute the steady states in the other compartments analytically. This approach was taken by, among others, Lassiter et al. (1979) who developed an effective computer program (EXAMS) for predicting the fate of a chemical compound in natural waters. With the aggregation analysis we found that, at least for fenitrothion, a detailed three-compartment model is necessary when a clear understanding of all chemical processes involved is required, as in this case. In fact, a three-compartment model requires more data to be developed and validated. The aggregation analysis showed that for prediction purposes in water, a one-compartment model is quite adequate and that the total relative error was quite small. A two-compartment model is also quite adequate to represent most processes if the surface microlayer can be neglected. In this paper more emphasis has been given to the complete model since we are interested in understanding as well as predicting the behavior of the fenitrothion.

Laboratory experiments were performed to measure rates of hydrolysis and volatilization (Maguire and Hale, 1980; see also Table 1, footnote b). When the search for the "best fit" parameter values was performed, it was found that some estimates were too low (k_1, k_4, k_5, k_7), or too high (k_3, k_6) and that a model run with these estimates would not fit the data: the global decay rate would be too slow. Care must be taken when applying laboratory data to the modeling of field conditions, especially when not all environmental conditions are taken into account. The sensitivity analysis also showed that a careful determination of field conditions, in particular sediment volume, was very important to the understanding of the behavior of fenitrothion in the pond. Since the volume of actively absorbing sediment is difficult to measure, some uncertainty remains, which presently cannot be eliminated. However, we have found that if we are willing to ignore the relative importance of fenitrothion in the sediments, we can still obtain relatively good estimates of the time needed for the chemical to disappear completely. Prediction capability is less influenced by noisy data and lack of knowledge than is the understanding of the chemistry. Therefore we conclude that, for fenitrothion in a given pond in New Brunswick, the problems of understanding and prediction under uncertainty are weakly coupled and each can be approached separately. Future work will seek confirmation of this hypothesis with other toxic substances in other environments.

REFERENCES

Halfon, E. (1979). The effects of data variability in the development and validation of ecosystem models. In B.P. Zeigler, M.S. Elzas, G.J. Klir, and T. I. Ören (Editors), Methodology in Systems Modelling and Simulation. North-Holland, Amsterdam, pp. 335–343.

Lassiter, R.R., Cline, D.M., and Burns, L. (1979). Research and development on an exposure analysis modeling system (EXAMS). US Environmental Protection Agency, Athens, Georgia (unpublished prospectus).

Maguire, R.J. and Hale, E.J. (1980). Fenitrothion sprayed on a pond: kinetics of its distribution and transformation in water and sediment. Journal of Agricultural and Food Chemistry, 28: 372–378.

National Research Council of Canada (1975). Fenitrothion: Effects of its Use on Environmental Quality and its Chemistry. NRCC Publication No. 14104, Ottawa, Ontario.

National Research Council of Canada (1977). Proceedings of a Symposium on Fenitrothion: Long-term Effects of its Use in Forest Ecosystems. NRCC Publication No. 16073, Ottawa, Ontario.

O'Neill, R.V. and Gardner, R.H. (1979). Sources of uncertainty in ecological models. In B.P. Zeigler, M.S. Elzas, G.J. Klir, and T.I. Ören (Editors), Methodology in Systems Modelling and Simulation, North-Holland, Amsterdam, pp. 447–463.

O'Neill, R.V. and Rust, B. (1979). Aggregation error in ecological models. Ecological Modelling, 7: 91–105.

Price, W.L. (1977). A controlled random search procedure for global optimisation. Computer Journal, 20: 367–370.

Ralston, A. and Wilf, H.S. (1960). Mathematical Methods for Digital Computers. Wiley, New York, pp. 95–109.
Theil, H. (1970). Economic Forecasts and Policy. North Holland, Amsterdam.
Tiwari, J.L. and Hobbie, J.E. (1976). Random differential equations as models of ecosystems: Monte Carlo simulation approach. Mathematical Bioscience, 38:25–44.

INPUT DATA UNCERTAINTY AND PARAMETER SENSITIVITY IN A LAKE HYDRODYNAMIC MODEL

L. Somlyódy*
International Institute for Applied Systems Analysis, Laxenburg (Austria)

1   INTRODUCTION

Hydrodynamic models are often used to calculate the magnitude and direction of the wind-induced motion of water in lakes, in both engineering and water quality problems. The one- and two-dimensional model versions most frequently employed have two major parameters, the wind drag coefficient and the bottom friction coefficient. Although a number of important experiments have been performed in relation to the drag coefficient (for example Wu, 1969; Graf and Prost, 1980) and some information is also available to define a feasible range of values for the bottom friction, both parameters should be the subjects of model calibration as they are lumped in character.

The reliability of a well-established hydrodynamic model depends primarily on two factors:

(i) parameter sensitivity, which indicates how the model simulation is distorted by a given error in the parameter vector; and
(ii) the influence of input data (in this case the wind velocity vector) uncertainty, in other words, input data sensitivity.

If the parameters have any meaningful physical interpretation (as is the case here), factor (i) is more related to research on the general subject concerned, while factor (ii) pertains to data collection for the specific system studied. Since no general rules are available to decide which issue is the more important but the consequences — whether to concentrate on research or on data collection — are quite different, both factors should be analyzed separately and their influence compared and contrasted.

There are numerous papers in the literature on parameter sensitivity (for example Halfon, 1977; Rinaldi and Soncini-Sessa, 1978; Kohberger et al., 1978; van Straten and de Boer, 1979; Gardner et al., 1981; Beck, 1983) but far fewer on input data uncertainty

---

*On leave from the Research Centre of Water Resources Development, VITUKI, Budapest, Hungary.

(Somlyódy, 1981; Scavia et al., 1981; Fedra, 1983; among others); joint studies of *both* factors are very rare. This is especially true for hydrodynamic models where even parameter sensitivity is seldom incorporated in the analysis and very little interest is shown in input data uncertainty.

Our objective here is as follows: for the example of a one-dimensional lake hydrodynamic model we wish to study both parameter and input data sensitivity, and compare and contrast the two. A trial and error method is used for model calibration and deterministic parameter sensitivity analysis is performed numerically. An order of magnitude analysis and a Monte Carlo simulation are performed to investigate input data uncertainty. The 1-D hydrodynamic model, and also more comprehensive model versions, accounting for more than just longitudinal movement (Shanahan and Harleman, 1981; Shanahan et al., 1981), were developed in the wider framework of a eutrophication study of Lake Balaton. For more details of the Lake Balaton study the reader is referred to van Straten and Somlyódy (1980) and Somlyódy (1981).

The structure of the paper is as follows. The model is presented in Section 2, while Section 3 gives some background information and a description of the wind data for Lake Balaton. Section 4 deals with calibration and parameter sensitivity and Section 5 discusses validation. The influence of input data uncertainty is considered in Section 6, while Section 7 gives an extension for multidimensional models and other lakes. Finally, the main conclusions are summarized in Section 8.

2 MODEL FORMULATION

2.1 Governing Equations

The motion of water along the lake's axis x (see Figure 1) is described by the one-dimensional equations of motion and continuity often adapted to river flow situations (Mahmood and Yevjevich, 1975; Kozák, 1977)

$$\frac{\partial U}{\partial t} = -g\frac{\partial z}{\partial x} - \frac{1}{2}\frac{\partial}{\partial x}(U^2) + \frac{1}{H\rho}(\tau_s + \tau_b) \quad (1)$$

$$\frac{\partial A}{\partial t} = -\frac{\partial Q}{\partial x} \quad (2)$$

where the latter can be rewritten as

$$B\frac{\partial z}{\partial t} = -\frac{\partial}{\partial x}[UB(H_1 + z)] \quad (2a)$$

Here

$U = Q/A$ = longitudinal flow velocity averaged over the cross-section,
$A = B(H_1 + z)$,
Q = stream flow rate,

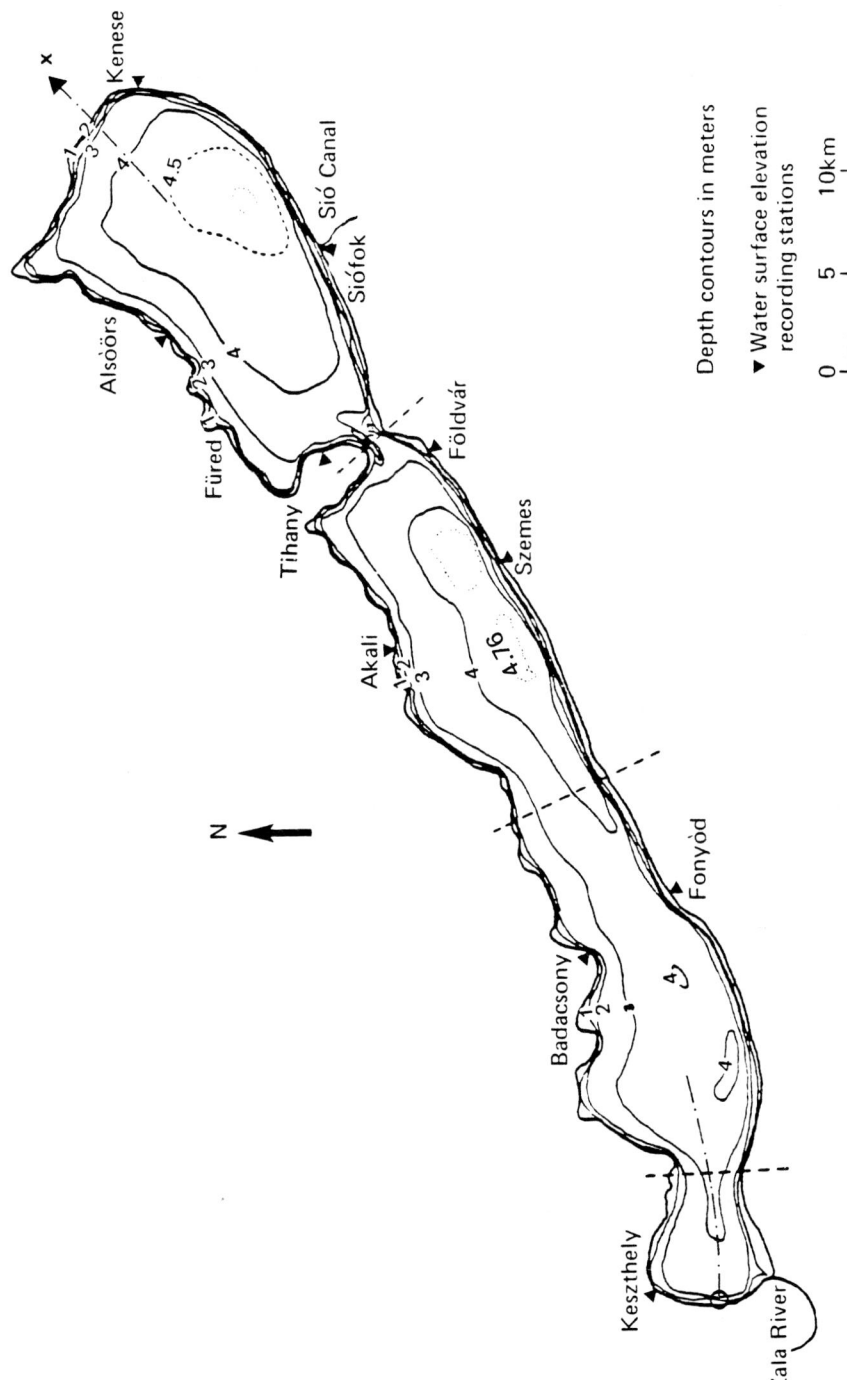

FIGURE 1 Map of Lake Balaton, showing depth contours and the locations of the stations measuring water surface elevation.

z = water level elevation (e.g., due to wind),
B = width of lake,
H_1 = nondisturbed water depth,
$H = H_1 + z$ = real depth,
τ_s = wind shear stress at the water surface,
τ_b = frictional shear stress at the lake bottom,
ρ = density of water, and
g = acceleration due to gravity.

The shear stresses are described by introducing the drag coefficient C_D and bottom friction coefficient λ (see for example Lick, 1976; Virtanen, 1978)

$$\tau_s = \rho_a C_D W_x |W| \qquad (3)$$

$$\tau_b = -\rho \lambda U |U|^n = -\rho \lambda Q |Q|^n / A^{1+n} \qquad (4)$$

where

ρ_a = density of air,
W = wind speed,
W_x = longitudinal component of wind speed, and
n = bottom friction exponent ($0 \leq n \leq 1$).

Here the quadratic law will be used, so that $n = 1$. As can be seen from eqn. (4), τ_b is related to the cross-sectional average velocity rather than to the local velocity in the vicinity of the bed; consequently λ is a lumped parameter.

In subsequent stages the equations listed above are rearranged for z and Q as unknown variables and dimensionless quantities are introduced; for details see Somlyódy and Virtanen (1982). Boundary conditions for one of the variables should be defined at the two ends of the lake, where $x = 0$ and $x = L$, respectively. For typical lake problems $Q(t, 0)$ and $Q(t, L)$ are generally given. If $Q(t, 0) = Q(t, L) = 0$, no inflow or outflow takes place, a situation which will be considered here.

2.2 Numerical Solution

An implicit finite difference scheme (Mahmood and Yevjevich, 1975) is selected and coupled to a matrix sweep technique. Time derivatives are approximated by differences centered in both space and time, while for space derivatives the differences are centered in space but weighted in time. Space-centered, forward-time approximations are employed for all coefficients and nonderivative terms except the bottom shear term (eqn. 4) where a more detailed approach, centered in space but weighted in time, is used; for details see Somlyódy and Virtanen (1982).

The resulting scheme is unconditionally stable in time. For z and Q, $(2N - 2)$ linear algebraic equations are generated which are then closed by the boundary conditions to the $2N$ unknown variables (N is the number of grid points). In order to make computation more economical a matrix sweep method was developed for the solution of the system of equations; using this method the number of elementary operations (and thus the execution time) is proportional to N rather than N^3 (which is the case for conventional matrix inversion methods). The price of this computational advantage is that spatial stability should be assured. As was shown by Somlyódy and Virtanen (1982), "weak" stability can be maintained either for relatively small or for relatively large time steps (e.g., $\Delta t <$ 100–200 s or $\Delta t > 900$ s), with $\Delta x = 2000$ m and $H = 3$–4 m which are typical values for the Lake Balaton problem.

3 CHARACTERISTICS OF LAKE BALATON: THE WIND FIELD AND ITS ASSOCIATED UNCERTAINTIES

Lake Balaton is long and narrow (78 km × roughly 8 km, see Figure 1) and extremely shallow. The average depth is 3.14 m, and the lake is less than 5 m deep everywhere except in one small area near the Tihany peninsula which divides the lake into two. In this latter region a river-type water motion is observed which changes direction depending on wind conditions and the associated water level oscillation. The velocity in this region sometimes exceeds 1 m s^{-1} (corresponding to a flow rate of around 4000 m^3 s^{-1}) — a very high value (Muszkalay, 1973) — while in other areas of the lake it is generally less than 0.1–0.15 m s^{-1}. The shallowness of the lake permits a water motion response to even mild winds and because of the fluctuations in the wind field a steady state never exists. This is well illustrated by the velocity measurements of Shanahan et al. (1981).

The prevailing wind direction lies between northwest and north. This is particularly apparent if strong wind (>8 m s^{-1}) events and summer periods are considered (Béll and Takács, 1974). The monthly average wind velocity ranges from 2 to 5 m s^{-1}; however the maximum may reach 30 m s^{-1}. The hourly average wind velocity exceeds 8 m s^{-1} at Siófok (see Figure 1) during approximately 15% of the year (Béll and Takács, 1974). The number of seiche-type events (strictly speaking seiche is the lake's response to a single wind impulse) is about 1000 per year (Muszkalay, 1979).

The temporal and spatial variations in wind are strongly influenced by the surrounding hills of the northern shoreline. The sequence of hills results not only in a nonuniform wind distribution along the northern shoreline itself but also in a highly variable velocity field above the lake due to sheltering, channelling, deviating, and separating effects. The average wind characteristics at various points clearly demonstrate these phenomena. For instance, at the eastern end of the lake the prevailing wind direction is northwesterly, at the other end northerly, while at the middle of the lake on the southern shoreline it is northeasterly (Béll and Takács, 1974). The magnitude of the spatial changes in direction is well illustrated by comparison of the records for Keszthely and Siófok (see Figure 1) during 1977: for $W > 3$ m s^{-1}, the mean value of the difference in wind direction is 39°, while its standard deviation is 36°, suggesting a relatively wide range within which the wind direction can fluctuate above the lake.

Due to sheltering effects the yearly average wind speed is 40–60% higher at Siófok than at Keszthely. This marked spatial variation in behavior is observed for nearly all individual storms. Transverse inhomogeneity in wind speed (i.e., at sites on opposite *sides* of the lake rather than opposite *ends*), due to the presence of mountains and the relative "smoothness" of the water surface compared to the surrounding terrain, has also been indicated by some observations (Béll and Takács, 1974). The number of wind recording stations (at Siófok, Szemes, and Keszthely, together with some temporary gauges (see Somlyódy, 1979)) is insufficient for accurate specification of the wind field described above (although an acceptable estimate can be made for the longitudinal distribution of W; see previous sections and Somlyódy and Virtanen (1982)). Consequently, uncertainty plays an important role and should be explicitly accounted for in the course of any modeling effort.

Out of the two variables defining the velocity vector, the absolute value of wind speed and the wind direction, the latter is far more important (see Section 6). Therefore further discussion of the uncertainties associated with wind direction is now necessary.

The most important sources of error are as follows:

(i) Incorrect registration and time averaging of the direction as a stochastic variable. As a result of turbulent fluctuations, the continuous records often define a domain 40–60° wide rather than a single line.
(ii) The discrete resolution of many of the measuring instruments, involving steps of 22.5° or sometimes even 45°, rather than continuous measurement.
(iii) The nonuniformity (randomness in space) of the wind field. Based on the example given previously (the comparison of records for Keszthely and Siófok) this may exceed 90°.

Each of these factors must be dealt with in a different way. Concerning item (i), for example, a Gaussian distribution can be hypothesized. For item (ii), the angle α defining wind direction can randomly take three discrete values (the mean, ±22.5°, or ±45°). No information is available concerning the character of spatial randomness and therefore the assumption of a uniform distribution is the most feasible. The corresponding strategies used in the course of the Monte Carlo simulation are given in Section 6.

The most detailed study to date on the motion of water in the lake was performed by Muszkalay (1973) who collected a set of water surface-elevation observations for ten years at up to ten stations around the lake (see Figure 1). Simultaneous measurements of wind speed at one or two of the stations and occasional measurements of water current in the Strait of Tihany completed his data set, which will serve as the basis for our analysis.

From his observations Muszkalay selected typical stormy events and looked for empirical relationships between wind parameters and I (which is the difference of the extreme water levels at the two ends of the lake observed during a storm, divided by the length of the lake), and between wind parameters and the maximum water velocity in the Strait. Some of the results based on the regression equations he developed (Muszkalay, 1973) are illustrated in Figures 2 and 3, for storm durations of 2 and 12 h and for $|\alpha^*| < 22.5°$ (α^* is the angle defined by the wind velocity vector and the longitudinal axis of the lake). Muszkalay gave I as a function of the instantaneous peak wind speed, W'_{max}, which is essentially higher than the maximum for a reasonable averaging period (e.g., an hour), $\bar{W}_{max}$; the ratio $W'_{max} : \bar{W}_{max}$ varies in the range 1.2–1.3. Since W'_{max} will not be used in the calibration stage and a rectangular wind input will be employed, the range given in Figure 2 corresponds to the factor 1.2–1.3. It should be stressed that this range does not incorporate the complete scatter of the original data. As is apparent from the figure, I depends linearly

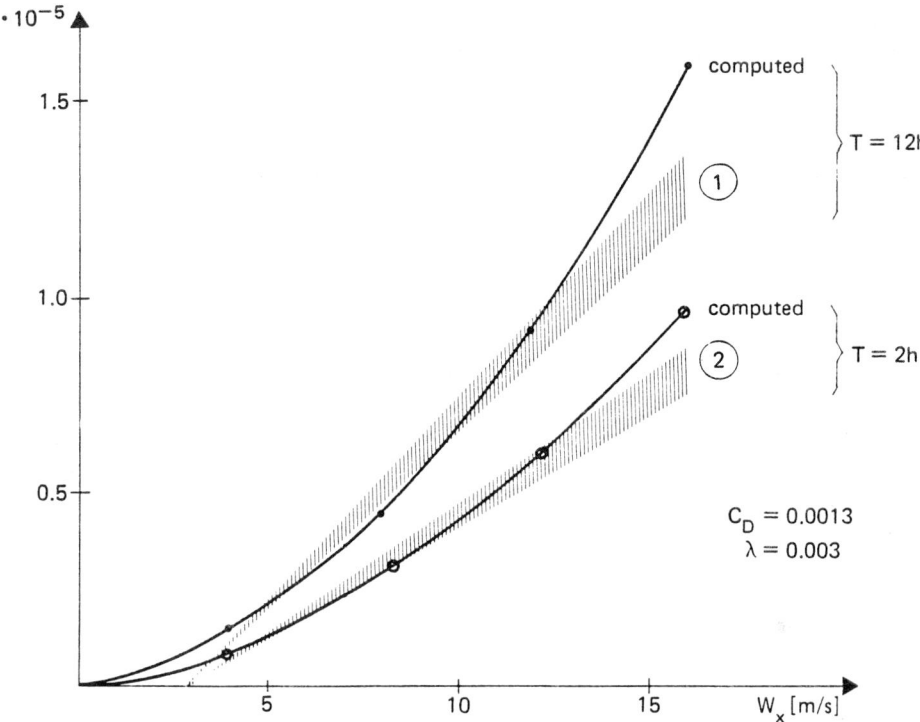

FIGURE 2 Maximum water level differences along the lake for wind events of (1) 12-h and (2) 2-h duration. The shaded areas are based on the results of Muszkalay (1973).

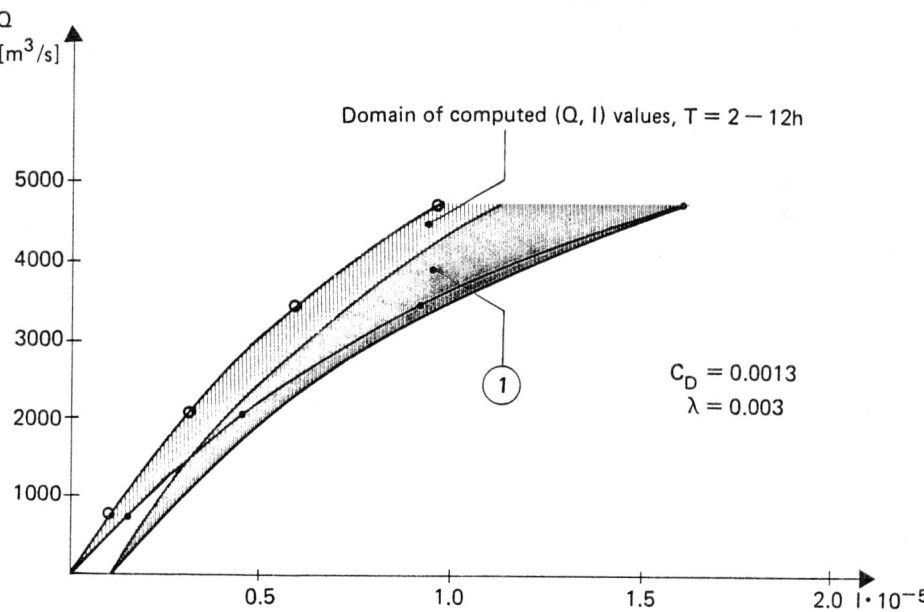

FIGURE 3 Stream flow rate at Tihany. The shaded area (1) is based on the results of Muszkalay (1973).

on W_x although theory suggests a quadratic relation. It is noted, however, that most of the data utilized lay in the 5–12 m s^{-1} domain and a quadratic fitting could also have been performed. As can be seen from Figure 2, the maximum value of I (in absolute terms) can exceed 1 m, a very high value compared to the average depth of the lake.

A domain is also given for the stream flow in Figure 3. Originally Muszkalay derived an empirical equation for the velocity 1 m below the free surface in a typical vertical cross-section. From this the stream flow rate can be calculated as a function of I but only approximately, as suggested in the figure.

One more essential finding of Muszkalay (1973) is mentioned here. He concluded from the observations that $I > 0$ even if $|\alpha^*| = 90°$ ($W_x = 0$). This is due to some deterministic effects (e.g., the deviating role of the hills) and to the stochastic nature of various other spatial nonuniformities (see item (iii) above). The deterministic effects were accounted for by a slight transformation of the wind speed vector; for details see Somlyódy and Virtanen (1982).

The results for I and Q (Figures 2 and 3) will be used next for model calibration, while the uncertainties in the wind field description and in the wind data will be discussed later in Section 6.

4 PARAMETER SENSITIVITY AND CALIBRATION

Realistic ranges for the two essential parameters C_D and λ (see eqns. 3 and 4, respectively) can be defined on the basis of literature values. The drag coefficient*, C_D, moves approximately between 0.001 and 0.0015 (Wu, 1969; Bengtsson, 1978; Graf and Prost, 1980). For λ, the bottom friction coefficient, no direct observations are available for lakes. For channel flows (in the turbulent domain for small roughness coefficients) it varies between 0.007 and 0.03; this range can serve as a guideline for lake situations (values near to or below the lower end of the range are generally expected).

During the calibration no effort was made to define loss functions or to use recently-developed versions of Monte Carlo procedures (see elsewhere in this volume for several examples of the use of these techniques). Instead, a straightforward trial and error fitting was performed. The empirical findings shown in Figures 2 and 3 incorporate the major features of the system's behavior so these plots were employed as a basis, rather than the more complex approach of using historical data. For the computations a time step Δt of 1800 s and a space step Δx of 2000 m (giving 40 grid points) were used, with geometric data from VITUKI (1976). The wind input profile was rectangular and characterized by the duration and one speed value; a sensitivity analysis on the shape of the profile (see Somlyódy and Virtanen, 1982) showed that the application of this simple distribution is fully acceptable for the present purpose.

As an example, Figure 4 shows the oscillation of the water level at the western end of the lake. As can be seen the dynamics of the system are very fast, characterized by a

---

*Note that the assumption $C_D(W) =$ constant was used as the slight wind dependence sometimes introduced is overruled by the uncertainty in wind data.

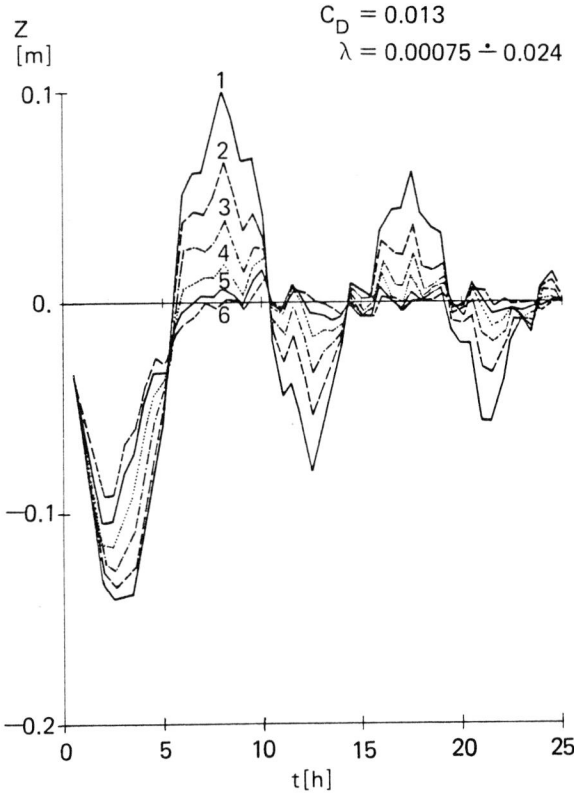

FIGURE 4 Sensitivity of the lake system to bottom friction, as illustrated by the oscillation of the water level at Keszthely ($W_x = 8$ m s^{-1}, $T = 2$ h).

seiche period of around 10 h. The bottom shear coefficient influences the peak amplitude but has an even stronger effect on the damping — an important feature which will be utilized later.

With increasing duration of wind input the duration of any negative (or positive) elongation also increases and in the case of a step-like wind input the water level approaches a steady state via several small oscillations. The stream flow through any given cross-section shows a similar pattern to that given in Figure 4 but the dynamics are even more rapid and the oscillation obviously decays for small and long durations, as well.

Next, the sensitivity of the system to λ and C_D is shown in terms of effects on I and Q (the latter measured at Tihany). Figure 5 illustrates the influence of the bottom shear coefficient as compared to the calibrated situation ($\lambda_0 = 0.003$, $C_{D0} = 0.0013$, $I_0 \approx 0.3 \times 10^{-5}$, and $Q_0 \approx 2100$ m^3 s^{-1}; these values are discussed later). The figure shows that the maximum water level difference is quite insensitive to λ and over the entire domain ($\lambda_{max}/\lambda_{min} \approx 30$) its influence moves in the range +15% to −27%. For values larger than $\lambda/\lambda_0 = 8$, β_1 is practically constant. Similar conclusions can be drawn for the stream flow although the sensitivity is slightly higher, especially if λ/λ_0 is small. The most sensitive behavior — in accordance with Figure 4 — is shown by β_3 (Figure 5),

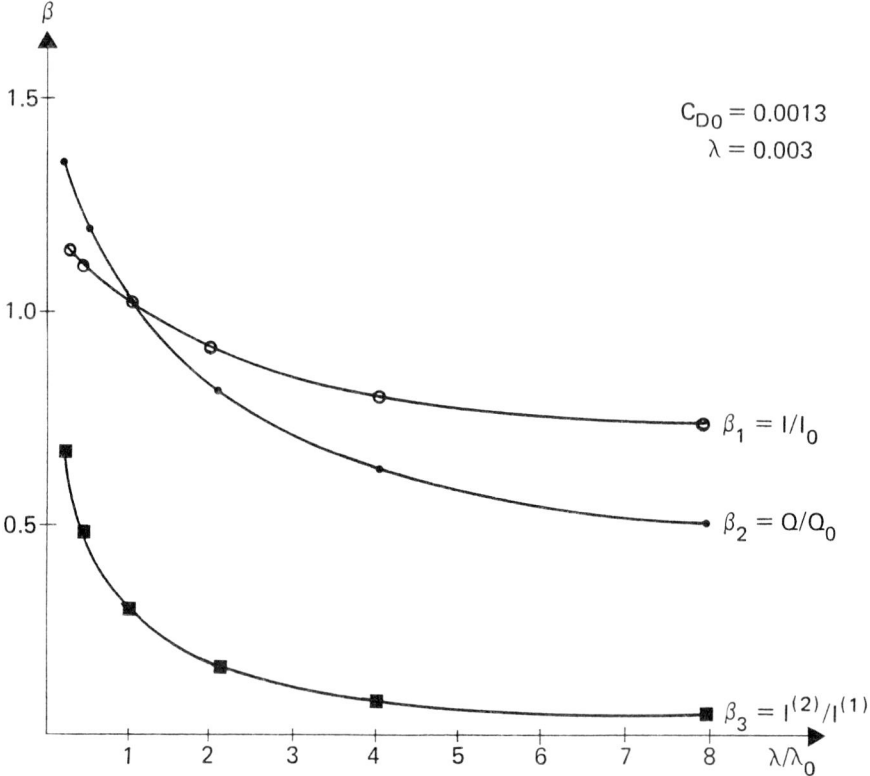

FIGURE 5 Parameter sensitivity: the influence of the bottom friction coefficient ($W_x = 8$ m s^{-1}, $T = 2$ h).

the ratio of the peak water level differences during the first and second elongation periods. Similar conclusions were also drawn for wind inputs other than those given in Figure 5.

Figure 6 illustrates the influence of the drag coefficient. As expected, both β_1 and β_2 depend approximately linearly on C_D. The model output is more sensitive to this parameter than to λ because the drag coefficient directly influences the energy input to the system.

From the mutually opposing influences of λ and C_D on both I and Q, it follows that no unique, "best" parameter combination can be found for the model without having further knowledge of the system. In the ranges $C_D = 0.0011$–0.0014 and $\lambda = 0.002$–0.008, fittings of approximately the same quality can be arrived at for I and Q. At this stage the damping properties of the system can be utilized for further information. From study of the historical data it is apparent that the damping is quite fast due to the shallowness of the lake; the amplitude of the second oscillation is around 30% of the first one, while the amplitude of the fourth oscillation is negligible. Based on this observation λ was fixed at 0.003 (see Figures 4 and 5), corresponding to $C_D = 0.0013$. Both values are realistic for lake situations (λ corresponds to a Chézy coefficient $C = (g/\lambda)^{1/2} \approx 60$). Comparisons with empirical results for I and Q are given in Figures 2 and 3. In the light

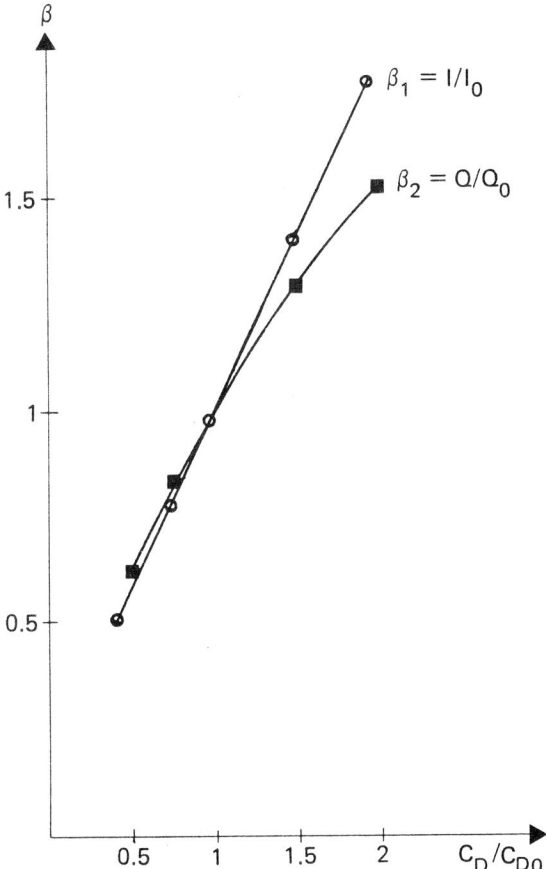

FIGURE 6 Parameter sensitivity: the influence of the wind drag coefficient ($W_x = 8$ m s^{-1}, $T = 2$ h).

of the explanation in Section 3, any closer agreement cannot be expected. Note that the calibration of a vertically integrated two-dimensional model for Lake Balaton resulted in the same parameter set (Shanahan and Harleman, 1981).

5 VALIDATION

In contrast to the calibration procedure, in which aggregate knowledge on the lake's behavior was utilized, historical data on typical stormy events (selected from among Muszkalay's (1973) observations) were employed for validation. Altogether more than ten events of different nature were simulated (Somlyódy and Virtanen, 1982) without changing the parameter values found. Hourly wind data measured at Szemes (see Figure 1) were used as input. The time step of the computation was 3600 s. For comparison water levels observed at the two ends of the lake (Keszthely and Kenese) and discharge values derived from velocity measurements at the Tihany peninsula (where available) were used. Three examples will now be discussed. The first is characterized by wind directions coinciding approximately with the long axis of the lake and the second by wind directions perpendicular to the long axis, while the third example concerns a set of consecutive but different events.

Example 1. (Date and starting time of storm: 16/11/1966, 08:00.) The period studied involved a relatively long-lasting storm with the wind blowing from the East, followed by three smaller storms of various natures (Figure 7a). The corresponding wind shear stress pattern is illustrated as a plot of $F = W_x|W|$ in Figure 7b. The shape of the water level curve (Figure 7c) is quite similar to that of the wind force and from this single example a linear relationship between the two could be hypothesized. The maximum water level difference is 0.7 m, approximately one fourth of the average depth and one of the highest values historically observed. No second peak can be observed, mainly due to the gradual decay in the wind shear. The agreement between simulated and observed water levels is excellent. The discharge, ranging from -2000 to $+3000$ m^3 s^{-1}, shows a highly fluctuating character. The mean value of the time series is negligible compared to the absolute values simulated.

Figure 8 shows the entire solution $z(t, x)$ in three-dimensional space, making the fluctuation of the free surface much more visible. The drastic change near to the Tihany Strait is particularly apparent; this is an obvious consequence of the Venturi-type structure encountered here. □

Example 2. (Date and starting time of storm: 8/7/1963, 08:00.) This example represents a fairly typical situation for the lake, with a strong wind perpendicular to the long axis (Figure 9a) resulting in a relatively small longitudinal shear stress component. The behavior of the water level is very complex (the changes are small and random) and the observed flow rate exhibits much larger fluctuations than the simulated value. This is a case where the model fails as a consequence of the uncertainties in the wind direction (see Section 6). One could argue that the inaccurate simulation is partly due to the one-dimensional treatment, but the 2-D model gave equally unsatisfactory results (Shanahan, 1981). □

Example 3. (Date and starting time of storm: 18/4/1967, 14:00.) This is the most comprehensive example studied: within the eight-day observation period more than five different situations covering the wind speed range 0–25 m s^{-1} and the complete directional domain of α can be distinguished, resulting in the irregular $F(t)$ pattern shown in Figures 10a and 10b. On comparing the observed and simulated water levels the "noisy" character of the latter becomes apparent. When moving averages are used for the computations reasonable agreement is achieved for the eastern end of the lake, but this is not true for Keszthely at the western end. The model gives an over-prediction around $t = 100$ h when the wind blows from the North. This is probably due to spatial nonuniformities in the wind field causing a strongly curved water surface not adequately characterized in the model. The discharge reflects the noisy character of the water level and shows the largest oscillations among the three examples discussed here. □

In summary we can state that the model has been satisfactorily calibrated. The validation is acceptable for situations where the wind blows along the main axis of the lake but inadequate for situations where the wind blows across the lake. This problem is discussed in Section 6.2.

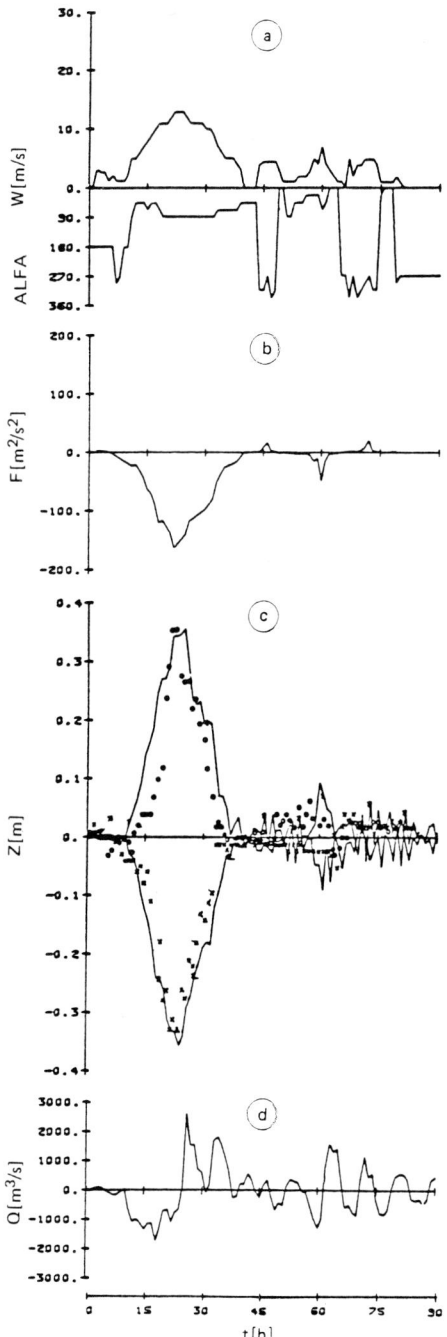

FIGURE 7 Simulation of a historical storm, 16/11/1966: (a) wind speed and direction at Szemes; (b) $F = |W|W_x$; (c) water level at Keszthely (1) and Kenese (2), respectively, together with observed values; (d) flow rate at Tihany.

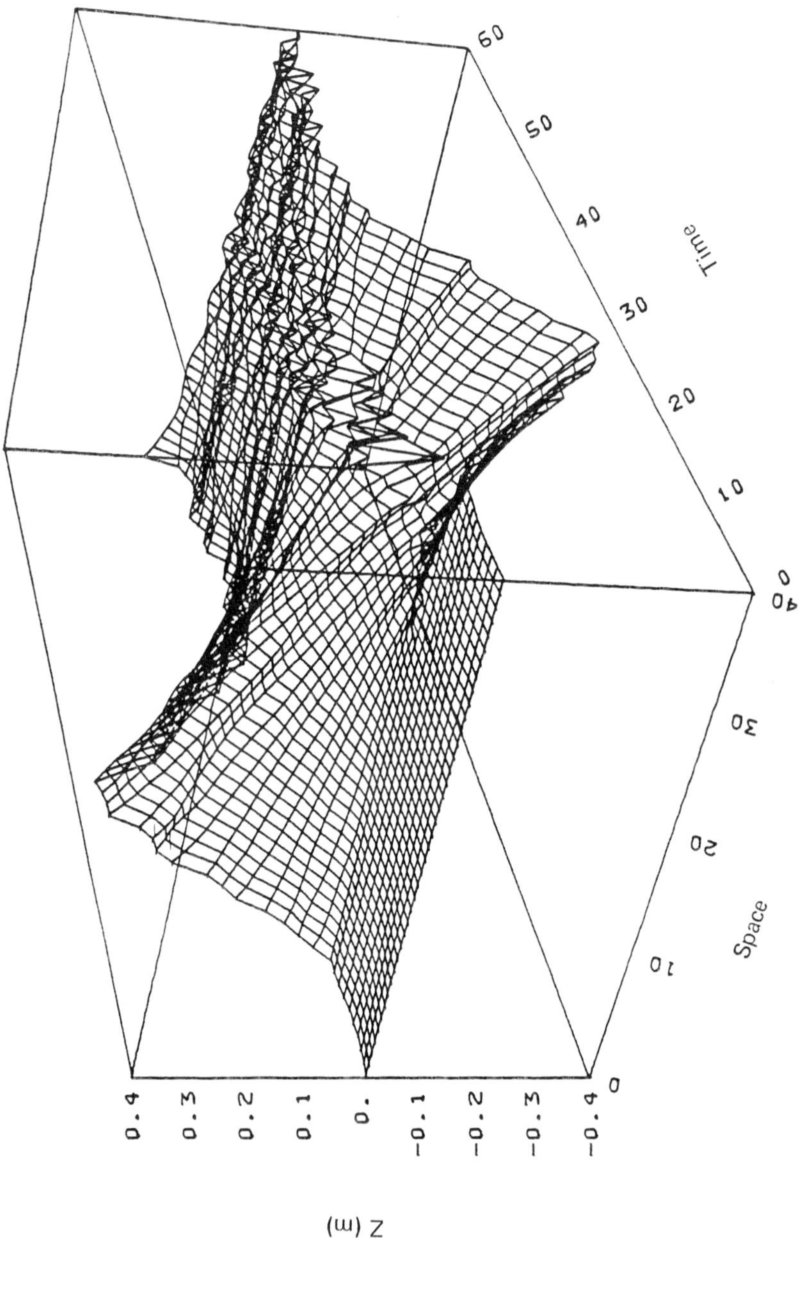

FIGURE 8 $z(t, x)$ solution surface for the event of 16/11/1966. Space coordinates 0 and 40 correspond to Keszthely and Kenese, respectively.

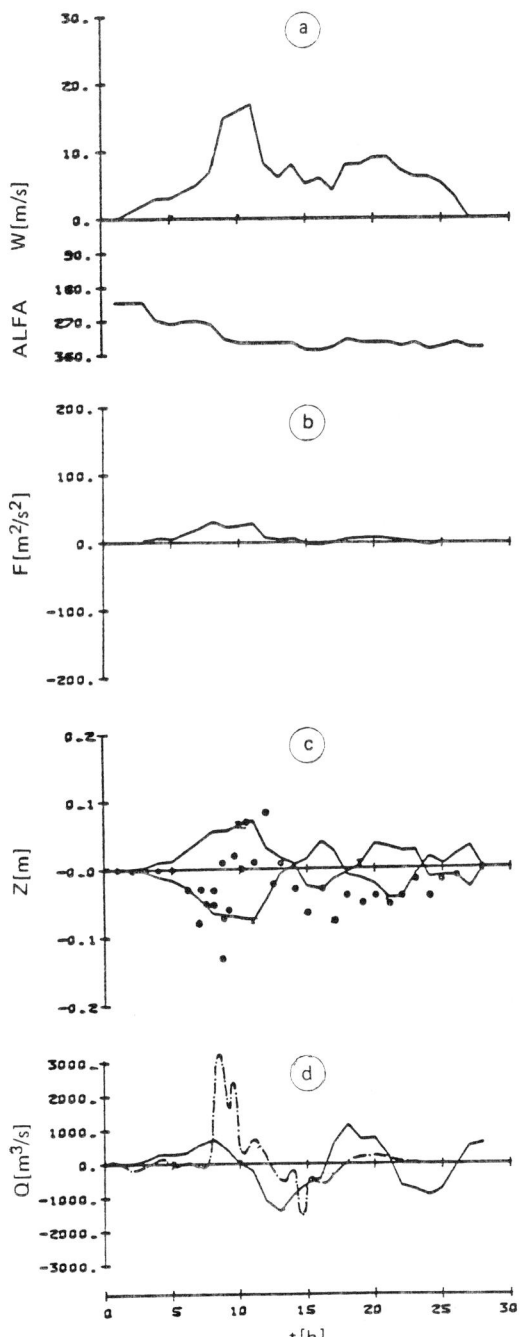

FIGURE 9 Simulation of a historical storm, 8/7/1963: (a) wind speed and direction at Szemes; (b) $F = |W|W_x$; (c) water level at Keszthely (1) and Kenese (2), respectively, together with observed values for Keszthely; (d) flow rate at Tihany (the broken line is derived from observed values).

6 WIND INPUT UNCERTAINTY

6.1 An Order of Magnitude Analysis

To begin this analysis we return to eqn. (3) and assume small errors, ΔC_D, ΔW, and $\Delta \alpha$, respectively, in the variables C_D, W, and α. The error propagation in the wind shear stress (related to $\tau_{smax0} = \rho_a C_D W^2$) in the vicinity of the nominal point 0 is then characterized by

$$\frac{\tau_{s0} + \Delta \tau_s}{\tau_{smax0}} \approx \left(\frac{\Delta C_D}{C_{D0}} + 2 \frac{\Delta W}{W_0} \right) \cos \alpha_0^* + (\cos \alpha_0^* - \Delta \alpha \sin \alpha_0^*) \tag{5}$$

derived from eqn. (3) after linearizing it. The term $\Delta C_D/C_{D0}$ indicates parameter sensitivity while the remaining terms relate to input data sensitivity. The errors ΔC_D and ΔW are similar in nature. For instance errors of ±20% in C_D or ±10% in W result in errors of roughly ±15% in I and Q, respectively (see Figure 6). These errors influence the magnitudes involved but not the dynamic characteristics of the system; therefore they can be handled reasonably well and require no further discussion here.

The situation regarding the second term in eqn. (5) is different because $\Delta \alpha$ can change the sign of τ_s and through this completely distort the time-dependent flow field. This can be shown by the following order of magnitude analysis.

Based on simulation results (see Figures 2 and 3) for wind durations of $T = 2$ h, the maximum water level difference and the flow rate at the Tihany peninsula can be expressed approximately as functions of τ_s. The error term α^* can be also introduced, thus leading to the following equations

$$I \approx 2.8 \times 10^{-5} \rho_a C_D W^2 (\cos \alpha_0^* - \Delta \alpha \sin \alpha_0^*) \tag{6}$$

and

$$Q \approx 1.1 \times 10^6 |I|^{1/2} \tag{7}$$

where I is dimensionless and Q is in units of m³ s⁻¹. Here α^* is zero if $x = 0$ (Figure 1) and, in contrast to the previous definition, I can be positive or negative. The sign of Q is the same as that of I.

Equation (6) clearly shows the main features of error propagation. There is no effect if $\alpha_0^* = 0$ or 180° (corresponding to longitudinal winds), while the largest effect occurs for $\alpha_0^* = 90°$ or 270° (corresponding to transverse winds). The case of $\alpha_0^* = 90°$ corresponds fairly closely to the prevailing wind direction and thus is of major importance. In this case – depending on the sign of $\Delta \alpha$ – positive or negative first amplitudes at the same end of the lake and both flow directions are all possible. For example, relatively small variations in input conditions ($\Delta \alpha = \pm 22.5°$ if $W = 10$ m s⁻¹, or $\Delta \alpha = \pm 10°$ if $W = 15$ m s⁻¹) correspond to strikingly wide error ranges in z (±13 cm) and Q (±1500 m³ s⁻¹).

When we recall some other features mentioned previously, such as the poor resolution in the directional data, the inappropriate time averaging of the wind data (often, only three-hour averages are available), and the fast response of the system, it is obvious

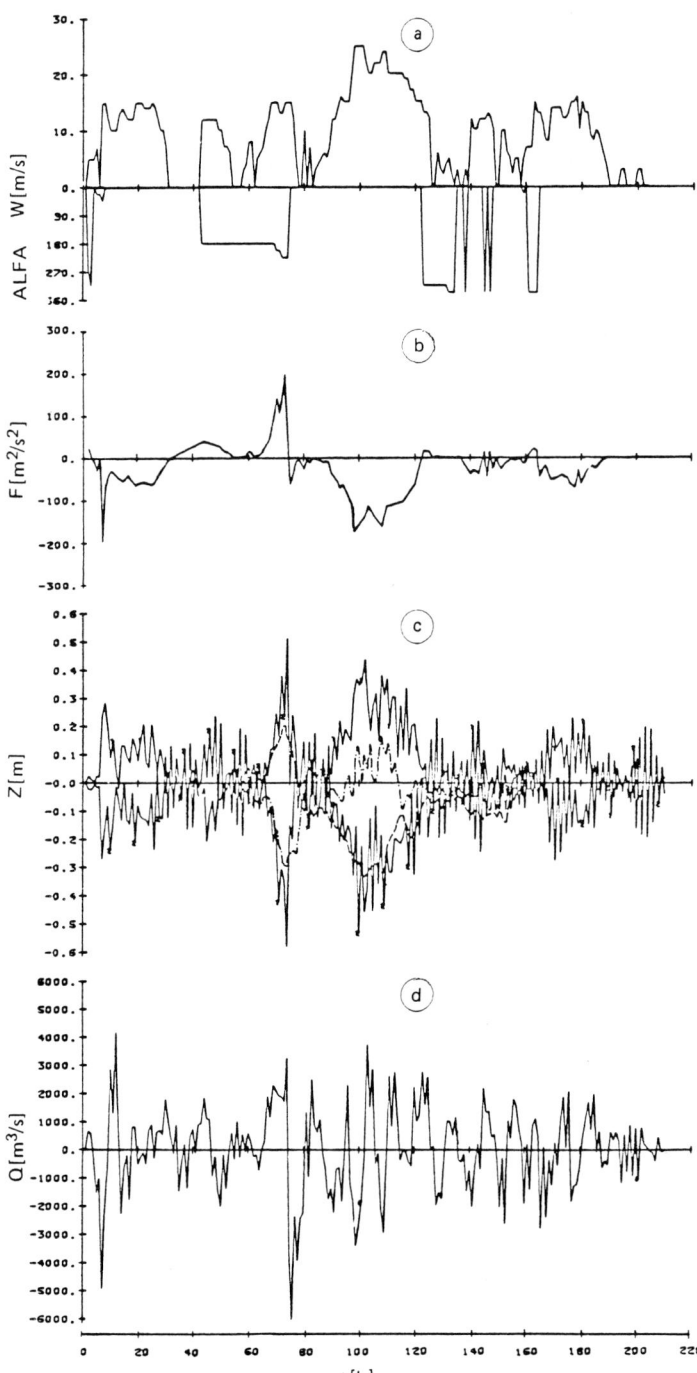

FIGURE 10 Simulation of a historical storm, 18/4/1967: (a) wind speed and direction at Szemes; (b) $F = |W|W_x$; (c) water level at Keszthely (1) and Kenese (2), respectively, together with observed values (broken line); (d) flow rate at Tihany.

that in the vicinity of $\alpha_0^* = 90°$ an error in a single direction datum can result in the sort of drastic change in the simulation described above. From this simple argument it follows that any better agreement between simulation and observation for the last two examples in Section 5 would, in fact, be surprising. For these reasons the more accurate Monte Carlo simulation procedure described next was employed.

6.2 Monte Carlo Simulations

In the course of the Monte Carlo procedure a random component $\Delta\alpha(t)$ is added to the mean scenario $\bar{\alpha}^*(t)$. The generation of $\Delta\alpha$ takes place numerically according to the three types of error sources and their respective distributions as outlined in points (i)–(iii) of Section 3. In this way a large number of $\alpha_i^* = \bar{\alpha}^* + \Delta\alpha_i$ scenarios can be calculated and computer simulations performed with them. Finally the statistics of the model output are evaluated. The number of simulations required was initially tested on Example 2 (see below); experiments were made with between 50 and 1000 Monte Carlo runs and it was eventually decided to use 100 runs. The values 16.8°, 22.5°, and 33.8° were assumed as basic, realistic values for the standard deviations (or half ranges) of the Gaussian, "discrete", and uniform distributions (see Section 3), respectively, but other values were also employed in order to check sensitivity. The transformations of the original distributions with respect to wind force, water level, and flow rate, respectively, were also studied. Some of the results obtained are now discussed.

Example 1. The Monte Carlo simulations depicted in Figure 11 correspond to parts b–d of Figure 7 and show the effects of considering the means, standard deviations (±), and extreme values of each parameter (uniform distribution, half range 33.8°). In agreement with the findings of Section 6.1, the uncertainty in α influences the wind shear to only a small extent (for wind conditions near to longitudinal). This is also true for z: for example, around $t = 25$ h the variance is practically zero. Again in accord with Section 6.1, the uncertainty range for Q is essentially wider indicating at the same time that model validation for discharge (or velocity) is more difficult than that for water level (see also Figure 5). Note that the mean trajectories agree reasonably well with the deterministic simulations (Figure 7), and that the discrete generator (option (ii)) with a half range of 22.5° gave practically the same results as those illustrated in Figure 11.

In conclusion we may state that in this case uncertainty is not too important, this being one reason why the validation in Section 5 was successful. □

Example 2. Results for the uniform distribution are given in Figure 12. In contrast to the previous example, the uncertainty produced in the wind shear stress by the same range in α is much wider due to the cross-wind conditions (see Section 6.1). In practice the water level variation can range between 0.15 and −0.15 m, thus including all the observed values. The results explain the "noisy" character of the water level well and show that under such conditions a model with deterministic input cannot be validated. For the discharge a strikingly broad domain was obtained, covering most of the measured values. Although it was stated in Section 5 that the model failed for this event, from the present example it follows that the statement is only true if uncertainties are not accounted for. Note that the order of magnitude estimate given in Section 6.1 coincides well with the results described in Figure 12.

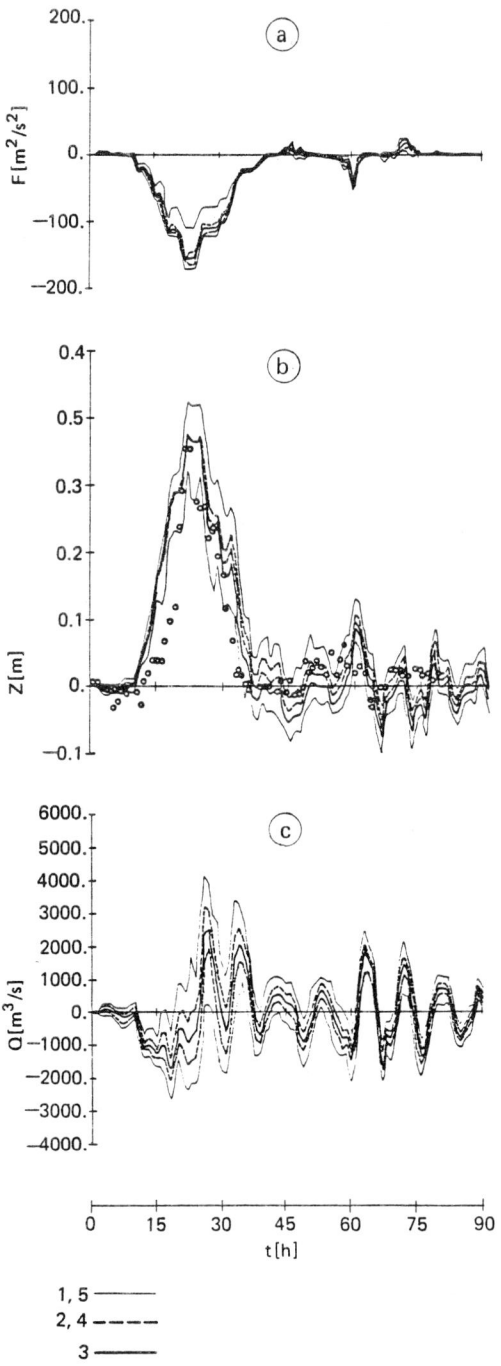

FIGURE 11 Monte Carlo simulation of the storm of 16/11/1966: (a) $F = |W|W_x$; (b) water level at Keszthely, together with observed values; (c) flow rate at Tihany ((3) mean value, (4) and (2) mean value ± standard deviation, (1) and (5) extreme values).

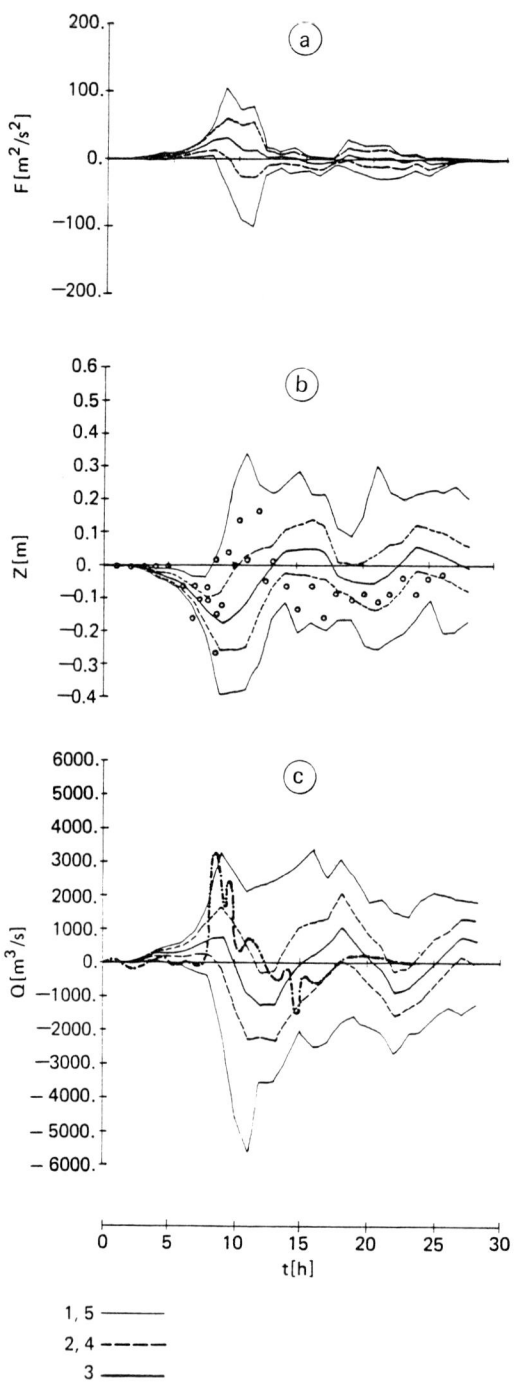

FIGURE 12 Monte Carlo simulation of the storm of 8/7/1963: (a) $F = |W|W_x$; (b) water level at Keszthely, together with observed values; (c) flow rate at Tihany ((3) mean value, (4) and (2) mean value ± standard deviation, (1) and (5) extreme values; the dot-dash line is derived from observed values).

Figure 13 summarizes the results obtained using various distributions for the discharge at Tihany. It is stressed that the situation described in Figure 13c represents the smallest uncertainty range possible in practice since the resolution of the standard wind-direction data is 22.5°.

As compared to Example 1, the mean trajectories depend more strongly on the distribution assumed for $\Delta\alpha$ and differ from the deterministic simulation. □

Example 3. Results for the water level at the eastern end of the lake are given in Figure 14 (uniform distribution, half range 33.8°). The very wide uncertainty domain (see also Section 6.1) is due to the cross-wind conditions around $t = 100$ h and the high wind speed value (Figure 10). More attention is paid here to the mean trajectory which is essentially different to the deterministic result; it is less noisy and agrees better with the observations. □

Two main conclusions may be drawn from this section: (i) except in the case of longitudinal wind conditions the model is far more sensitive to directional data than to other model parameters, and this should be taken into account in model development and use; (ii) the model was successfully validated in a stochastic fashion, which is a considerable achievement as compared to deterministic simulations.

7 MULTIDIMENSIONAL MODELS AND OTHER SYSTEMS

It is of some interest to analyze first, whether the behavior of multidimensional models as regards uncertainty propagation is similar to that of the 1-D model and second, what conclusions can be drawn for other lakes.

7.1 Multidimensional Models

For multidimensional approaches (3-D or horizontal 2-D models in the present case) the wind shear stress vector as model input is given by the following equation

$$\tau_s = \rho_a C_D W^2 (\cos\alpha^* e_x + \sin\alpha^* e_y) \tag{8}$$

where e_x and e_y are unit vectors for directions x and y, respectively. In contrast to eqn. (3), a second term now appears in parentheses representing the transverse shear stress. The sensitivity of this equation to $\Delta\alpha$ can be characterized by the relationship

$$(\tau_{s0} + \Delta\tau_s)/\tau_{smax0} \approx (\cos\alpha_0^* - \Delta\alpha \sin\alpha_0^*) e_x + (\sin\alpha_0^* + \Delta\alpha \cos\alpha_0^*) e_x \tag{9}$$

In the 1-D model an error in the wind direction influences the absolute value (and also the sign) of the shear stress appearing in the input. In contrast, the absolute value is unaffected for 2-D or 3-D situations; only the direction and through this the components are modified (eqn. 9). The sensitivity structure of eqn. (9) is, however, not very suitable. For wind directions nearly perpendicular to the lake's axis τ_{sx} is characterized by the same (large) sensitivity as in the 1-D model while for τ_{sy} the sensitivity is negligibly small (cos $\alpha_0^* \approx 0$). As the cross-sectional average discharge in a 2-D model is primarily dependent on τ_{sx} we may conclude that the sensitivity of the 2-D model to input data

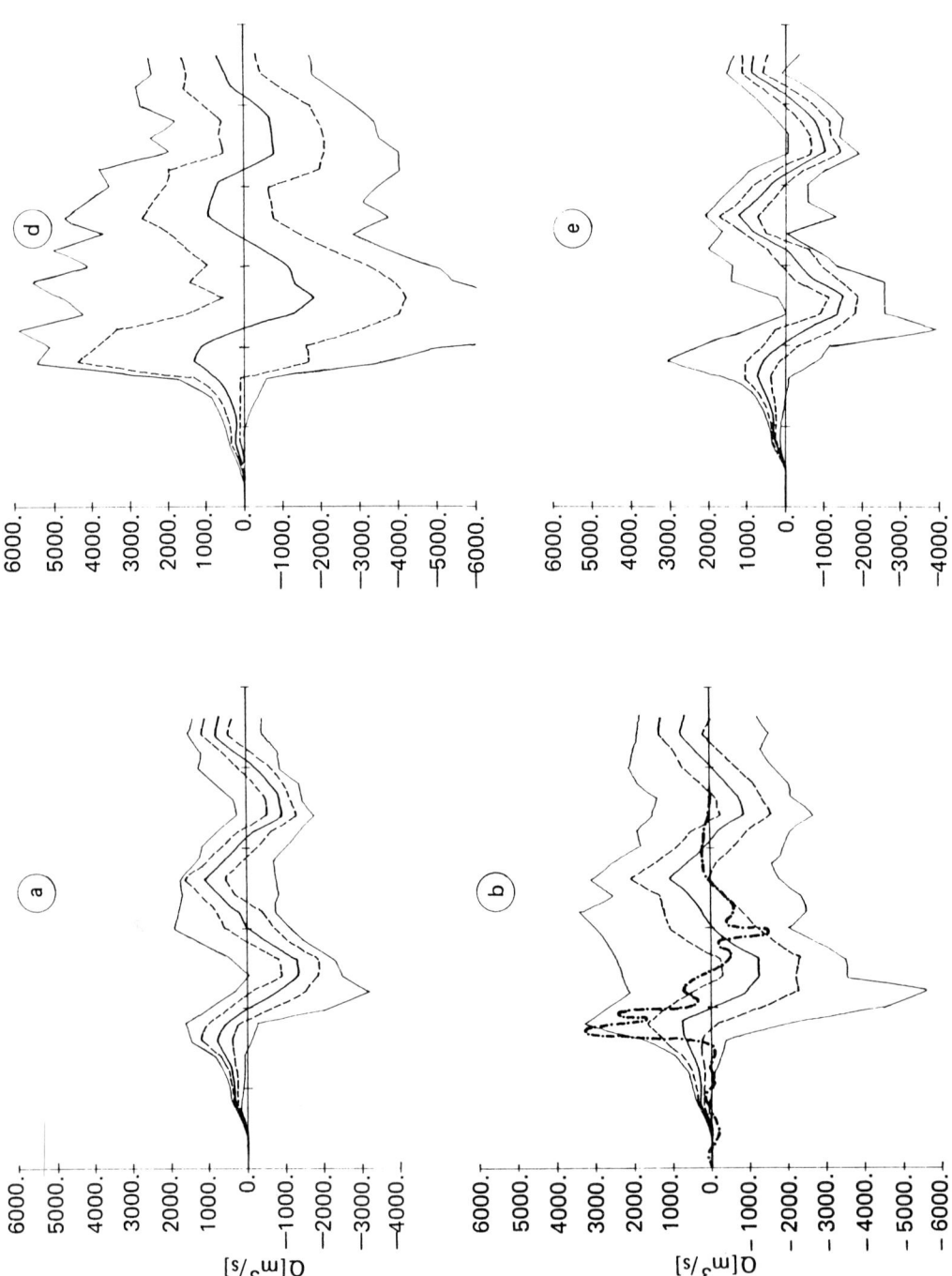

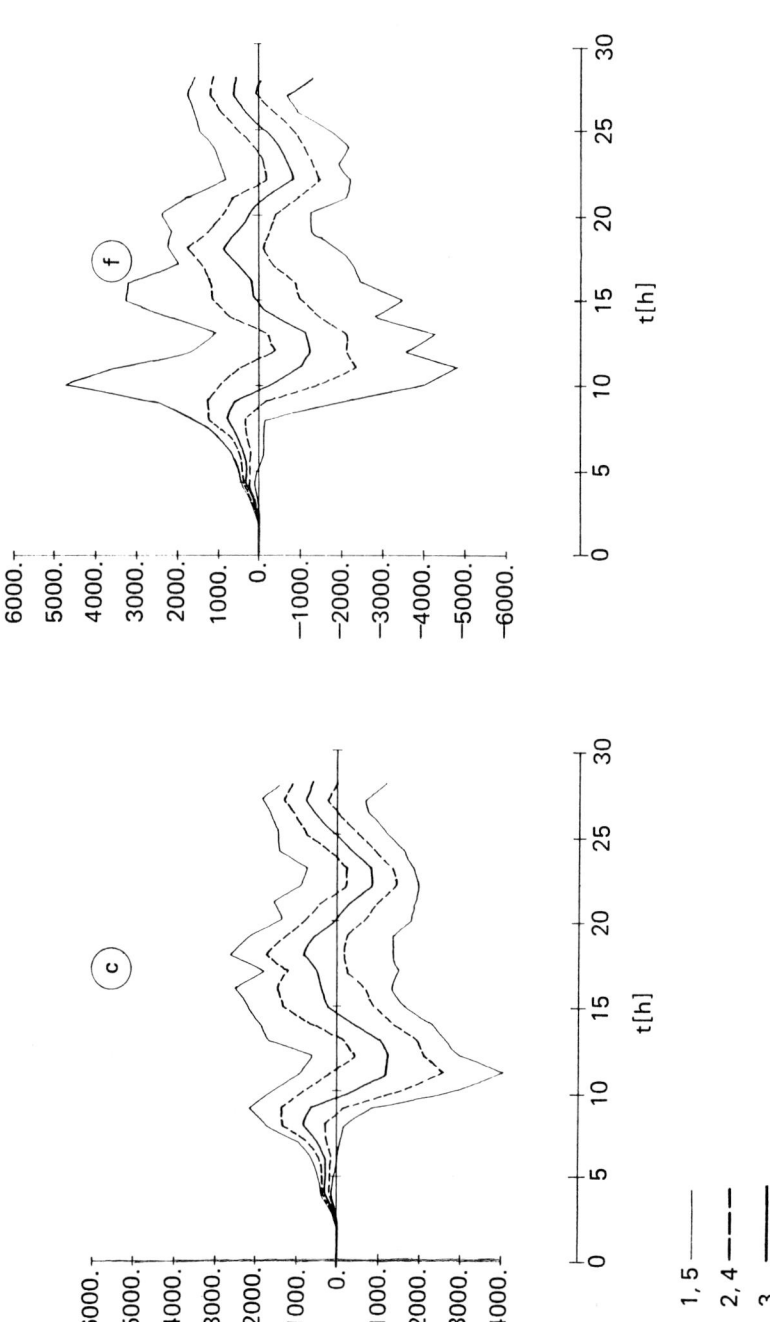

FIGURE 13 Influence of various types of distribution and distribution parameters on the Monte Carlo simulation, as illustrated by the flow rate at Tihany during the storm of 8/7/1963. (a) uniform distribution, half range 22.5°; (b) uniform distribution, half range 33.8° (the dot-dash line is derived from observed values); (c) discrete distribution, half range 22.5°; (d) discrete distribution, half range 45°; (e) Gaussian distribution, standard deviation 11.3°; (f) Gaussian distribution, standard deviation 16.8° ((3) mean value, (4) and (2) mean value ± standard deviation, (1) and (5) extreme values).

uncertainty has a similar character to that encountered with the 1-D approach. This explains why Shanahan and Harleman (1981) failed to simulate the discharge adequately for the two storms discussed in their report.

It is more difficult to reach conclusions on the possible behavior of a 3-D model as an additional important parameter, the vertical eddy viscosity, would appear in such a model. This would influence first of all the shape of the vertical velocity profile (excluded from both the 1-D and 2-D approaches), and through this also the cross-sectional average stream flow. Thus it is suspected that the input data sensitivity might be smaller for a 3-D model, but that it would still be significant.

The sensitivity in τ_{sy} (see eqn. 9) is highest for longitudinal wind conditions. This will, however, primarily influence the circulation pattern, but only slightly affect Q and I.

7.2 Other Lake Systems

Among the most characteristic features of Lake Balaton are the lake's geometry, the prevailing wind direction which is approximately perpendicular to the lake's axis, and the relatively fast response of the system. The wind field also shows specific patterns of temporal and spatial change. However, the methods of data collection used do not in fact follow satisfactorily either the features of the wind field or those of the lake system (due to insufficiently exact or inappropriate resolution in space, time, and direction). These are all reasons for the large uncertainty found and for the dominant position of input data sensitivity rather than parameter sensitivity.

Certainly other lakes can and do essentially differ from Balaton in their major characteristics. However all the systems are specific in their own different ways: for example, no regular lake of circular shape has yet been found in nature for which uncertainties in wind direction would equally influence the x and y components of a model simulation. Moreover, for most lakes, typical values of length, width, and prevailing wind directions can be straightforwardly defined. From these major characteristics the ranges and types of possible uncertainties follow. More importantly, the major characteristics contain clues on how the various uncertainties can be diminished by developing an appropriate monitoring strategy.

8 CONCLUSIONS

Parameter sensitivity and the influence of input data uncertainty has been studied for a one-dimensional model of Lake Balaton. The major characteristics of the Balaton system are the long, narrow shape of the lake and the prevailing wind direction which is approximately transverse to the lake's axis. In addition to detailed simulations, two aggregated parameters, the maximum water level difference I along the lake and the volume flow rate Q at the smallest cross-section, were used to describe the major features of the system. Our conclusions are as follows:

(i) The wind field exhibits characteristic temporal and spatial changes. The response time of the lake is very short: a typical measure is the longitudinal seiche period of around ten hours according to the model. Storms of short duration (1–2 h) induce considerable

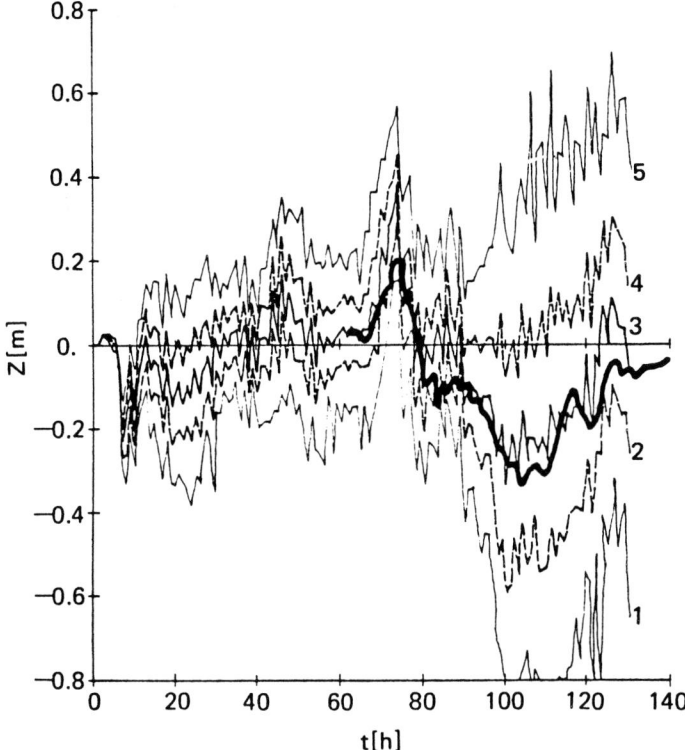

FIGURE 14 Monte Carlo simulation of the storm of 18/4/1967: water level at Kenese ((3) mean value, (4) and (2) mean value ± standard deviation, (1) and (5) extreme values; the bold line is derived from observed values).

movement so that a steady state is practically impossible to define. The instrumental resolutions of wind data in time, space, and direction are inappropriate for the known properties of the wind field and the fast dynamics of the system. This inappropriate monitoring strategy causes large uncertainties in the wind input data and should be accounted for in the course of model development.

(ii) The influences of the two major parameters, the wind drag and bottom friction coefficient, on model performance are opposite. Thus, it is difficult to find a unique, "best" parameter combination. In the ranges $C_D = 0.0011-0.0014$ and $\lambda = 0.002-0.008$, fittings of approximately equal quality can be arrived at for I and Q. Based also on the damping properties of the system a parameter vector (0.0013, 0.003) was calculated. Essentially the same results were obtained from the independent development and calibration of a horizontal 2-D model for Lake Balaton.

(iii) The model's behavior is obviously very sensitive (in fact, almost directly proportional) to the drag coefficient which influences the energy input to the system. In marked contrast, the model output is quite insensitive to the bottom friction coefficient (the only parameter directly associated to internal variables). In the range of λ related to the nominal calibrated value (0.25–8), I varies in the range $+15\%$ to -27%. For Q the sensitivity is slightly higher, but both I and Q are practically independent of λ for $\lambda/\lambda_0 > 8$.

(iv) The model was successfully validated for longitudinal wind conditions. However, this was not possible for winds closer to transversal and particularly not for the stream flow, one of the two variables. The 2-D model showed the same properties.

(v) An order of magnitude analysis clearly indicated that the failure in validation was due to uncertainties in the wind direction, which has almost no influence on the model performance for longitudinal winds but a very major effect for transversal conditions. Errors of ± 15 cm and ± 1500 m^3 s^{-1} can easily occur for I and Q, respectively (the nominal value would be zero for both variables). While errors in the wind speed are of a similar nature to those in the drag coefficient, in that they do not influence the direction of the flow or the sign of a water level amplitude, an error in the wind direction can completely distort the time pattern of the simulation. The behavior of the 2-D model is expected to be similar.

(vi) By introducing an error component solely in the wind direction, Monte Carlo simulations were performed. This more accurate approach justified the findings of the order of magnitude analysis. Assuming realistic uncertainties in the wind direction (in this case, a $\pm 22.5°$ error domain) the model was successfully validated for all the historical storms simulated. The mean trajectories of the Monte Carlo runs are close to the deterministic simulations for longitudinal wind directions with increasing deviation observed on approaching cross-wind conditions.

(vii) In this particular case we can conclude that input data sensitivity is more dominant than parameter sensitivity for the 1-D model; the situation would probably be similar for the 2-D model version. However, the pattern of course can and will be different for other lakes. Nevertheless, it is generally of great importance to work out a proper wind monitoring network, knowing the major features of the system studied (see conclusion (i) above), in order to reduce the possible influence of various input data uncertainties.

REFERENCES

Beck, M.B. (1983). Sensitivity analysis, calibration, and validation. In G.T. Orlob (Editor), Mathematical Modeling of Water Quality: Streams, Lakes and Reservoirs. Wiley, Chichester, pp. 425–467.
Béll, B. and Takács, L. (Editors) (1974). The climate of the Lake Balaton region. Publications of the Hungarian Meteorological Institute, Vol. XL. (In Hungarian.)
Bengtsson, L. (1978). Wind induced circulation in lakes. Nordic Hydrology, 9:75–94.
Fedra, K. (1983). A Monte Carlo approach to estimation and prediction. In M.B. Beck and G. van Straten (Editors), Uncertainty and Forecasting of Water Quality. This volume, pp. 259–291.
Gardner, R.H., O'Neill, R.V., Mankin, J.B., and Carney, J.H. (1981). A comparison of sensitivity analysis and error analysis based on a stream ecosystem model. Ecological Modelling, 12:173–190.
Graf, W.H. and Prost, J.P. (1980). Aerodynamic drag and its relation to the sea state: with data from Lake Geneva. Archiv für Meteorologie, Geophysik und Bioklimatologie, Series A, 29:67–87.
Halfon, E. (1977). Analytical solution of the system sensitivity equation associated with a linear model. Ecological Modelling, 3:301–307.
Kohberger, R.C., Scavia, D., and Wilkinson, J.W. (1978). A method for parameter sensitivity analysis in differential equation models. Water Resources Research, 14(1):25–29.
Kozák, M. (1977). Computation of the Unsteady Flow in Open Channels. Akadémiai Kiadó, Budapest. (In Hungarian.)
Lick, W. (1976). Numerical modeling of lake currents. Annual Review of Earth and Planetary Sciences, 4:49–74.

Mahmood, K. and Yevjevich, V. (Editors) (1975). Unsteady Flow in Open Channels. Water Resource Publications, Vol. 1. Fort Collins, Colorado.

Muszkalay, L. (1973). Characteristic Water Motions in Lake Balaton. VITUKI, Budapest. (In Hungarian.)

Muszkalay, L. (1979). Water motions in Lake Balaton. In S. Baranyi (Editor), Summary of Research Results for Lake Balaton. VIZDOK, Budapest, pp. 34–128. (In Hungarian.)

Rinaldi, S. and Soncini-Sessa, R. (1978). Sensitivity analysis of generalised Streeter–Phelps models. Advances in Water Resources, 1(3):141–146.

Scavia, D., Canale, R.P., Powers, W.F., and Moody, J.L. (1981). Variance estimates for a dynamic eutrophication model of Saginaw Bay, Lake Huron. Water Resources Research, 17(4):1115–1124.

Shanahan, P. and Harleman, D.R.F. (1981). Linked Hydrodynamic and Biogeochemical Models of Water Quality in Shallow Lakes. Technical Report No. 268. Ralph M. Parsons Laboratory for Water Resources and Hydrodynamics, Department of Civil Engineering, Massachusetts Institute of Technology, Cambridge, Massachusetts.

Shanahan, P., Harleman, D.R.F., and Somlyódy, L. (1981). Modeling Wind-Driven Circulation in Lake Balaton. CP-81-7. International Institute for Applied Systems Analysis, Laxenburg, Austria.

Somlyódy, L. (1979). Hydrodynamical Aspects of the Eutrophication Modeling in the Case of Lake Balaton. CP-79-1. International Institute for Applied Systems Analysis, Laxenburg, Austria.

Somlyódy, L. (1981). Modeling a Complex Environmental System: The Lake Balaton Case Study. WP-81-108. International Institute for Applied Systems Analysis, Laxenburg, Austria.

Somlyódy, L., and Virtanen, M. (1982). Application of a One-Dimensional Hydrodynamic Model for Lake Balaton. Working Paper. International Institute for Applied Systems Analysis, Laxenburg, Austria, forthcoming.

van Straten, G. and de Boer, B. (1979). Sensitivity to Uncertainty in a Phytoplankton–Oxygen Model for Lowland Streams. WP-79-28. International Institute for Applied Systems Analysis, Laxenburg, Austria.

van Straten, G., and Somlyódy, L. (1980). Lake Balaton Eutrophication Study: Present Status and Future Program. WP-80-187. International Institute for Applied Systems Analysis, Laxenburg, Austria.

Virtanen, M. (1978). Computation of Two-Dimensional Unsteady Flow in Shallow Water Systems. Report prepared on the Ninth International Post-Graduate Course on Hydrological Methods for Developing Water Resources Management. UNESCO–VITUKI, Budapest.

VITUKI (1976). Standard Cross-Sections of Lake Balaton. VITUKI, Budapest.

Wu, J. (1969). Wind stress and surface roughness at air–sea interface. Journal of Geophysical Research, 74(2):444–455.

MAXIMUM LIKELIHOOD ESTIMATION OF PARAMETERS AND UNCERTAINTY IN PHYTOPLANKTON MODELS

G. van Straten
*Department of Chemical Engineering, Twente University of Technology,
P.O. Box 217, Enschede (The Netherlands)*

1 PROBLEM STATEMENT

In modeling large complex systems, estimates for the model parameters cannot always be obtained by controlled experimentation or independent measurements. Moreover, most parameters are lumped parameters in the sense that they represent a wealth of underlying processes for which separate modeling is undesirable or impractical, so that their numerical value has a well-defined physical meaning only for the system under study within the context of the model specified. Consequently, some form of model calibration, achieved by adjusting the parameters in some way, is inevitable.

Generally, modeling is used to enhance our understanding of the behavior of a system as a whole, preferably in quantitative terms, from the action of each of the components separately. Of course the eventual aim is to use models as a tool to decide upon the effects of possible control actions or alternative management strategies. Obviously, the success of such applications depends critically upon the quality of the calibration. Here it is not only essential to have parameter estimates that fit the data well, but also important to have some idea about the accuracy of the estimates, and about the uncertainty in the model predictions resulting from the uncertainty in the parameters.

Formal calibration, or, if preferred, parameter estimation, based on minimization of the sum of squared differences between model results and data (least-squares methods) does allow for the simultaneous estimation of the parameter variance-covariance matrix (see, for example, Draper and Smith, 1966). However, least-squares methods as such ignore knowledge that might exist about the error structure of data, inputs, or model. If something is known or can be assumed about the error structure, improved estimates with lower variances can in principle be obtained by employing maximum likelihood estimation.

In a previous paper maximum likelihood estimation was applied to a model for phytoplankton dynamics (DiToro and van Straten, 1979), using data obtained for Lake Ontario during the International Field Year of the Great Lakes (IFYGL, 1972–1973).

In this case, *a priori* knowledge about the data error was available in the form of spatial variances because the data were lake-wide means computed from numerous local samples. The present paper reports on a second application of the same approach using the same model for the same lake, but now for a ten-year lumped-data set. Here, the uncertainty in the data arises from the variability among the different years.

First, the theory of the method as developed by DiToro and van Straten will be restated, but cast in a more general framework. Next, the results of the application to the IFYGL data will be summarized and supplemented with experience gained with the ten-year lumped-data set. The key section is the discussion, in which the mathematical treatment is scrutinized to explain some of the difficulties encountered in the practical application. The principal aim of this contribution is to show what can be learnt about maximum likelihood estimation from practical experience, and how this leads to suggestions for improvement. The actual implementation of these improvements will be the subject of future research.

2 THE MODEL

To help the reader in the rather abstract derivations to follow, the model is first briefly discussed, although detailed knowledge is not essential for the development in the subsequent sections. The model was developed originally by the Manhattan College (Thomann et al., 1975). State variables are phytoplankton carbon, herbivorous zooplankton carbon, and carnivorous zooplankton carbon; organic nitrogen, ammonia nitrogen, and nitrate nitrogen; and organic phosphorus and orthophosphorus. Each of these variables is modeled for both the epilimnion and the hypolimnion, but in the horizontal plane it is assumed that the lake is homogeneous. Driving variables are total daily solar irradiation, day length, extinction in the water column, water temperature in the epilimnion and hypolimnion, mixing over the thermocline, and inputs of phosphorus and nitrogen compounds. Data are available from the regular sampling program on a roughly ten-times-a-year basis from 1967 to 1976, and in addition from the IFYGL program for a one-year period from 1972 to 1973. The measurements comprise chlorophyll-a, total zooplankton biomass, total Kjeldahl nitrogen, ammonia, nitrate, and total and soluble reactive phosphorus. A comparison with model results is made possible by a suitable linear combination of the model state variables.

Previously, the model has been calibrated using data for 1967–1970 (Thomann et al., 1975), thus providing an initial guess for the parameter vector in the present application. In what follows, the forcing functions representative of the period 1967–1970 have also been used, rather than the IFYGL or ten-year averaged forcings. Although year-to-year differences for Lake Ontario are not excessive, some error will result from this simplification, which is not accounted for here.

3 METHODOLOGY

3.1 Estimation

In this section the methodology as originally developed by DiToro and van Straten (1979) is restated in a slightly more general form.

Let f_j be the column vector of model results at time instant t_j ($j = 1, 2, \ldots, n$). Thus

$$f_j = \{f_1(t_j; \hat{\beta}), \ldots, f_k(t_j; \hat{\beta}), \ldots, f_s(t_j; \hat{\beta})\}^T$$

where $f_k(t_j; \hat{\beta})$ denotes the model outcome for state variable k at time t_j given the parameter vector $\hat{\beta}$ (dimension p). The dimension of f_j is s, the number of state variables. Further let c_j be the s-dimensional vector of the observed lake-wide (average) concentration for the state variables:

$$c_j = \{c_1(t_j), \ldots, c_k(t_j), \ldots, c_s(t_j)\}^T$$

Since f_j is a model for c_j, one may postulate that

$$c_j = f_j(\beta) + v_j \tag{1}$$

where v_j denotes the sum of all errors at time t_j, which may be composed of measurement errors, spatial errors, and model structural errors (including propagated input errors).

Under the additional assumption that the total error $\hat{v}_j$ is normally distributed with variance–covariance matrix R_j (dimension $s \times s$), the multivariate probability density function of c_j given the process parameters b is

$$p(c_j; b) = (1/((2\pi)^{s/2}(\det R_j)^{1/2})) \exp[-(1/2)(c_j - f_j(b))^T R_j^{-1}(c_j - f_j(b))] \tag{2}$$

The multivariate likelihood function for c_j given the model parameters $\hat{\beta}$, $l(c_j; \hat{\beta})$, has the same form as eqn. (2). If the disturbances $\hat{v}_j$ are not correlated in time, the likelihood function for the full time series can be written as

$$L(\hat{\beta}) = \prod_{j=1}^{n} l(c_j; \hat{\beta}) \tag{3}$$

or

$$-\ln L(\hat{\beta}) = (sn/2) \ln(2\pi) + \frac{1}{2} \sum_{j=1}^{n} \ln(\det R_j) + \frac{1}{2} \sum_{j=1}^{n} [(c_j - f_j(\hat{\beta}))^T R_j^{-1}(c_j - f_j(\hat{\beta}))] \tag{4}$$

Note that in eqn. (1) it is tacitly assumed that f_j has the same dimension as the observation vector. Thus, linear transformations of the actual model state variables are understood to have been performed before eqn. (1) is applied, and consequently, f_j should, in fact, be viewed as the data-oriented model results vector rather than as a state vector as such. It should also be noted that the summations in eqn. (4) must be taken over all data points excluding missing data, so that in the case of missing data sn should read N, the total number of actual observations made ($N \leq sn$).

In eqn. (4) both the elements of R_j and the elements of the parameter vector $\hat{\beta}$ are unknown. The estimation problem can now be formulated as: find R_j, $\hat{\beta}$ that minimize the right-hand side of eqn. (4) (these are then the estimates $\hat{R}_j$ and $\hat{\beta}$). At this point DiToro and van Straten (1979) make the tacit additional assumption that the estimate of $\mathbf{R}_j$ does not depend upon the parameter estimate $\hat{\beta}$. Under this assumption the parameters can be found by differentiating eqn. (4) with respect to $\hat{\beta}$, to yield

$$\sum_{j=1}^{n} (c_j - f_j)^T \mathbf{R}_j^{-1} (\partial f_j / \partial \hat{\beta}) = 0 \tag{5}$$

The result of eqn. (5) is further simplified by assuming that no correlations exist between the disturbances among states, so that $\mathbf{R}_j$ is a diagonal matrix:

$$\mathbf{R}_j = \begin{bmatrix} \sigma_{\nu1}^2(t_j) & & 0 \\ & \sigma_{\nu k}^2(t_j) & \\ 0 & & \sigma_{\nu s}^2(t_j) \end{bmatrix} \tag{6}$$

The log likelihood is now reduced to

$$-\ln L(\beta) = (sn/2) \ln(2\pi) + \frac{1}{2} \sum_{j=1}^{n} \sum_{k=1}^{s} \ln(\sigma_{\nu kj}^2) + \frac{1}{2} \sum_{j=1}^{n} \sum_{k=1}^{s} (1/\sigma_{\nu kj}^2)(c_{kj} - f_{kj})^2 \tag{7}$$

where $\sigma_{\nu kj}^2$, c_{kj}, and f_{kj} are shorthand notations for $\sigma_{\nu k}^2(t_j)$, $c_k(t_j)$, and $f_k(t_j; \hat{\beta})$, respectively. Differentiation with respect to β yields, in a similar way to the derivation of eqn. (5):

$$\sum_{j=1}^{n} \sum_{k=1}^{s} \{(c_{kj} - f_{kj})/\sigma_{\nu kj}^2\}(\partial f_{kj}/\partial \hat{\beta}) = 0 \tag{8}$$

This is equivalent to a weighted least-squares problem with weights

$$w_{kj} = 1/\sigma_{\nu kj} \tag{9}$$

i.e., equivalent to minimizing

$$S = \sum_{k=1}^{s} \sum_{j=1}^{n} w_{kj}^2 (c_{kj} - f_{kj}(\hat{\beta}))^2 \tag{10}$$

Equation (10) demonstrates the well-known fact that maximum likelihood gives a probabilistic justification for the use of weighted least-squares, and, in addition, provides a method to choose the weights, which must otherwise be chosen by engineering judgment.

At this stage of the development the weights are not known. The values of $\sigma_{\nu kj}^2$ must also be estimated. In order to do this, ν_{kj} is split into a known, time-dependent part μ_{kj}, i.e., the variance of the spatial mean of state variable k observed at time t_j, and an unknown part η_{kj}, expressing all other errors, mainly the model error. If both errors are independent and normally distributed, and if the model error is supposed to be time-invariant in lack of information, then

$$\sigma_{vkj}^2 = \sigma_{\eta k}^2 + \sigma_{\mu kj}^2 \tag{11}$$

Differentiation of eqn. (7) with respect to the unknown $\sigma_{\eta k}^2$ together with some algebraic rearrangement leads to

$$\sigma_{\eta k}^2 = \sum_{j=1}^{n} \{(c_{kj} - f_{kj})^2 - \sigma_{\mu kj}^2\}/(\sigma_{\eta k}^2 + \sigma_{\mu kj}^2)^2 \bigg/ \sum_{j=1}^{n} 1/(\sigma_{\eta k}^2 + \sigma_{\mu kj}^2)^2 \tag{12}$$

This is an implicit formula for $\sigma_{\eta k}^2$, but in this form is amenable to solution by successive substitution. Note that if the variance of the spatial mean concentration is small, i.e., if $\sigma_{\mu kj}^2 \ll \sigma_{\eta k}^2$, then $\sigma_{\eta k}^2 + \sigma_{\mu kj}^2$ is practically time-invariant and eqn. (12) transforms into

$$\sigma_{\eta k}^2 = (1/n) \sum_{j=1}^{n} (c_{kj} - f_{kj})^2 - (1/n) \sum_{j=1}^{n} \sigma_{\mu kj}^2 \tag{13}$$

so that the model error variance is the residual variance less the average of the spatial heterogeneity contributions.

3.2 Parameter Variance–Covariance

A lower bound for the parameter variance–covariance is provided by the Cramér–Rao inequality (see, for example, Eykhoff, 1974):

$$\text{cov}(\hat{\boldsymbol{\beta}}) = E\{(\hat{\boldsymbol{\beta}} - \boldsymbol{b})(\hat{\boldsymbol{\beta}} - \boldsymbol{b})^T\} \geq \boldsymbol{J}^{-1} \tag{14}$$

where $\boldsymbol{J}$ is the Fisher information matrix:

$$\boldsymbol{J} = E\{(\partial \ln L/\partial \boldsymbol{b})(\partial \ln L/\partial \boldsymbol{b})^T\} \tag{15}$$

Following DiToro and van Straten

$$E\{(\partial \ln L/\partial b_\alpha)(\partial \ln L/\partial b_\beta)\} = \sum_k \sum_j \sum_{k'} \sum_{j'} (1/\sigma_{vkj}^2)(1/\sigma_{vk'j'}^2)(\partial f_{kj}/\partial \beta_\alpha)(\partial f_{k'j'}/\partial \beta_\beta)$$

$$\cdot E\{(c_{kj} - f_{kj})(c_{k'j'} - f_{k'j'})\} \tag{16}$$

But, because it is assumed that the residuals are not correlated in time and among states

$$E\{(c_{kj} - f_{kj})(c_{k'j'} - f_{k'j'})\} = \sigma_{vkj}^2 \delta_{jj'} \delta_{kk'} \tag{17}$$

where δ_{mn} is the Kronecker delta (equal to 1 if $m = n$, and 0 otherwise), so that the final result is (asymptotically)

$$\text{cov}(\hat{\beta}_\alpha, \hat{\beta}_\beta) = \left[\sum_{k=1}^{s} \sum_{j=1}^{n} (1/\sigma_{vkj}^2)(\partial f_k(t_j; \hat{\beta})/\partial \beta_\alpha)(\partial f_k(t_j; \hat{\beta})/\partial \beta_\beta) \right]^{-1} \tag{18}$$

This is the conventional nonlinear weighted least-squares expression for the covariance–variance matrix of the parameter estimates.

3.3 Prediction Error

When the parameter estimates $\hat{\beta}$ and the associated variance–covariance matrix are available one may also estimate the error in the prediction resulting from the parameter uncertainty:

$$V\{f_{kj}(\hat{\beta})\} = E\{[f_{kj}(\hat{\beta} + \Delta\beta) - f_{kj}(\hat{\beta})]^2\} \tag{19}$$

Linearization of $f_{kj}(\hat{\beta} + \Delta\beta)$ around $\hat{\beta}$ by a Taylor-series expansion yields

$$V\{f_{kj}(\hat{\beta})\} = E\{(\partial f/\partial \beta_\alpha)(\partial f/\partial \beta_\beta)\Delta\beta_\alpha \Delta\beta_\beta\} = \sum_{\alpha=1}^{p}\sum_{\beta=1}^{p} (\partial f/\partial \beta_\alpha)(\partial f/\partial \beta_\beta)\, \mathrm{cov}(\beta_\alpha, \beta_\beta) \tag{20}$$

This result is reasonable only if the linearization around the parameter values is reasonable. In cases of pronounced nonlinear behavior, eqn. (20) may be expected to yield misleading results.

4 APPLICATION

4.1 IFYGL Data

In the application to the IFYGL data $\sigma^2_{\mu kj}$ is the variance of the spatial mean, computed as

$$\sigma^2_{\mu kj} = (1/N_{kj})\sigma^2_{\epsilon kj} \tag{21}$$

$$\sigma^2_{\epsilon kj} = \{1/(N_{kj}-1)\}\sum_{i=1}^{N_{kj}} (c_{ijk} - \bar{c}_{kj})^2 \tag{22}$$

$$\bar{c}_{kj} = (1/N_{kj}) \sum_{i=1}^{N_{kj}} c_{ijk} \tag{23}$$

where c_{ijk} is the observation of variable k at time t_j at location i, N_{kj} the total number of locations sampled for variable k at time t_j, $\bar{c}_{kj}$ the spatial average concentration, and $\sigma^2_{\epsilon kj}$ the sample variance of the individual c_{ijk} values.

The parameter-estimation procedure involves the solution of eqn. (8), which can be done by conventional nonlinear least-squares routines. The only difference is that the weights in this application depend upon the parameters chosen, and a continuous update of the weights according to eqn. (12) is necessary during the optimization.

The results can be summarized as follows.

(a) The objective function (eqn. 10) using a continuous weight update turned out to be very insensitive to the parameter choice (see discussion below); consequently, convergence was slow. Since the principle interest of DiToro and van Straten was in the parameter uncertainty, they decided to skip the optimization step in this preliminary application. Instead, the nominal values resulting from the 1967–1970 hand calibration were used as reasonable approximations of the maximum likelihood estimates.

(b) Coefficients of variation for the parameters resulting from the analysis are shown in Table 1, column a. Because no direct calibration with the 1972 data was done (and, moreover, forcings representative of 1967–1970 rather than the actual 1972 data were used), the fit is not as good as would otherwise have been possible. Consequently, the model errors as shown in Table 2 are definitely larger than expected. This, in turn, inflates the parameter covariances, allowing for only qualitative conclusions. Generally, the kinetic parameters and stoichiometric ratios associated with the phytoplankton, phosphorus, and nitrogen cycles can be estimated with fair accuracy. On the basis of the field data less can be said about the value of the Michaelis–Menten coefficients, especially for nitrogen. This is not surprising because, in Lake Ontario, nitrogen is only limiting for a relatively short period. Perhaps most striking in column a of Table 1 is the relatively large uncertainty in the zooplankton parameters. Apparently, zooplankton kinetic parameters cannot be estimated with confidence from the available field data, at least for herbivorous zooplankton. It is somewhat surprising that the uncertainty for the carnivorous zooplankton seems to be smaller, especially because its behavior has only a secondary effect on the phytoplankton and nutrient concentrations.

(c) Table 3, column a, gives an impression of the prediction error due to parameter uncertainty, calculated from eqn. (20). Prediction errors tend to be largest in summer and smallest in winter but only yearly averages are shown in Table 3. Clearly, the prediction of zooplankton is without much meaning in this case. The numbers in parentheses indicate the prediction error when the covariance structure of the parameters is ignored. Since several parameters are strongly correlated the effect is rather dramatic. Correlations arise from the structure of the model. For example, since

$$dA/dt = (G-D)A \tag{24}$$

where A is phytoplankton, G the specific growth rate, and D the death rate, G and D have a strong positive correlation upon estimation, because it is in fact the difference that determines the behavior pattern of phytoplankton rather than the individual G and D associated parameters. Consequently, a large uncertainty in each of the parameters individually does not necessarily imply a large prediction error, because of the mitigating effect of the covariance terms in eqn. (20).

4.2 Ten-Year Data Set

In the application of this paper the ten-year data were used to construct a lumped-data set for one "average" year. For this purpose the measurement data for a particular month were averaged over ten years (as far as data were available). Since the year-to-year

TABLE 1 Parameter uncertainty in coefficients of variation (%)[a].

	a	b	c	d
Phytoplankton				
Growth rate	63	48	71	83
Respiration rate	81	26	27	38
N, Michaelis–Menton	479	267	>1000	>1000
PO_4, Michaelis–Menton	411	115	174	173
N/chlorophyll-a	53	19	28	29
PO_4/chlorophyll-a	63	16	16	20
Settling velocity	76	>1000	86	102
Nitrogen				
Organic N–NH_3 transformation rate	31	42	57	57
NH_3–NO_3 transformation rate	25	33	51	51
N settling	85	38	53	57
Phosphorus				
Unavailable P–PO_4 transformation rate	113	43	48	48
P settling	237	32	41	45
Zooplankton				
C/chlorophyll-a	235	131	247	515
Herbivorous grazing	170	86	228	428
Herbivorous assimilation efficiency	800	210	–	771
Herbivorous respiration	>1000	79	–	920
Herbivorous grazing saturation	>1000	280	–	>1000
Carnivorous grazing	95	49	90	317
Carnivorous assimilation efficiency	116	78	–	785
Carnivorous respiration	160	51	–	896

[a] a: 1972 IFYGL data. Nominal parameter set without optimization. b, c, d: 1967–1976 data. Optimization with fixed weights; b, all parameters, c, five parameters fixed at nominal value, and d, full uncertainty with parameter set found under c.

variability was much larger than the spatial variance the latter was ignored, and the variance assigned to each of the twelve data points was simply taken as the sample variance about the ten-year mean. (Note that this variance may be up to ten times larger than the variance of the mean. This choice implies that the ten-year mean is seen as the most likely value for any individual year, while the sample variance relates to the interval in which the datum is expected for any individual year rather than the interval in which the mean is expected.) The variance obtained in this way somehow comprises the year-to-year variability in inputs and forcing functions, and thus compensates more or less for the fact that these variations have not been explicitly taken into account. Next, the lumped-data set was substituted for the IFYGL set and the estimation procedure repeated in the same fashion as before. However, because of the insensitivity of the objective function, as mentioned previously, it was decided to optimize the parameters with a fixed weight, based on the data variance alone. Thus, no weight update incorporating the model error was performed during the optimization. However, once the optimum point was found, an estimate of the model error was made using eqn. (12), and the parameter variance–covariances were calculated with the updated weights according to eqns. (11) and (18). Because in this case the data variance is much larger

TABLE 2 Model error versus data error[a].

	a		b		c/d	
	Model	Data	Model	Data	Model	Data
Epilimnion						
Chlorophyll-a	89	7	2	22	6	22
TKN[b]	52	4	–	47	–	47
NH_3-N	260	6	10	21	21	21
NO_3-N	44	2	13	34	13	34
Total P	10	1	–	42	–	42
PO_4-P	87	3	–	19	5	19
Hypolimnion						
Chlorophyll-a	70	18	7	59	9	59
TKN[b]	17	4	–	50	–	50
NH_3-N	390	6	–	15	–	15
NO_3-N	8	1	20	40	16	40
Total P	16	1	1	41	1	41
PO_4-P	91	2	–	17	–	17

[a] See footnote to Table 1.
[b] Total Kjeldahl nitrogen.

TABLE 3 Annually averaged prediction error (as % coefficient of variation) for some of the epilimnion variables[a,b].

	a	b	c	d
Chlorophyll-a	56 (705)	10 (295)	8 (110)	9 (150)
Zooplankton	350 (2050)	43 (1200)	14 (130)	27 (460)
Ortho-P	74 (430)	9 (90)	9 (150)	10 (150)
Total P	6 (36)	5 (35)	4 (18)	5 (21)

[a] See footnote to Table 1.
[b] Values in parentheses show the prediction error when the covariance structure of the parameters is ignored.

than in the IFYGL case, calculation of the model error with eqn. (12) was not always possible. That is, the data error encompasses the span of all possible errors including model errors, and there is simply not enough information to separate the individual errors.

In contrast to the IFYGL application a parameter optimization was performed prior to uncertainty calculations, rather than using the nominal parameter set. First, a full unconstrained 20-parameter search was made. No difficulty was met in finding a parameter set that produced a better fit than the nominal one. Consequently, as can be seen from Table 1, column b, the parameter error is much less than in the case of no optimization (the high relative error for the settling velocity is artificial, because its value turns out to be estimated as nearly zero, in which case a coefficient of variation is not a suitable error measure). With the optimized set, apparently, a lower model error (as far as this could be computed, see Table 2, column b) and a lower prediction error (Table 3, column b) are obtained. However, when looking at the parameter values arrived at, it became clear immediately that the improved fit was obtained at a physically nonfeasible point in parameter space. For example, the carnivorous assimilation efficiency was raised from 0.6 in the nominal set to a value much larger than 1 in the optimized set, which is, of course, impossible. And although most other parameters remained within reasonable ranges, some of the suggestions made by the optimization, such as a five-fold larger carbon-to-chlorophyll ratio, are highly unlikely. These results serve to illustrate that multiparameter optimizations in complex systems with relatively few data tend to yield nonsensical answers if no constraints are imposed on the parameters.

To overcome these difficulties a second run was made in which some of the most uncertain and some of the physically most constrained parameters were fixed at their nominal values (all associated with the zooplankton cycle). A new optimization was carried out with these constraints, and a new optimum was found with values quite different from those of the full parameter case. As expected, the model error was larger in this case (see Table 2, column c/d) because fixing some of the parameters restricts the number of degrees of freedom to adapt the state trajectory to the data. The parameter covariances were computed in two ways. Firstly, the fixed parameters were treated as constants, so that they do not appear in the variance–covariance calculations (see Table 1, column c). This is equivalent to assuming that these parameters are known with certainty. In the second method, uncertainty was also assigned to the fixed parameters by including them in the variance–covariance computations. This is equivalent to assuming that the fixed parameters belong to the optimal set. The results are given in Table 1, column d. As can be seen, the fixed parameters are associated with very large uncertainties, which is not unexpected because uncertainty was one of the criteria for selecting them to be fixed. In the case of the optimal point, a large uncertainty would have meant that these parameters could hardly be estimated from the data. Away from the optimal point this conclusion is not necessarily justified (compare columns a and d). On the other hand, a large uncertainty does indicate that at the actual parameter point the model is not very sensitive to these parameters. This is why it is perhaps more appropriate to speak of sensitivities rather than uncertainties when the estimates are not close to the "true" values (see Schweppe, 1973).

It is also interesting to note from Table 1 that the assumption that some of the parameters are known with certainty decreases the uncertainty in the remaining parameter estimates. This agrees with the intuitive feeling that the availability of such a strong piece of external information permits a better estimation to be made. This point is further illustrated in Figure 1, in the simple case of two parameters. Suppose that the point (C, D) is found after optimization with p_1 fixed at C. Now the uncertainty of the parameters resulting from the Cramér–Rao inequality (eqn. 14), assuming both parameters to be unknown, is represented by the ellipse around (C, D), which is a linear approximation of the true nonlinear confidence contour represented by the line drawn. Thus, the uncertainty in parameter p_2 is given by σ_3. However, if C is considered to be known with certainty, the true point cannot be everywhere within the confidence ellipse but must be on the line $p_1 = C$. Then, the uncertainty in p_2 under the condition $p_1 = C$ is the distance σ_2 only, and is therefore smaller than before.

Assuming the parameters to be known with certainty reduces the uncertainty of the remaining parameters only if we stay in the same point of parameter space. It is not true in general, as can be seen from a comparison of columns b and c in Table 2. Allowing

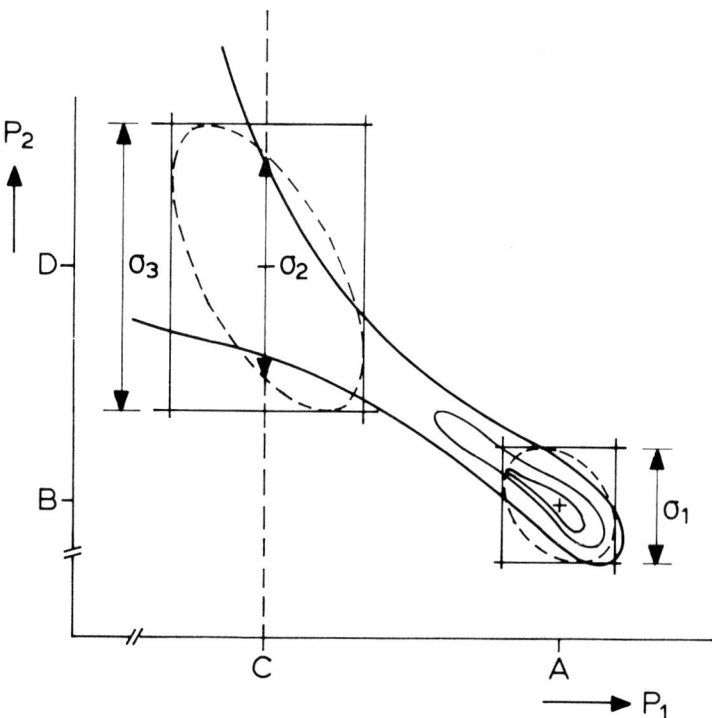

FIGURE 1 Effect on parameter uncertainty of fixing a parameter off the optimal point (two-parameter example, for explanation see text).

for a full optimization does decrease the parameter variances. Again, reference is made to Figure 1. If, instead of fixing p_1 at C, a full optimization is carried out, one would find the true minimum (A, B). Since this is a true minimum the variance σ_1 is expected to be lower than at other points somewhere "up the hill". It may, of course, be that the point (A, B) is physically not feasible (as is the case with column b in Table 1).

Finally, a word must be said about the prediction error in the three cases. As expected, the average prediction error is smaller in case c than in case d because the parameters are known with less uncertainty. However, it is surprising that the prediction errors in the full parameter case (column b) are somewhat larger than in cases c and d, despite the better fit in this point. It may be that this has something to do with the non-linearities in the model. One might expect the model to be particularly nonlinear around nonfeasible points in parameter space (as in case b), and this might have an inflating effect on the prediction uncertainties. The result confirms that care must be exercised in applying eqn. (20).

5 DISCUSSION

The theme of central importance in maximum likelihood estimation is what is known or can be assumed about the structure of the errors. The results are reasonable inasmuch as the assumptions are reasonable. In this application the assumption was that model errors are additive with the state variables, so that they can be treated in essentially the same fashion as observation errors. It was also implicitly assumed that input errors are either absent or propagate linearly with the state estimates, which is perhaps not very likely in the case of an essentially nonlinear model. It may also be that system disturbances do not lead to additive state noise. In such situations a stochastic differential equation might be a more appropriate model than a deterministic description (see McLaughlin, 1978). Further, the initial conditions were assumed to be known with certainty. In reality, of course, they are not known exactly and this will, in principle, lead to a nonwhite state error because the initial condition error vanishes as time proceeds in this kind of model. A possible solution to this problem could be to treat the initial conditions as additional parameters and estimate them simultaneously.

With these limitations in mind, some of the problems encountered can now be discussed. First consider the question of the observed insensitivity of the objective function of eqn. (10) to the parameters. In both applications of this study the total error as given by eqn. (11) is only a weak function of time; in the IFYGL case because $\sigma^2_{\mu kj} \ll \sigma^2_{\eta k}$, and in the ten-year data case because the error in the data estimates is not much different throughout the year. Then $\sigma^2_{\eta k}$ is approximated by eqn. (13), and substitution back into eqn. (11) yields

$$\sigma^2_{\nu kj} \simeq (1/n) \sum_{j=1}^{n} (c_{kj} - f_{kj})^2 - (1/n) \sum_{j=1}^{n} \sigma^2_{\mu kj} + \sigma^2_{\mu kj} \qquad (25)$$

Obviously the two latter terms almost cancel out so that

$$\sigma^2_{\nu k} \simeq (1/n) \sum_{j=1}^{n} (c_{kj} - f_{kj})^2 \qquad (26)$$

Thus, the objective function given by eqn. (10)

$$S = \sum_{k=1}^{s} \sum_{j=1}^{n} (1/\sigma^2_{\nu kj})(c_{kj} - f_{kj}(\hat{\beta}))^2$$

becomes

$$S \simeq \sum_{k=1}^{s} n = ns \qquad (27)$$

In other words, S is almost constant, irrespective of the parameter choice. Indeed, in the applications the numerical value of the objective function was always approximately equal to the total number of observations as predicted by eqn. (27).

The situation can be analyzed further in more general terms. Consider the maximization of the likelihood function (eqn. 4) in two cases: first where $\mathbf{R}_j$ is known *a priori*. In this case the ln terms in eqn. (4) do not depend on the parameters and the parameter estimation is done by minimizing

$$S = \sum_{j=1}^{n} [(c_j - f_j(\beta))^T R_j^{-1}(c_j - f_j(\beta))] \tag{28}$$

This is a weighted least-squares procedure with fixed weights. The optimization step in the ten-year data analysis falls in this category.

Now assume R_j is not known. In this case it is reasonable to assume that R_j is time-invariant. It can be shown that differentiation of the likelihood function with respect to the unknown element of R yields

$$\hat{R} = \frac{1}{n} \sum_{j=1}^{n} (c_j - f_j(\hat{\beta}))(c_j - f_j(\hat{\beta}))^T \tag{29}$$

in the optimum point $\hat{\beta}$. This result is easily verified for a diagonal R matrix, but it is true in general (see Schweppe, 1973). Thus, in the optimum the last term of the likelihood function (eqn. 3) is

$$\sum_{j=1}^{n} [(c_j - f_j(\hat{\beta}))^T \hat{R}^{-1}(c_j - f_j(\hat{\beta}))] = \mathrm{tr}\{\hat{R}^{-1} \sum_{k=1}^{n} (c_j - f_j(\hat{\beta}))(c_j - f_j(\hat{\beta}))^T\}$$

$$= \mathrm{tr}\{\hat{R}^{-1} n \hat{R}\} = n \, \mathrm{tr}\{I\} = ns \tag{30}$$

In other words, this term is virtually constant (in agreement with what was found in the derivation of eqn. 27), and optimization of the likelihood function must be performed by minimizing

$$S_2 = \ln |R| \tag{31}$$

or by minimizing

$$S_3 = \left| \sum_{j=1}^{n} (c_j - f_j(\beta))(c_j - f_j(\beta))^T \right| \tag{32}$$

This is not equivalent to a weighted least-squares procedure.

To get a better feeling for the meaning of eqn. (32) it is interesting to consider the special case where cross-correlations in the residuals among states are absent. In that case R is diagonal. The parameter estimation is then equivalent to

$$\text{minimize } S_4 = \prod_{k=1}^{s} \sum_{j=1}^{n} (c_{jk} - f_{jk}(\beta))^2 \tag{33}$$

Or in words: minimize the product of the sum of the squared differences of the states. Equation (33) has attractive properties in that it automatically solves the problem of different unit dimensions normally encountered in multistate least-squares. Also it puts automatically more weight on those state variables for which the model fits well.

But what about the case where R contains both unknown and known, time-variable parts, as was the situation in the applications in this paper? A full mathematical treatment is not so easy, but by analogy one might infer that minimization of

$$S_5 = \prod_{k=1}^{s} \sum_{j=1}^{n} (1/\sigma_{\mu jk}^2)(c_{jk} - f_{jk})^2 \tag{34}$$

is a reasonable pragmatic approach in this case. Weighting the residuals by the measurement variances reduces the influence of unreliable data points. This idea can be incorporated in eqn. (32) for the more general correlated case.

A final word about the prediction problem. The prediction error, eqn. (20), is not a true error propagation formula. It merely states the variance in the prediction that arises from parameter uncertainty at any point in time around the given, deterministic trajectory. Thus, errors do not build up as time proceeds. It may be that real error-propagation calculations based on recursive-filtering techniques (Beck et al., 1979) are more useful. Both prediction-error and error-propagation computations depend heavily upon the validity of the linearization around the optimum. When in doubt, the use of Monte Carlo simulation, employing the estimated parameter variance–covariance structure, may be preferable. An additional advantage of a Monte Carlo approach is the possibility of studying the effects of input uncertainty and uncertainty in initial conditions. Maximum likelihood estimation is still possible in these more general cases (Schweppe, 1973), but the equations are rather complex and their practical implementation is cumbersome.

6 CONCLUSION

Maximum likelihood estimation is a practical and useful procedure in cases involving additive process and measurement noise. The resulting weighted least-squares optimization can be successfully performed when the error statistics are known. For errors that are only partly known our experience has been less favorable. The least-squares objective function with continuous weight update during optimization is particularly insensitive to the choice of parameters, especially when the total error is only a weak function of time, as is frequently the case. The theoretical development for completely unknown error statistics shows that weighted least-squares is doomed to fail in this situation, and that the product of the sum of squared residuals is a more appropriate objective function. This may also be the proper function to use when the errors are partially known, in which case weighting based on the known error component, e.g., measurement error, may be a suitable modification. A forthcoming study will investigate these ideas in practical applications.

Experience from this study shows that unconstrained optimization of multi-parameter systems may easily lead to nonfeasible solutions. To prevent this undesirable behavior, constrained optimization is needed, but this has a definite effect on the parameter variance–covariance matrix. In the case of the most extreme constraint possible, namely fixing some of the parameters entirely, the overall parameter uncertainty increases. Moreover, these uncertainty estimates are still too low when the fixed parameters, though assumed to be known with certainty, are in reality uncertain. Thus, the practice of fixing some of the parameters must be viewed with caution. In fact one may question whether a model is actually well-structured if the use of parameter constraints is the only way to avoid nonfeasible solutions. Further research in this area is definitely needed.

Uncertainty in predictions because of parameter uncertainty is strongly mitigated by the parameter covariance structure. This must be taken into account when employing Monte Carlo simulation as an alternative to the error prediction formula derived from a linearization around the optimum. The results indicate that the linearized equation may lead to suspicious results, especially for points where the model is nonlinear, or for points located at a constraint. More insight can probably be obtained by comparison with results from error-propagation calculations based on recursive-filtering algorithms.

ACKNOWLEDGMENT

I am indebted to Gerard Golbach who created the ten-year lumped-data set and redesigned the computer program. His critical comments were highly appreciated and have led to considerable rethinking of some of the basic concepts of the approach.

REFERENCES

Beck, M.B., Halfon, E., and van Straten, G. (1979). The Propagation of Errors and Uncertainty in Forecasting Water Quality, Part I – Method. WP-79-100. International Institute for Applied Systems Analysis, Laxenburg, Austria.

DiToro, D.M. and van Straten, G. (1979). Uncertainty in the Parameters and Predictions of Phytoplankton Models. WP-79-27. International Institute for Applied Systems Analysis, Laxenburg, Austria.

Draper, N.R. and Smith, H. (1966). Applied Regression Analysis. Wiley, New York, pp. 263–299.

Eykhoff, P. (1974). System Identification, Parameter and State Estimation. Wiley, New York, pp. 410–414.

McLaughlin, D. (1978). Parameter estimation problems in water resource modeling. In G.C. Vansteenkiste (Editor), Modeling, Identification, and Control in Environmental Systems. North-Holland, Amsterdam, pp. 137–151.

Schweppe, F.C. (1973). Uncertain Dynamic Systems. Prentice-Hall Series in Electrical Engineering. Prentice-Hall, Englewood Cliffs, New Jersey, pp. 423–468.

Thomann, R.V., DiToro, D.M., Winfield, R.P., and O'Connor, D.J. (1975). Mathematical Modeling of Phytoplankton in Lake Ontario, 1. Model Development and Verification. EPA-660/3-75-005. US Environmental Protection Agency, Corvallis, Oregon.

IDENTIFICATION METHODS APPLIED TO TWO DANISH LAKES

Henning Mejer and Leif Jørgensen

Københavns Teknikum, Copenhagen 2200 N (Denmark)

1. NOMENCLATURE USED IN THE PAPER

Throughout this paper the following nomenclature will be used:

$\bar{\psi}$ = model state variables,
$\bar{\psi}^o$ = observed state variables,
t = time,
$\bar{r}$ = space coordinates,
$\bar{a}$ = parameter set,
$\bar{\theta}$ = normalized parameter set,
$\bar{\eta}$ = (weighted) residuals,
s = cubic splines,
$\bar{\bar{J}}$ = (weighted) sensitivity matrix (Jacobian),
λ = Marquardt parameter,
i = state variables ($i = 1, 2, \ldots, n$),
j = sampling times ($j = 1, 2, \ldots, m_i$),
k = estimatable parameters ($k = 1, 2, \ldots, p$),
r = iterations in parameter search ($r = 0, 1, 2, \ldots$),
$^-$ = (single bar) vector quantity,
$^=$ = (double bar) matrix,
^{T} = transposition,
^{0} = (zero) initial guess, and
^{o} = observed value.

2. INTRODUCTION TO THE PROBLEMS

Given a set of field data on a lake system $\{\psi_i^o(t_j)\}$ at sampling times $t_1 < t_2 < \ldots < t_j < \ldots < t_{m_i}$ and a proposed deterministic model

$$d\bar{\psi}/dt = \bar{f}(\bar{\psi}, t; \bar{a}) \tag{1}$$

at least three questions immediately arise:

(i) How should observations ψ_i^o (or linear combinations of ψ_i^o) be identified with model state variables ψ_i?
(ii) Which feasible parameter set(s) $\bar{a}$ will minimize deviations between $\bar{\psi}$ and $\bar{\psi}^o$, given a model structure $\bar{f}$?
(iii) How could submodel constructs be selected and modified in ways that both obey *a priori* biological and chemical knowledge in a qualitative sense and fit the observed data in a more quantitative sense?

Since the dimensions of both state space and parameter space are usually high (>10) in structured lake models, traditional parameter-search routines often fail to answer these questions unless the initial parameter guess is very close to an optimum.

This paper contains some down-to-earth methods for answering these questions. Most of the techniques described may be extended to apply to distributed models of the form

$$\partial \bar{\psi}/\partial t = \bar{f}(\bar{\psi}, \partial \bar{\psi}/\partial \bar{r}, \partial^2 \bar{\psi}/\partial \bar{r}^2, t; \bar{a}) \tag{2}$$

3 PARAMETER-SEARCH METHODS

Three iterative methods will be discussed: the Gauss–Newton, the steepest descent, and the Marquardt. To avoid scaling problems, both the parameters ($\bar{a}$) and the state variable residuals ($\bar{\psi} - \bar{\psi}^o$) are normalized in the following way

$$\theta_k \equiv a_k/a_k^0 \tag{3}$$

$$\eta_{ij} \equiv w_i[\psi_i(t_j) - \psi_i^o(t_j)]/\langle\psi_i^o\rangle, \text{ where } \langle\psi_i^o\rangle = (1/m_i) \sum_j \psi_i^o(t_j) \tag{4}$$

$\{a_k^0\}$ is an initial parameter guess and w_i are weights that may be chosen, for example, as inverse normalized elements of the diagonal variance–covariance matrix of the observed state variables (see also Di Toro and van Straten, 1979).

Introducing a Jacobian (sensitivity matrix) as

$$\bar{\bar{J}} = \partial \bar{\eta}/\partial \bar{\theta} \quad \text{(dimension: } \sum_{i=1}^{n} m_i \times p \text{)} \tag{5}$$

the three methods may be formulated as follows

Gauss–Newton: $\bar{\theta}_{r+1} = \bar{\theta}_r - (\bar{\bar{J}}^T \bar{\bar{J}})^{-1} \bar{\bar{J}}^T \bar{\eta}$

Steepest descent: $\bar{\theta}_{r+1} = \bar{\theta}_r - \bar{\bar{Y}}^{-1} \bar{\bar{J}}^T \bar{\eta}$

Marquardt: $\bar{\theta}_{r+1} = \bar{\theta}_r - (\bar{\bar{J}}^T \bar{\bar{J}} + \lambda \bar{\bar{I}})^{-1} \bar{\bar{J}}^T \bar{\eta}$

where r numerates the iterations ($r = 0, 1, 2, \ldots$), λ is the Marquardt parameter, $\bar{\bar{Y}}$ is a positive definite matrix, and $\bar{\bar{I}}$ is the $p \times p$ identity matrix.

$\overline{\overline{Y}}$ is usually set as an identity matrix for the first iteration and updated at each subsequent iteration, as shown by Fletcher and Powell (1963). λ is reduced each time $\overline{\eta}^T \overline{\eta}$ decreases and increased when $\overline{\eta}^T \overline{\eta}$ increases (Marquardt, 1963). As $\lambda \to 0$ the search direction in parameter space approaches Gauss–Newton directions. As $\lambda \to \infty$ the direction approaches steepest descent, i.e., the direction opposite to the gradient. It is therefore possible to shift "dynamically" between the three methods according to new information on the Jacobian gained during computations. The Jacobian is for the first iteration, when $r = 1$, evaluated as difference quotients, but for later iterations higher-order estimates may be used.

4 CSMPOPT – A PARAMETER ESTIMATION PACKAGE

A computer package using a combination of the three numerical methods mentioned in the previous section and including inequality constraints on the parameter set as a whole (i.e., $|\overline{\theta}| \leq$ a user-specified limit) and several other options was written by one of us (H.M.) in 1975. Since then the package has been tested on about a dozen different water-quality models. Practical experience has shown that this tool only yields feasible parameter estimates when the dimensions of the problem are low (<5) or when the package is used for fine-tuning an already well-calibrated model.

CSMPOPT is a program that is transparent to the user in the sense that his model should be coded as a CSMP source text. Except for a few statements to be added, there are no restrictions in the CSMP facilities (including the possibility that the entire source text might be called from a FORTRAN subprogram embracing the whole model). The supplementary statements are as follows:

- FIT <list of variable names>,
  specifying state variables to be fitted to observed variables;
- ADJUST <list of parameter names>,
  specifying parameters to be perturbed;
- STORAGE <obs(dim)>, <wg(dim)>, ...,
  specifying tables of observed time series ($\overline{\psi}^o$) and weights (w_i).

Additionally, an initial grid search may be required.

The output from the program includes iteration history and, if convergence has been achieved, final parameter values, residuals, and (optionally) an approximate sensitivity matrix and various measures of error statistics.

5 DERIVATIVE ESTIMATES

Originally intended to provide a good initial parameter guess for CSMPOPT, the derivative ($d\overline{\psi}/dt$ for lumped models, $\partial \overline{\psi}/\partial t$, $\partial \overline{\psi}/\partial \overline{r}$, and $\partial^2 \overline{\psi}/\partial \overline{r}^2$ in distributed models) were estimated leaving $\overline{a}$ as the only unknown in model equations (1) and (2). When this technique was applied, some useful side-effects were discovered, namely the possibility of state-variable identification and model structure validation (or rather, invalidation).

Cubic splines (and bicubic splines in the distributed two-dimensional case) have also been applied. The resulting algorithm for the lumped model case is shown in Table 1; the underlying reasoning is explained in Mejer et al. (1980a, b).

TABLE 1 Algorithm for computing cubic-spline estimates ds_j/dt of time derivatives $d\psi(t_j)/dt$, given $s_j = \psi(t_j), j = 1, 2, \ldots, m$.

Step 1 $(j = 2, 3, \ldots, m-1)$

$\alpha_j = (t_{j-1} - t_j)/2(t_{j+1} - t_{j-1})$

$\beta_j = (t_{j+1} - t_j)/2(t_{j+1} - t_{j-1})$

$\gamma_j = 3\{[(s_{j+1} - s_j)/(t_{j+1} - t_j)] - [(s_j - s_{j-1})/(t_j - t_{j-1})]\}/(t_{j+1} - t_{j-1})$

Step 2

$\tilde{\beta}_1 = 0$

$\tilde{\gamma}_1 = 0$

Step 3 $(j = 1, 2, \ldots, m-2)$

$\tilde{\beta}_{j+1} = \beta_{j+1}/(1 + \alpha_{j+1}\tilde{\beta}_j)$

$\tilde{\gamma}_{j+1} = (\gamma_{j+1} + \alpha_{j+1}\tilde{\gamma}_j)/(1 + \alpha_{j+1}\tilde{\beta}_j)$

Step 4

$d^2 s_m/dt^2 = 0$

Step 5 $(j = m-1, m-2, \ldots, 1)$

$d^2 s_j/dt^2 = \tilde{\gamma}_j - (\tilde{\beta}_j d^2 s_{j+1}/dt^2)$

Step 6

$ds_1/dt = [(s_2 - s_1)/(t_2 - t_1)] - [(d^2 s_2/dt^2)(t_2 - t_1)/6]$

Step 7 $(j = 1, 2, \ldots, m-1)$

$ds_{j+1}/dt = (ds_j/dt) + \{(t_{j+1} - t_j)[(d^2 s_j/dt^2) + (d^2 s_{j+1}/dt^2)]/2\}$

The spline functions $s_i(t)$ are forced through the observed values

$$s_i(t_j) = \psi_i^o(t_j) \quad (t_1 < t_2 < \ldots < t_j < \ldots < t_{m_i}) \tag{6}$$

and the outputs of the algorithm are estimates of $d\bar{\psi}/dt$ at sampling times t_j

$$d\psi_i(t_j)/dt \approx ds_i(t_j)/dt \tag{7}$$

Since the spline functions here are not used for interpolation — as is usually the case — the algorithm in Table 1 is very much simplified compared to other reported algorithms (see, for example, Greville, 1967).

It is vital to the accuracy of eqn. (7) that $t_{j+1} - t_j$ should be sufficiently small. To get an impression of how small, the plots presented in Figure 1 were prepared; the figure shows — albeit in a rather academic way — that the number of samples per peak should be at least four. Considering typical timescales in lake processes, this means as a rule of thumb that the sampling frequency should be about twice a week when studying the water body and about once per fortnight when sediment samples are taken.

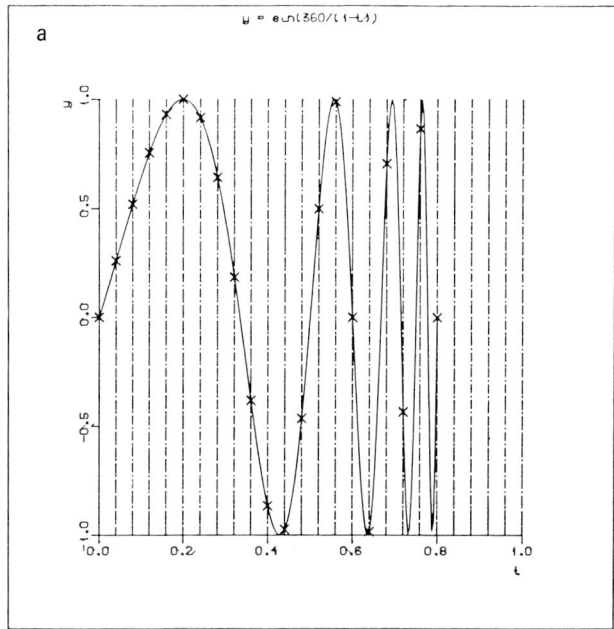

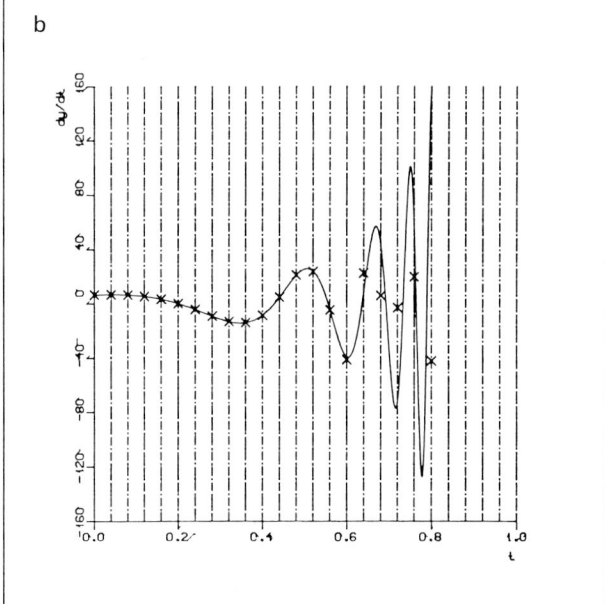

FIGURE 1 (a) A hypothetical state variable $y = \sin[2\pi/(1-t)]$ sampled at equal intervals $(t_{j+1} - t_j = 0.04)$. (b) The corresponding analytically known time derivative and the time-derivative estimates (×) found by using the algorithm in Table 1.

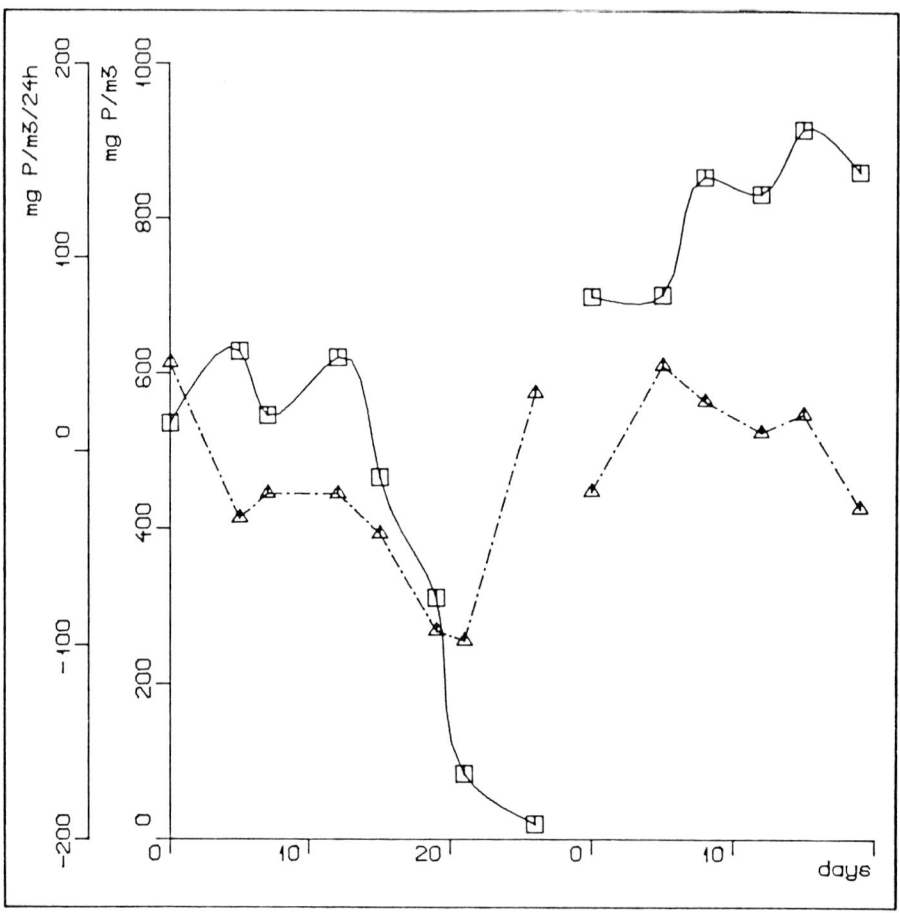

FIGURE 2 Time-derivative estimates (△) based on observed orthophosphate concentrations (□) for Lake Glumsø, Denmark. Note the time gap between the two sampling periods, in spring and autumn.

Figure 2 shows a more realistic case, namely the concentration of soluble orthophosphate in Lake Glumsø, Denmark, as measured in two distinct sampling studies.

6 EXAMPLES

6.1 State Variable Identification

The technique described above has been applied to a 17 state-variable/14 calibrated-parameter model for the shallow Lake Glumsø, Denmark (see Jørgensen et al., 1978). One of the state variables (*PHYT*) was intended to describe phytoplankton biomass

$$dPHYT/dt = [(CDR_{max} f_T f_N f_P f_C) - S_a - (Q/V) - (G_z/Y)] PHYT \qquad (8)$$

where the f's denote functions of temperature (T), intracellular nitrogen (N), phosphorus (P), and carbon (C); Q is the water outflow and G_z is a zooplankton grazing function. Other symbols represent constants: CDR_{max} = maximum cell division rate, S_a = settling rate of algae, V = lake volume, and Y = a yield factor.

At the outset, eight observed time series were possible candidates for identifying $PHYT$: chlorophyll-a, algal fresh weight, chemical oxygen demand (COD), glucose, organic phosphorus (1–80-μm fraction), organic nitrogen (1–80-μm fraction), dry matter, and Secchi-disk readings.

Rearranging eqn. (8) and neglecting the zooplankton term we obtain

$$CDR_{max} \approx [(d \ln PHYT/dt) + S_a + (Q/V)]/f_T f_N f_P f_C \qquad (9)$$

By substituting each of the eight candidate parameters one at a time for $PHYT$ in this equation it was possible to test which gave the most constant values of CDR_{max}. Note that any proportionality factor relating observed and modelled state variables is irrelevant here because of the logarithmic derivative. Since the last six candidate parameters are definitely known to include nonalgal condensed matter, it was expected that chlorophyll-a and algal fresh weight would most closely represent $PHYT$, and this was actually suggested by the analysis.

6.2 Local Parameter Estimation

Another equation in the model of Lake Glumsø mentioned above reads as follows

$$dN_c/dt = UN\, PHYT - [S_a + (Q/V) + (G_z/Y)] N_c \qquad (10)$$

where the new symbols N_c and UN represent intracellular nitrogen and nitrogen uptake rate, respectively. UN is calculated as

$$UN = UN_{max} g_N N_s/(KN + N_s) \qquad (11)$$

where g_N is a known function of N_c and $PHYT$, and N_s is soluble nitrogen in the water phase. The constants UN_{max} (maximum uptake rate) and KN (nitrogen Michaelis constant) remain to be estimated. Combining eqns. (10) and (11) and again neglecting the zooplankton term, we obtain

$$PHYT N_s g_N / \{(dN_c/dT) + [S_a + (Q/V)N_c]\} \approx (1/UN_{max})N_s + (KN/UN_{max}) \qquad (12)$$

which has the shape of a linear regression model

$$Y = ax + b \qquad (13)$$

Calculation of slope and intercept leads to estimates of UN_{max} and KN

$$\widehat{UN}_{max} = 0.043 \quad [\text{day}^{-1}] \quad (0.034-0.059)$$

$$\widehat{KN} = 0.26 \quad [\text{gN m}^{-3}] \quad (0.15-0.45)$$

Units are given in square brackets and the numbers in parentheses are 95% confidence limits.

A corresponding calculation for phosphorus only leads to an estimate of maximum uptake rate (UP_{max}); the Michaelis constant estimate (KP) fails, obviously because phosphorus was not limiting at any time during the measurement periods. Despite this overparameterization, KP was still retained in the model, mainly because phosphorus might become a limiting factor in this lake in the future.

6.3 Submodel Construct Determination

Measurement of primary production leads to an estimate of maximum cell division rate through the relation

$$\text{Production} = CDR_{max} \; PHYT \, f_T f_N f_P f_C \tag{14}$$

where

$$f_T = 1.17^{|T_{opt}-T|} \tag{15}$$

$$f_N = \max\{0, (N_c - N_c^{min})/N_c\} \quad \text{(and similar relations for } f_P \text{ and } f_C\text{)} \tag{16}$$

After a local calibration on N_c^{min}, etc., these parameters changed CDR_{max} as shown in Table 2.

TABLE 2 CDR_{max} estimated from primary production.

	Mean	Standard deviation	Coefficient of variation (%)
Original N_c^{min}, etc.	0.71	0.65	92
Local calibrated N_c^{min}, etc.	7.00	2.65	38
$f_T^{old} \to f_T^{new}$	4.71	1.00	21
Literature values	~4	~1	~25

By inspection of the factors f_T, f_N, f_P, and f_C, it was evident that f_T did not depict the observed response on growth rate very well, especially near the optimum temperature T_{opt}. Another construct was therefore suggested

$$f_T^{new} = e^{a(T-T_{opt})}[(T_{max} - T)/(T_{max} - T_{opt})]^{a(T_{max}-T_{opt})} \tag{17}$$

(see Lassiter, 1975). The constant a is chosen to make $f_T^{new} = f_T^{old}$ at $T = 0$, where f_T^{old} is given by eqn. (15). The resulting coefficient of variation decreased as shown in Table 2. Also, CDR_{max} now falls within the range of literature values (when corrected for the effects of light limitation on growth).

6.4 Distributed Models

In several versions of a sediment model for Lake Esrom, (see Kamp-Nielsen, 1978; Mejer, 1978; Mejer et al., 1980b) a significant discrepancy between observed and modelled exchangeable phosphorus (P_e) occurred for values for the month of May; otherwise the models behaved fairly well. One of the four partial-differential equations in the latest version is

$$\partial P_e/\partial t = -K(\tau)g(T, OX)P_e - (dZ/dt)(\partial P_e/\partial z) \tag{18}$$

where

$$dZ/dt = [(S-R)/10DM_0 P_{sed}] - \int_0^{z_{max}} K(\tau)g(T, OX)(P_e/P_{total})dz \tag{19}$$

and

$$K(\tau) = 1/[\tau + (1/<k>)] \tag{20}$$

There are two independent variables in the model, time (t) and depth in the sediment (z). Z is the displacement of the sediment/water interface determined mainly by the sedimentation rate (S), resuspension rate (R), and a compression term (the integral in eqn. 19). DM_0 denotes the dry matter content at the sediment surface and $g(T, OX)$ is a known function of temperature (T) and oxygen (OX). The decay rate $K(\tau)$ is a function of "effective" age $\tau = \tau(t, z)$ and an average rate constant of newly sedimented material $(<k>)$, and it should be strictly positive (cf. eqn. 20).

Applying bicubic splines to estimate $\partial P_e/\partial t$ and $\partial P_e/\partial z$, $K(\tau)$ was calculated at various sampling times and depths from eqn. (18). It turned out that $K(\tau)$ was negative for the May values, suggesting that the incorrect section of the model was located near the term $K(\tau)g(T,OX)P_e$ in eqn. (18). Unless data for this month are obsolete, a positive term is missing on the right-hand side of eqn. (18).

7 CONCLUSIONS

Traditional parameter-search methods usually fail when applied to structured lake models with more than about ten dimensions in state space and parameter space. Estimating derivatives from intensive measurement programs, e.g., by using spline functions, seems to improve the initial parameter guesses needed for these methods. Since low-order parameter subspaces are manipulated at each stage of the technique described, state variable identification and debilitated submodel diagnosis turn out to be valuable side-effects of this — admittedly somewhat simplistic — method.

ACKNOWLEDGMENTS

The cooperation of Sv. E. Jørgensen, ISEM (Denmark) on this study, and the facilities offered by IIASA, Laxenburg (Austria) to finish work on this paper are gratefully acknowledged.

REFERENCES

Di Toro, D.M. and van Straten, G. (1979). Uncertainty in the Parameters and Predictions of Phytoplankton Models. WP-79-29. International Institute for Applied Systems Analysis, Laxenburg, Austria.

Fletcher, R. and Powell, M.J.D. (1963). A rapidly convergent descent method for minimization. Computer Journal, 6:163–168.

Greville, T.N.E. (1967). Spline functions, interpolation and numerical quadrature. In A. Ralston and H.S. Wilfs (Editors), Mathematical Methods for Digital Computers. Vol. II. Wiley, Chichester.

Jørgensen, S.E., Mejer, H., and Friis, M. (1978). Examination of a lake model. Ecological Modelling, 4:253–278.

Kamp-Nielsen, L. (1978). Modelling the vertical gradients in sedimentary phosphorus fractions. Verhandlungen des internationalen Vereins für Limnologie, 20:720–727.

Lassiter, R.R. (1975). Modeling Dynamics of Biological and Chemical Components of Aquatic Ecosystems. EPA-660/3-75-012. US Environmental Protection Agency, Corvallis, Oregon.

Mejer, H. (1978). Interactions Between Sediment and Water. Proceedings of the Nordic Symposium on Sediments, 6th, Hurdal. ISBN-82-990528-0-7. University of Oslo, Blindern, Norway.

Mejer, H., Jørgensen, S.E., and Jørgensen, L.A. (1980a). On parameter estimation in lake modeling. Proceedings of the Task Force Meeting on Lake Balaton Modeling. Veszprém. 27–30 August, 1979. Vol. I, pp. 195–204.

Mejer, H., Jørgensen, S.E., and Kamp-Nielsen, L. (1980b). A sediment phosphorus model. Proceedings of the Task Force Meeting on Lake Balaton Modeling. Veszprém. 27–30 August, 1979. Vol. II, pp. 104–131.

ANALYSIS OF PREDICTION UNCERTAINTY: MONTE CARLO SIMULATION AND NONLINEAR LEAST-SQUARES ESTIMATION OF A VERTICAL TRANSPORT SUBMODEL FOR LAKE NANTUA

F. Chahuneau
Laboratoire de Biométrie, INRA-CNRZ Domaine de Vilvert, Jouy-en-Josas 78350 (France)

S. des Clers and J.A. Meyer
Laboratoire de Zoologie, Ecole Normale Supérieure, 46 rue d'Ulm, 75230 Cedex 05 Paris (France)

1 INTRODUCTION

Lake Nantua (see Figure 1 for lake characteristics) is a small eutrophic alpine lake undergoing frequent algal blooms (the blue-green alga *Oscillatoria rubescens*). The entire water body is thermally stratified from April to December and completely mixed after the late-autumn overturn (the lake is monomictic). Given the rather small lake area and its simple morphometry, the water body is considered horizontally homogeneous and a one-dimensional submodel was developed to describe vertical transport processes (eddy diffusion and advection).

The submodel (briefly described in this paper) accounting for density stratification is used here to compute evolution of temperature profiles due to heat transport and surface exchanges. This allows simulation of the thermal dynamics of Lake Nantua and subsequent modifications of vertical transport rates for the substances involved in chemical and ecological cycles. This transport submodel will be included in a larger water-quality/ecological model describing the lake ecosystem dynamics. This latter model, still under development, will help to assess the efficiency of different restoration techniques (such as hypolimnetic aeration or withdrawal) and to define optimal operational rules.

This paper describes how experimental temperature profiles are used in the nonlinear least-squares estimation of some submodel parameters. The calibrated model is then validated against another set of field data. Finally, Monte Carlo simulation, employing the parameter variance–covariance structure identified in the parameter estimation procedures, is used to investigate the prediction error variance (or uncertainty).

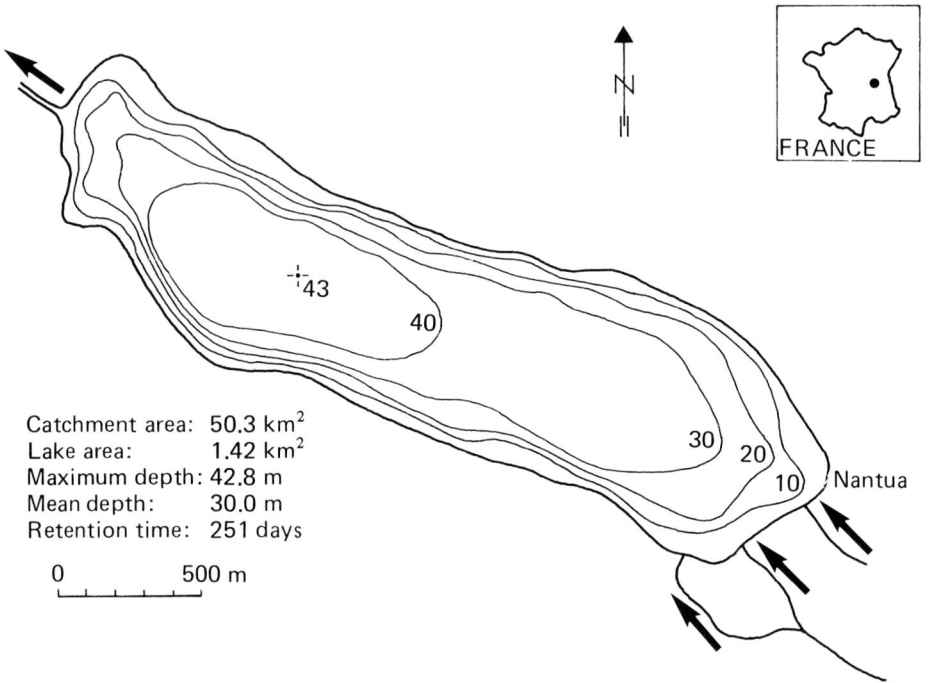

FIGURE 1 Lake Nantua: basic information.

This is done in order to avoid inaccuracies in the computation of prediction confidence arising when the usual linearization around the parameter estimates is used.

2 MODEL DESCRIPTION

2.1 Fundamental Equation

The water-quality model consists of a system of parabolic partial-differential equations describing vertical transport of dissolved substances or biota, and their interactions. For example, the mass-balance equation describing the temperature variations $T(z, t)$ with time (t) at a given depth (z) is

$$\frac{\partial T(z, t)}{\partial t} = \underbrace{\frac{1}{A(z)} \frac{\partial}{\partial z} \left(A(z) K(z, t) \frac{\partial T(z, t)}{\partial z} \right)}_{\text{transport by vertical dispersion}} - \underbrace{\frac{1}{A(z)} \frac{\partial}{\partial z} \left(A(z) W(z, t) T(z, t) \right)}_{\text{transport by vertical advection}}$$

$$+ \underbrace{Q_{\text{in}}(z, t) T_{\text{in}}(z, t)}_{\text{inflow}} - \underbrace{Q_{\text{out}}(z, t) T(z, t)}_{\text{outflow}} + \underbrace{SS(z, t)}_{\text{source-sink}} \qquad (1)$$

where

$A(z)$ = horizontal cross section at depth z (m²),
$K(z, t)$ = dispersion coefficient (m² s⁻¹),
$T(z, t)$ = temperature at depth z (°C),
$W(z, t)$ = vertical advection velocity (m s⁻¹),
$Q_{in}(z, t), Q_{out}(z, t)$ = rates of volume displacement corresponding to inflows and outflows, respectively (s⁻¹),
$T_{in}(z, t)$ = temperature of inflowing water (°C), and
$SS(z, t)$ = rate of change due to internal sources and sinks (°C s⁻¹).

Equation (1) is solved by a finite difference technique. The water body is vertically discretized; in each layer the concentration of a given substance is assumed to be homogeneous. In the case of Lake Nantua, 43 layers were considered (each 1.0 m thick). The Crank–Nicolson approximation, associated with central differences for spatial discretization, was applied. Nonlinearity in eqn. (1), due to the source–sink term and the dependence of $K(z, t)$ on the vertical temperature gradient, is treated by iteration at each time step (Remson et al., 1971). The time step (about 0.1 day) is automatically adjusted according to the convergence rate of the iterative process.

2.2 Description of Vertical Transport

2.1.1 Dispersion

Equation (1) includes a diffusional transport term. Given the spatial and temporal scale characterizing the model (Ford and Thornton, 1979), the "diffusion" coefficient expresses much more than eddy diffusion generated by local shear. The transport equation is spatially averaged over horizontal planes, and the characteristic time scale is of the order of one day (i.e., we are interested in day-to-day variations). Thus, in addition to the typical diffusional transport, the "diffusion" coefficient includes implicitly all transport mechanisms of an advective nature for which the fluxes through horizontal planes are balanced. This concerns such mechanisms as mixing by internal seiches, nocturnal thermal convection, local upwellings and downwellings generated by wind-driven circulation, and transient convection cells such as Langmuir circulations. This is expressed precisely in the concept of "effective dispersion" developed by Orlob and Selna (1970). The "effective dispersion" coefficient is an operational parameter which cannot be easily measured in the field, but can only be estimated by computing layer-to-layer mass or heat budgets.

Parametrization of the dispersion coefficient is based on the following qualitative considerations: vertical dispersion increases with wind speed (kinetic energy input), even in the thermocline (shear stresses) and in the hypolimnion (through large eddies associated with seiche oscillations), and decreases with the local degree of stability of the water column (density gradient), through buoyancy effects; vertical dispersion associated with large vertical eddies necessarily decreases near physical boundaries (including the free surface and thermocline). The vertical dispersion coefficient is calculated using the following formulae. In the epilimnion:

$$K(z) = (k/Pr)\, U^* l_{mix}(1 + SIGMA\, Ri)^{-POW}$$

where

$K(z)$ = vertical dispersion coefficient ($m^2\,s^{-1}$) at depth z (m),
k = von Karman constant (= 0.4),
Pr = turbulent Prandtl number (= 1.0),
U^* = friction velocity = $[(\rho_a/\rho_w)CD_{10}U_{10}^2]^{1/2}$ ($m\,s^{-1}$),
CD_{10} = drag coefficient for wind speed at 10 m (U_{10}, $m\,s^{-1}$),
ρ_a, ρ_w = densities of air and water, respectively,
l_{mix} = mixing length = distance (m) to the closest physical boundary with the additional constraint $l_{mix} \leqslant Z_{SCALE}$ (maximum eddy scale); for $z \leqslant l_{add}$ (surface layers), $l_{mix} = l_{add}$, where l_{add} = additional mixing length (expressing wave mixing and nocturnal convection effects),
$SIGMA, POW$ = empirical, nondimensional parameters expressing the sensitivity of turbulence to stratification effects, and
Ri = local Richardson number = $(g/\rho_w)(\partial\rho_w/\partial z)|\partial u/\partial z|^{-2}$, where $|\partial u/\partial z|$ is the vertical gradient in horizontal flow velocity, and is computed from a stratified boundary-layer approximation (see Tucker and Green, 1977).

In the metalimnion and hypolimnion:

$$K(z) = \max(K_{th}, ALPHA(N_{th}^2/N^2(z))^{POW2}K_{th}l_{mix})$$

where

$N(z)$ = Brunt–Väisälä frequency (s^{-1}) = $[(g/\rho_w)(\partial\rho_w/\partial z)]^{1/2}$,
K_{th}, N_{th} = values at the thermocline of the dispersion coefficient and the Brunt–Väisälä frequency, respectively,
l_{mix} = mixing length (m), and
$ALPHA, POW2$ = empirical parameters.

The additional constraint $K(z) \leqslant 10K_{th}$ is introduced in the hypolimnion.

A detailed justification of the semiempirical equations used will be given elsewhere (Chahuneau and des Clers, in preparation). The classical concept of mixing length (distance to boundaries), modulated by wind speed and local density gradient (see for instance Leonard et al., 1978), was used. Near the free surface, the dispersion coefficient is increased to account for wave mixing and nocturnal thermal convection.

The whole formulation involves a total of six parameters, namely, CD_{10}, $SIGMA$, POW, Z_{SCALE}, $ALPHA$, and $POW2$.

2.2.2 Advection

Computation of vertical advective transport is based on the empirical formulae used by the US Army Corps of Engineers in 1974 (see Tetra Tech., 1978) for reservoir modeling. The inflow zone is centered on the depth where the density of incoming water matches lake water density. The zone extends over and under this level; its total thickness depends on both local density gradient and the inflow rate. A uniform distribution of inflow is assumed (i.e., an equal proportion of total inflow is added to every layer in this zone). Water withdrawal is restricted to the well-mixed surface zone, where it is uniformly distributed.

More detailed approaches, for example, Gaussian distribution of inflows (Ryan and Harleman, 1971), were not attempted, since the uncertainty on loadings is large and Lake Nantua is a natural lake with little through-flow.

2.2.3 Convection

Thermal convection is not introduced in eqn. (1). It operates separately, in the discrete framework defined by the finite difference grid.

At the end of each time step, the algorithm checks the computed temperature profile for density instabilities. The checking procedure starts from the surface downwards. Whenever a density instability between two adjacent layers is detected, mixing starts and works upwards until the instability is removed (on some occasions, it may reach the surface). Starting again from the level where mixing was initiated, it checks the profile further down, and so on, until overall stability is obtained. Density is used here rather than temperature, so that the mixing algorithm is valid even in the case of inverse winter stratification.

Thus, thermal convection is modeled as an instantaneous process, which is in accordance with the characteristic time scale of interest.

2.3 Boundary Conditions for Thermal Simulation

It is assumed that no heat transfer occurs through the lake bottom. The equation that defines heat transfer at the air–water interface is a nonlinear function of surface and air temperatures, atmospheric pressure, vapor pressure, etc., taking into account explicitly the various heat-transfer processes illustrated in Figure 2.

The global heat budget is expressed by

$$Q_{net} = Q_{sw} + Q_{at} \pm Q_c \pm Q_e - Q_{br}$$

where

Q_{net} = net surface heat flux,
Q_{sw} = net short-wave radiation (only a fraction is absorbed at the surface),

Q_{at} = incoming long-wave radiation from the atmosphere,
Q_c = flux of convected heat (net convected energy),
Q_e = flux of latent heat (heat loss because of water evaporation), and
Q_{br} = long-wave back-radiation from the lake surface,

all measured in units of $W\,cm^{-2}$.

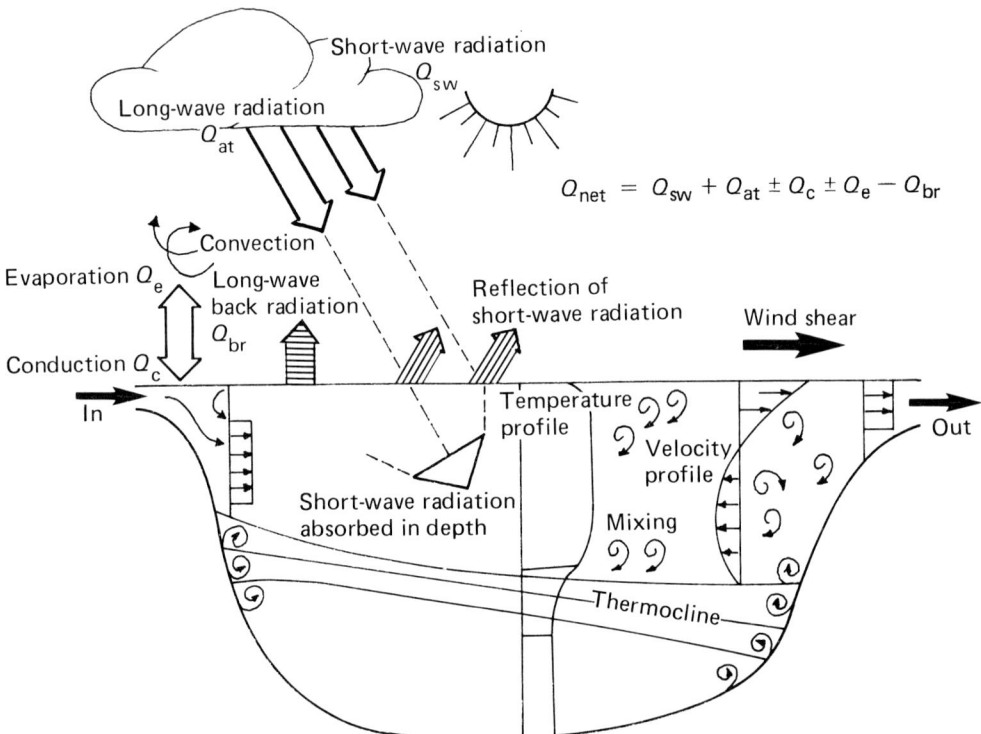

FIGURE 2 Physical processes accounted for in the thermal submodel (adapted from Svensson, 1978).

2.4 Forcing Variables for Thermal Simulation

The forcing variables used in the model are listed in Table 1 together with their measurement frequencies. Some forcing variables were not measured directly but were estimated from other field data. For example, global solar radiation was estimated from astronomic computations and daily sunshine duration data. Inflow temperature was linearly interpolated between monthly measurements. These estimations are the major sources of uncertainty in the forcings for the simulated period.

TABLE 1 Matrix of the forcing variables, showing basic field data used in their estimation (Lake Nantua, 1972–1973).

Forcings variables (daily averages)	Measurement frequencies and field data used						
	Continuous measurements			Daily average Wind speed	Daily sums		Monthly measurements Inflow temperature
	Air temperature	Relative humidity	Atmospheric pressure		Sunshine duration	Hydraulic inflows and outflows	
Measured							
Hydraulic inflows and outflows						×	
Inflow temperature							×
Air temperature	×						
Wind speed				×			
Estimated							
Global radiation					×		
Vapor pressure	×	×					
Atmospheric radiation	×	×			×		

3 MODEL CALIBRATION

3.1 Fixing the Values of Some of the Parameters

The estimation of the six parameters occurring in the formula for the vertical dispersion coefficient is not an easy task. To our knowledge, there is no unique, straightforward, and guaranteed strategy to organize parameter estimation for complex models. All that can be done is to fix those parameter values that are more certain than others, and concentrate on the latter in order to reduce the dimensions of the parameter space. One can also test the sensitivity of the model to various parameters and discard those parameters for which model sensitivity is low. Examination of the correlation between parameter estimates can also provide a guideline in the parameter-fixing procedure.

Reasonable estimates for the parameters CD_{10} and $POW2$ are available in the literature. For moderate winds and for small lakes, values of CD_{10} seem to vary between 0.7×10^{-3} and 2.3×10^{-3} (Bengtsson, 1978), and therefore CD_{10} was fixed at an average value of 1.7×10^{-3}, given the average wind speed at Nantua. For $POW2$, which appears in the relation between the eddy diffusion coefficient $K(z)$ and the Brunt–Väisälä frequency (N) in the hypolimnion, values ranging from 0.25 to 1.0 (Lerman, 1979) are reported in the literature. Thus, this parameter was fixed at a value of 0.6.

The next step was to evaluate the orders of magnitude of the remaining parameters. Values reported for $SIGMA$ vary widely, according to the way in which the vertical velocity gradient, and hence the Richardson number, is approximated. A range from 1.76×10^{-3} to about 10 can be found in the literature. Associated values for POW vary from 0.5 to 2.0 (Munk and Anderson, 1948; Newbold and Ligget, 1974; Bowden and Hamilton, 1975; Leonard et al., 1978; Walters et al., 1978).

The physical meaning of Z_{SCALE} is that of an upper limit to the size of wind-induced eddies (mixing length). In Lake Nantua (30-m average depth), a value of 10 m can be reasonably adopted as an order of magnitude (the range 5–10 m was defined as "physically acceptable").

The value of $ALPHA$ is unknown. However, when one considers the values of N below the thermocline, it can be seen that values of $ALPHA$ greater than 0.1 will produce a rapid increase in $K(z)$ under the thermocline, a feature which is not observed in the empirically obtained $K(z)$ profiles presented in the literature (Bella, 1970; Orlob and Selna, 1970; Li, 1973). Values ranging between 0.001 and 0.1 can be anticipated.

Once some parameter values have been fixed, or "physically acceptable" ranges have been defined, one can use visual fitting to get a first idea of model sensitivity to parameter changes. In this step, observation of animated sequences on a television raster display proved very useful (see below).

It was observed that changes in $ALPHA$ modify the $K(z)$ profile under the thermocline, but have little influence on the temperature profile. The thermal simulation is not very sensitive to $ALPHA$ so that the latter may be regarded as almost fixed. A value of 0.025 gives $K(z)$ profiles quite similar to the experimental profiles presented in the literature. It should be noted that $ALPHA$ may become important when materials consumed (such as dissolved oxygen) or produced (such as phosphate) at the sediment–water interface are included in the model.

3.2 Calibration

The three parameters *SIGMA* and *POW*, which express sensitivity of the dispersion coefficient to the local Richardson number, and Z_{SCALE} (maximum mixing length), which strongly influences thermocline depth, remain to be identified. The objective function to be minimized was defined as the sum of the squared differences between simulated and observed values (least-squares criterion). For the year 1972, temperature data were available for eight unevenly-spaced depths: 0.0 m, 2.5 m, 5.0 m, 7.5 m, 10 m, 20 m, 30 m, and 43 m, at roughly monthly intervals.

The optimization routine used is a slightly modified version of the Marquardt algorithm (Meeter, 1968), combining the steepest-descent method (far from the minimum) and the Gauss–Newton method (close to the minimum) (Marquardt, 1963).

The three initial parameter values were estimated from previous sensitivity tests, in the range suggested by literature values. The initial values were chosen as *SIGMA* = 10, *POW* = 1.8, and Z_{SCALE} = 9 m. Convergence of the algorithm led to *SIGMA* = 13, *POW* = 1.6, and Z_{SCALE} = 7.4 m. The sum of the squared deviations was reduced to 26% of its initial value. The uneven distribution of observed temperature data along depth (five points out of eight are in the first 10 m) favors transfer of residual deviation between model and data to the hypolimnion. This uneven weighting could have been avoided by interpolating the observed temperature profile at equally spaced depths.

From linear approximation around the minimum, the optimization routine provides an estimation of the correlation coefficients between the parameter estimates and the confidence limits for these estimates. Confidence intervals for model predictions are also computed with this linear approximation. Owing to the strongly nonlinear nature of the model (with respect to the parameters), these confidence intervals must be considered as rough estimates only. The Monte Carlo technique was used as an alternative method to estimate these confidence intervals (see Section 7). The correlation matrix for this first calibration is given in Table 2. Examination of the correlation matrix obtained at the end of the three-parameter calibration shows a strong negative correlation between *SIGMA* and *POW*: the parameter estimation problem is ill-conditioned.

TABLE 2 Parameter-estimate correlation matrices.

	SIGMA	POW	Z_{SCALE}
Calibration 1			
Correlation matrix	1		
	−0.894	1	
	−0.030	0.417	1
Final parameter values	13.0	1.6	7.4 m
95% confidence limits for estimates	16.87	1.72	9.01 m
	9.21	1.45	5.76 m
Calibration 2			
Correlation coefficients		1	
		0.834	1
Final parameter values		1.7	7.6 m
95% confidence limit for estimates		1.75	9.08 m
		1.64	6.07 m

Because of the antagonistic effect between *SIGMA* and *POW*, many (*SIGMA*, *POW*) pairs can produce equally good fits, based on the sum-of-squares criterion. However, it was found that values greater than 10.0 modify the shape of the thermal profile in the metalimnion, by shifting the point of inflexion of the profile close to the bottom of the mixed layer, a feature which is not observed in the experimental data. Therefore, we decided to perform a second calibration, on two parameters (*POW* and Z_{SCALE}), with *SIGMA* fixed at a value of 10.

Convergence was obtained with $POW = 1.7$ and $Z_{SCALE} = 7.6$ m (starting from values of 1.3 and 4, respectively), which was quite close to the previous results. The residual sum of squares ($69.8°C^2$) was quite close to its value for the three-parameter calibration ($67.4°C^2$). The correlation matrix, estimated parameters, and confidence limits for this second calibration are given in Table 2. It should be noted that fixing *SIGMA* narrows the confidence interval of the estimate of *POW*.

Convergence required 23 runs. Note that several runs of the model are necessary for each iteration in the calibration procedure, since the partial derivatives of the model with respect to the parameters are computed numerically and several trials may be necessary before a new search direction is determined. As a model run requires 2 min CPU time on an IBM 370/168, this last calibration used 46 min of CPU.

It should be noted that the estimated parameter values were very close to our initial guess, even in the second calibration where the initial values were intentionally

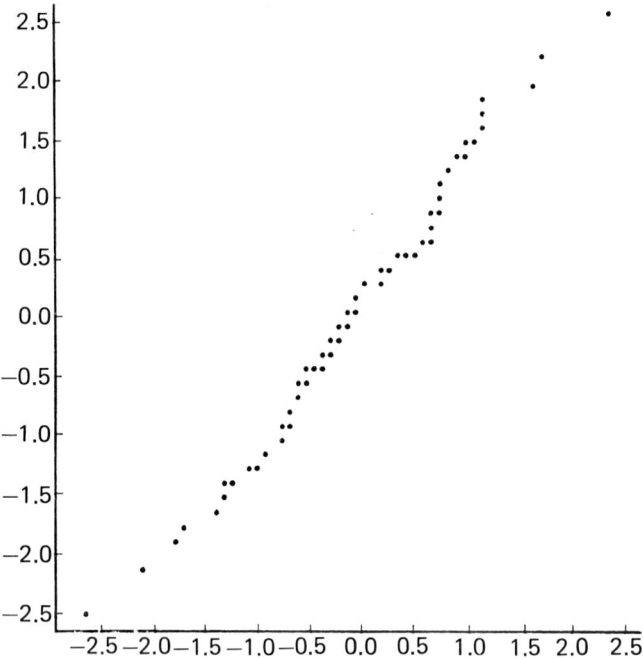

FIGURE 3 Normal probability plot of residuals.

Analysis of prediction uncertainty by Monte Carlo simulation 193

shifted to check the uniqueness of the minimum in the neighborhood of the solution obtained for the first calibration.

4 ANALYSIS OF RESIDUALS

The use of the least-squares criterion corresponds to the maximum likelihood criterion if model residuals are independent, normal, random variables with zero mean and equal variances. This property was checked *a posteriori* using a normal probability plot (probit test) of residuals (Figure 3). Fitting a straight line by eye through the set of points shows that the distribution mean (intercept with $y = 0$) does not depart very much from zero.

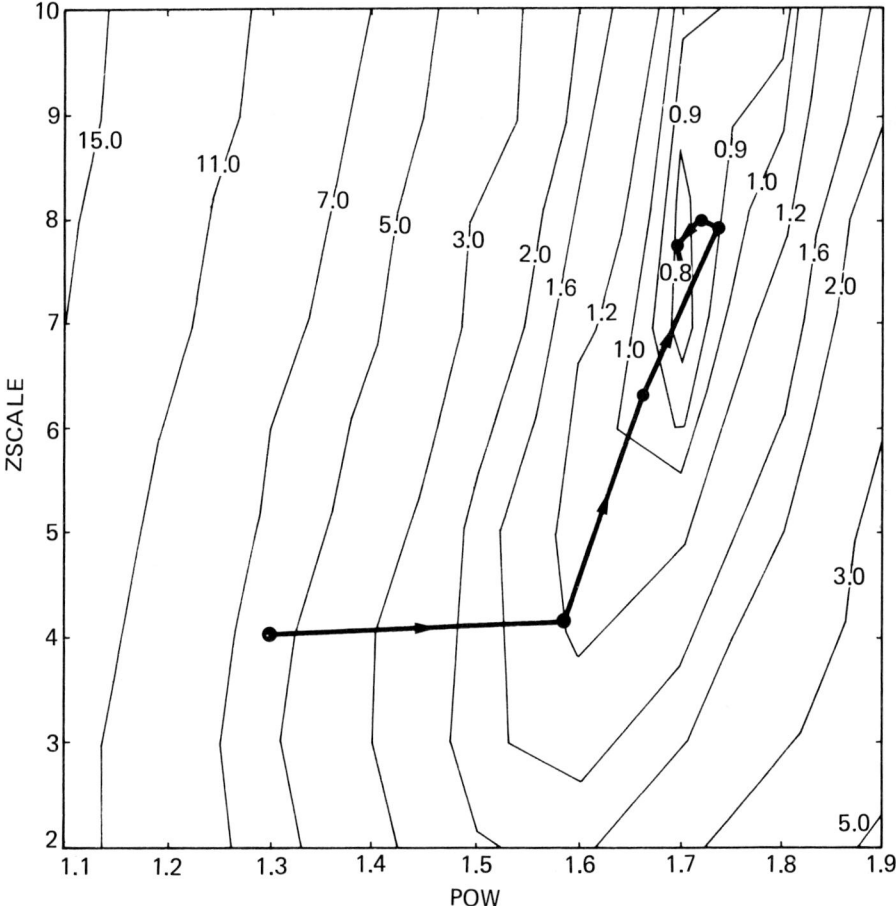

FIGURE 4 Contours of objective function (mean of squared deviations).

5 MAPPING THE PARAMETER SPACE

As only two parameters were left for calibration, a visual mapping of the parameter space was possible, for checking the uniqueness of the minimum in the range of physically acceptable values. The mapping also provides some visual impression of the nonlinearities in the model and the correlation between parameters, and shows the efficiency of the Marquardt algorithm.

The model was run for 81 different pairs of POW (varying from 1.1 to 1.9 at 0.1 intervals) and Z_{SCALE} (from 5.5 to 9.5 at 0.5 intervals) values. The least-squares criterion was computed and normalized (that is, the sum was divided by 12×8, which is the number of observed data). Contour lines were linearly interpolated on the 81-point grid (Figure 4). The path followed by the optimization routine between two iterations has been drawn *a posteriori*.

6 MODEL VALIDATION

Model validation was carried out for the year 1973, all parameters being held constant. Some calibration (1972) and validation (1973) results are given in Figure 5, which shows data points plotted and four predicted profiles. These pictures are snapshots taken from an animation sequence, which was generated on an experimental television raster display and stored on a videocassette (designed by P. Matherat of the Centre de Calcul de l'Ecole Normale Supérieure, Paris).

The agreement between experimental and predicted profiles was equally good for calibration and validation runs, which shows that goodness of fit for the year 1972 was not just an artifact of the calibration procedure (Figure 6).

7 MONTE CARLO SIMULATION

A Monte Carlo simulation (100 runs) was performed with two parameters (POW and Z_{SCALE}) randomly chosen from a bivariate correlated Gaussian distribution (Lehman, 1977). The distribution mean, standard deviations, and correlation coefficient estimates (Table 2) were those computed by the last calibration.

The use of the Monte Carlo technique enables the simulation outputs to be presented in terms of a mean (stochastic mean) (Tiwari et al., 1978) and associated variance. The Monte Carlo runs are then compared to the calibration run (deterministic run) to identify dates and depths critically affected by parameter uncertainty. Finally, Monte Carlo results are compared to the confidence intervals estimated from linear approximation, and to observed data points.

Monte Carlo simulation results are given in Figure 7 for four different depths. In the epilimnion layers (at 0.0 m, 2.5 m, and 7.5 m) there is relatively no error accumulation (in terms of prediction variance) due to parameter uncertainty. This is explained by the feedback mechanisms acting on the heat balance at the air–water interface and their characteristic time scale. In contrast, once the lake is thermally stratified, model error may accumulate in the hypolimnion predictions (at 30.0 m). This illustrates the

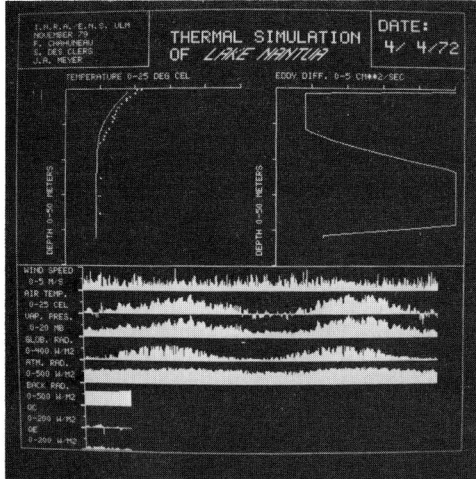
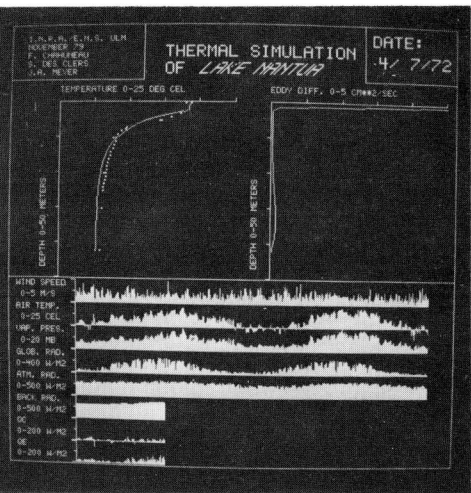
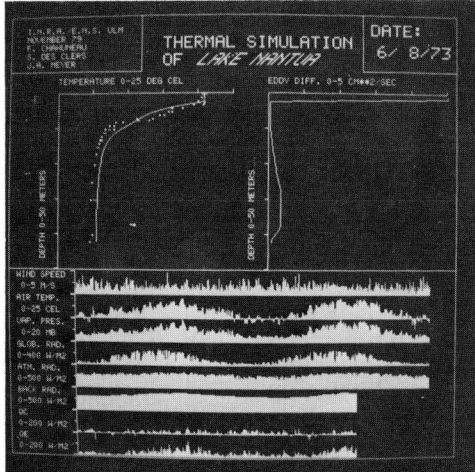
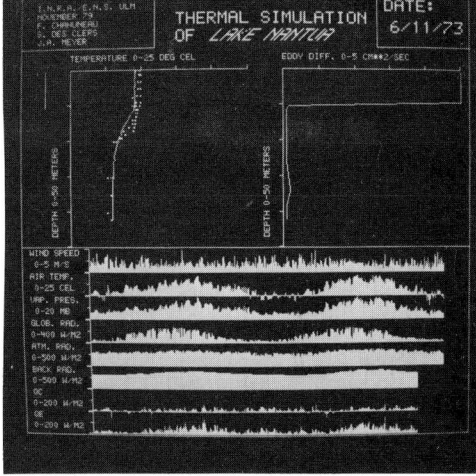

FIGURE 5 Thermal and vertical dispersion coefficient profiles at four different dates. Histograms are given for the forcing variables (wind speed m s^{-1}, air temperature °C, vapor pressure mb, global solar radiation W m^{-2}, and atmospheric radiation W m^{-2}) as well as for some fluxes calculated by the model (back-radiation, Q_c flux of sensible heat, and Q_e flux of latent heat, in W m^{-2}), in the lower part of the pictures. On the lefthand upper corner of the temperature profiles, the two lines show the zones of hydraulic inflow (left) and outflow (right). 1972 is the calibration year, 1973 the validation year.

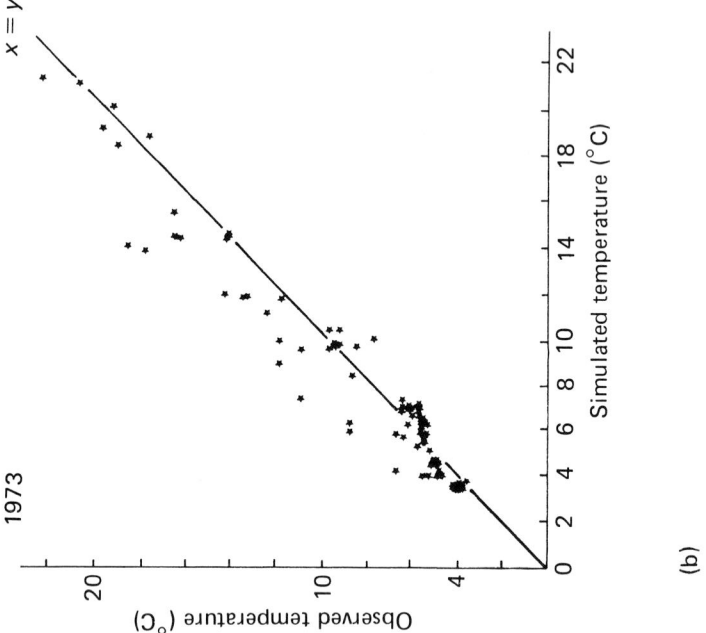
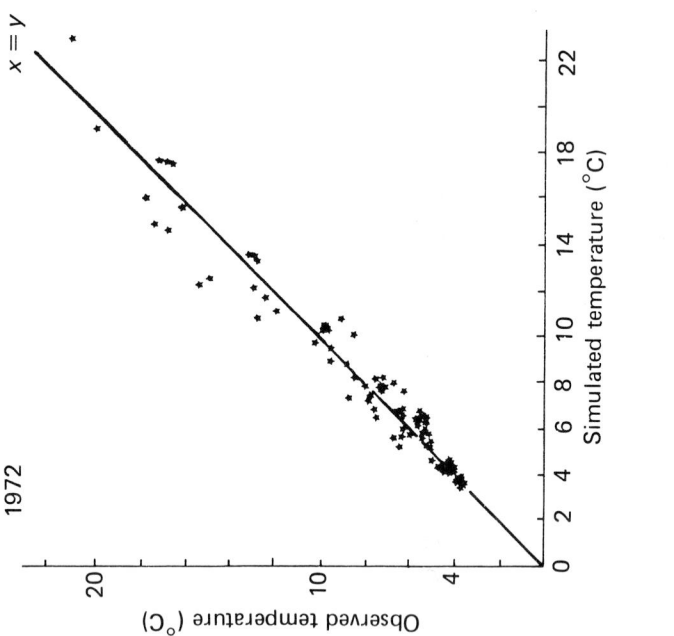

FIGURE 6   Observed versus simulated temperature.

sensitivity of the hypolimnetic equation for vertical dispersion to the two parameters, as well as the relatively high inertia of hypolimnetic temperature variations. By the end of a yearly cycle very little uncertainty remains that could be transferred to a second year of simulation, because of the complete mixing of the lake at that time.

Another representation of the Monte Carlo simulation results is given in Figure 8 where, following O'Neill and Gardner (1979), error accumulation over the year is given for four observation depths (0.0 m, 2.5 m, 7.5 m, and 30.0 m). The percentage of runs that remain "valid", given an error tolerance, at any particular day is plotted against time for the year 1972. The error criterion is given as a coefficient of variation around the deterministic value. For instance, up to day 219 at 0.0 m, about 10 runs from 100 satisfy the 2% error criterion but more than 80 runs from 100 satisfy the 10% error criterion. Concerning error accumulation over time at depth 0.0 m, the key dates — day 74 (14 March 1972) and day 350 (18 December 1972) — correspond to the triggering of thermal stratification and autumn mixing, respectively. Simulated hypolimnion temperatures are again seen to be more affected by parameter uncertainty (only 70% of the Monte Carlo runs remained within 20% error up to day 366).

The limited number of field data available for calibration makes comparison between Monte Carlo prediction variance and linear confidence interval estimations difficult, since these intervals are computed here only around the data points. Nevertheless, it can be seen (Figure 7) that the linearly estimated (95%) confidence intervals are always smaller than the Monte Carlo ± one standard-deviation intervals. Thus, in the case of this model, linearization around estimated-parameter values strongly underestimates prediction uncertainty. This would probably be the case for most highly nonlinear models, and thus shows the suitability of the Monte Carlo technique for uncertainty analysis.

The observed data nearly always lie within a ± one standard-deviation interval around the stochastic mean (and always within the ± two standard-deviation interval, as expected).

Finally, it is also noticeable that the "stochastic mean" and the deterministic run are sometimes quite different.

8 CONCLUSION

This study shows that some techniques which are currently used by statisticians in nonlinear curve-fitting problems can be successfully applied to complex, spatially distributed simulation models. Examination of error accumulation over time reveals some critical periods for the system's dynamics. This information could be used to optimize data collection, by concentrating on these critical periods. Such unevenly distributed data over time would favor information transfer from data to model parameters during the calibration phase. Other techniques, using model-sensitivity analysis for optimal sampling design (Vila, 1980), could probably be rewardingly applied to this kind of problem.

It is expected that statistical analysis on model behavior will become an important research field in environmental modeling. Such methodology should reinforce feedback from model-building to the design of field-data collection procedures.

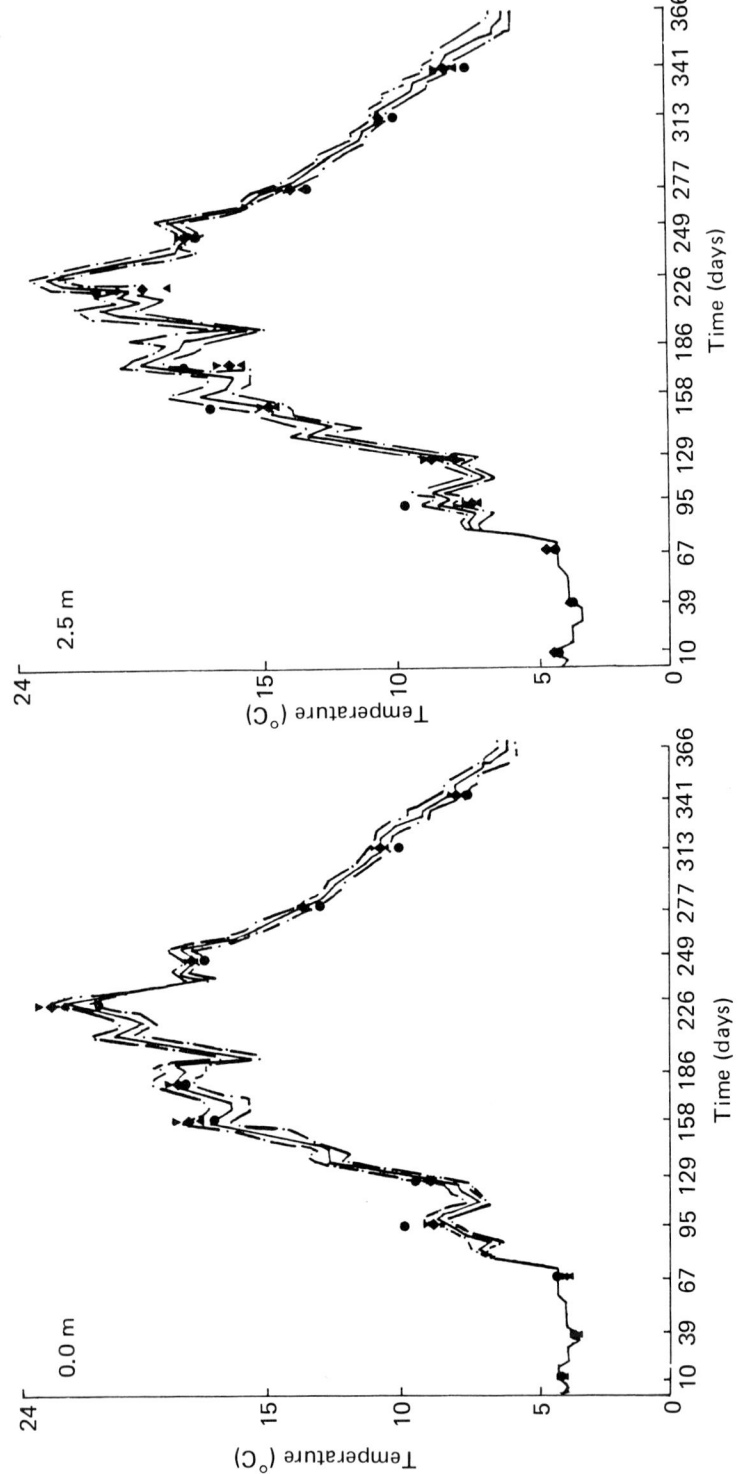

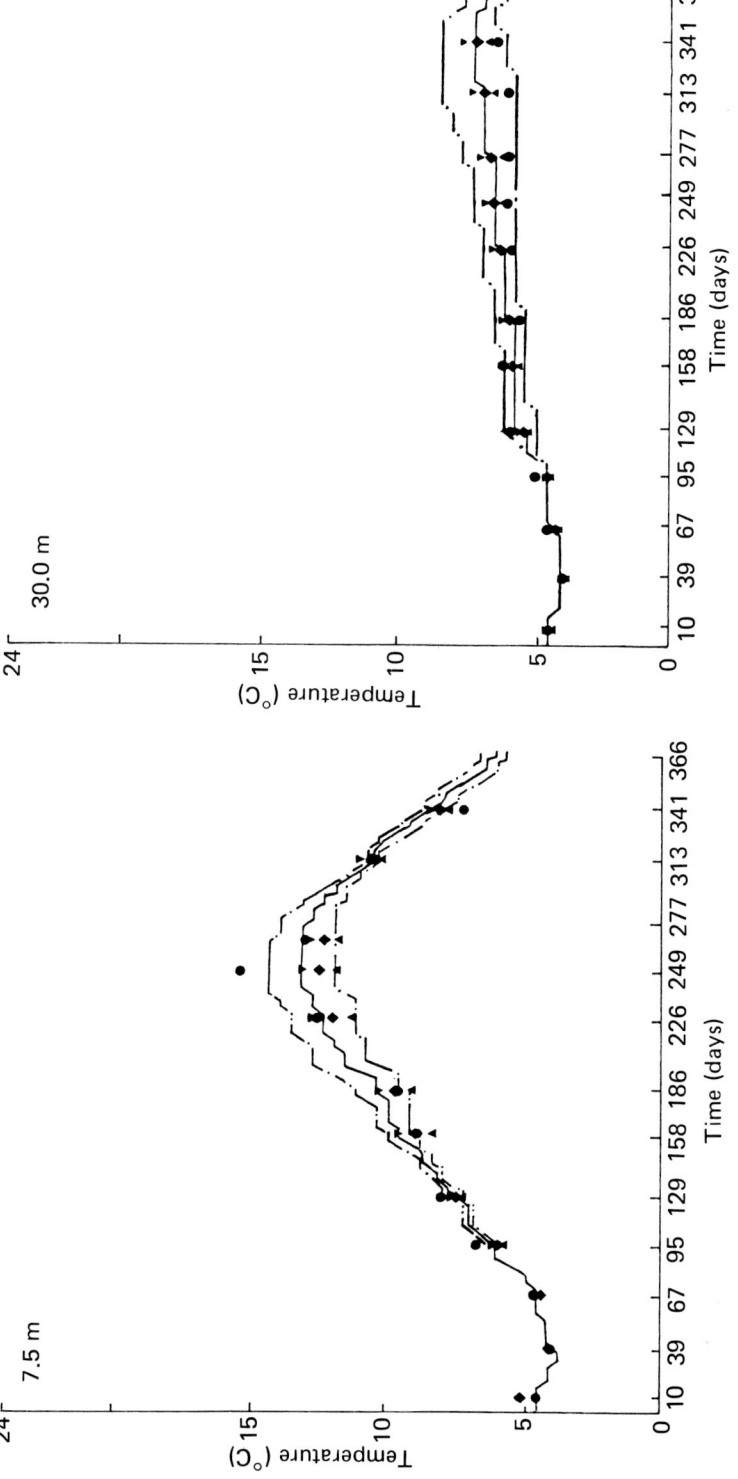

FIGURE 7 Temporal behavior of water-temperature predictions at different depths (0.0 m, 2.5 m, 7.5 m, and 30.0 m) in 1972. Symbols used in the figure have the following meanings: ♦, deterministic run; ▶◀, estimated confidence interval; ●, observed data; —, stochastic mean; —·—, ± one standard deviation.

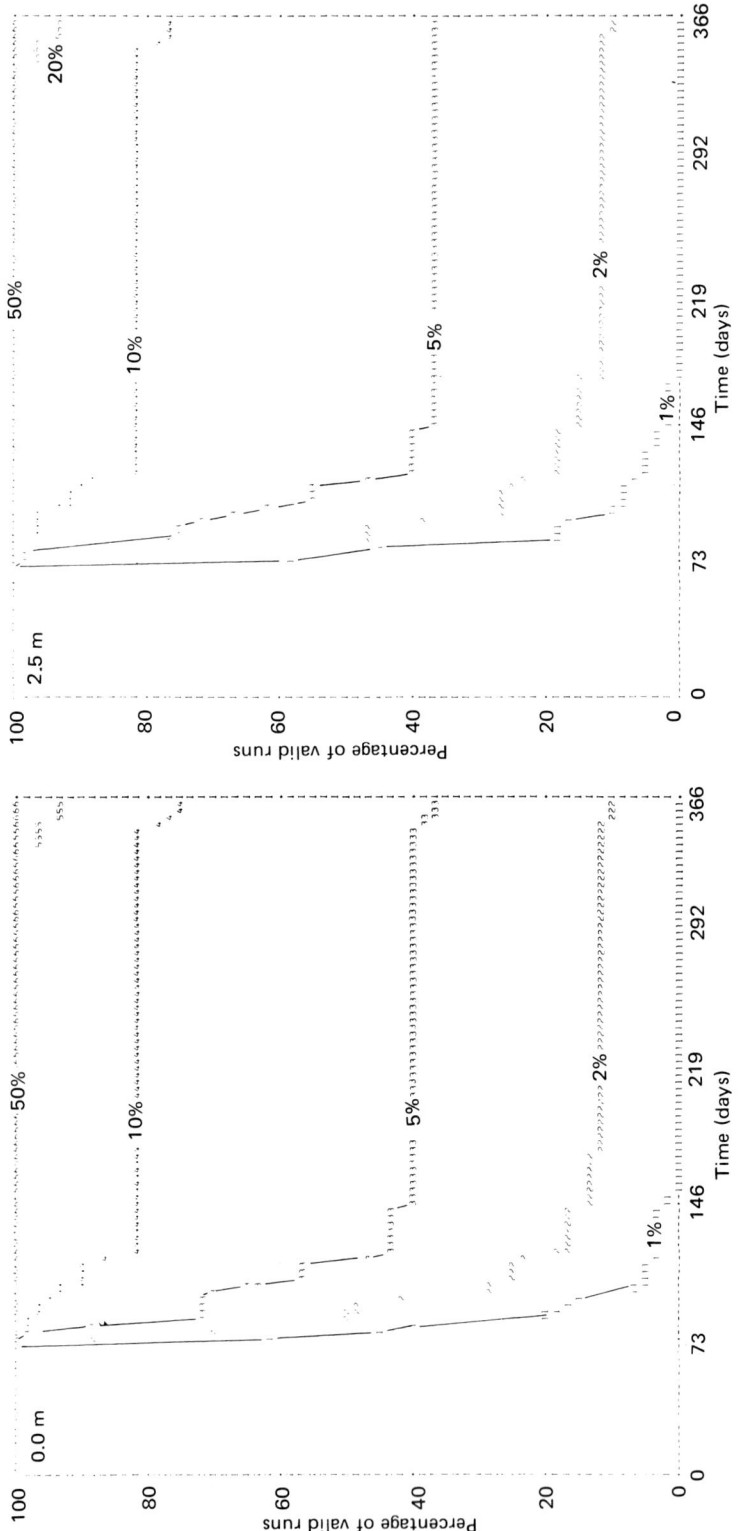

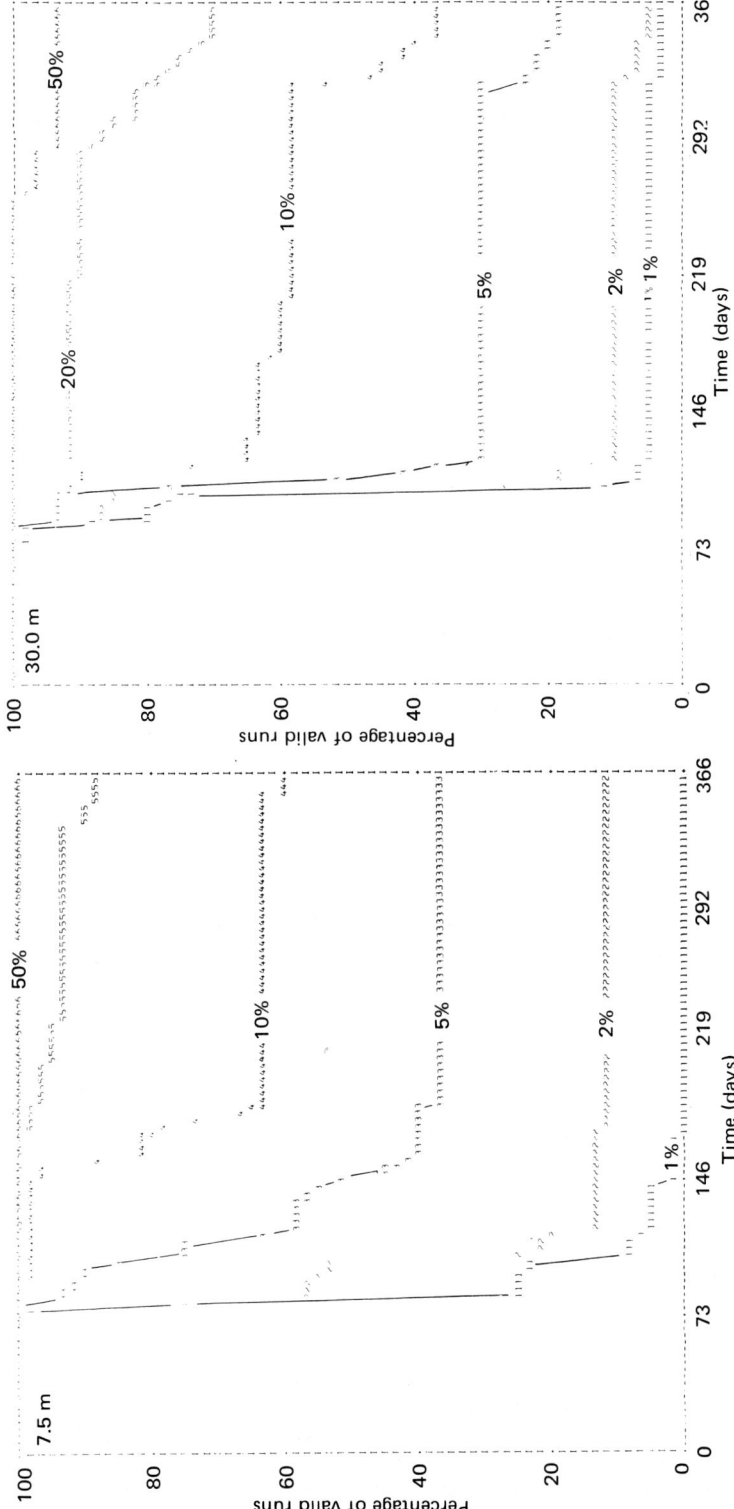

FIGURE 8 Error accumulation during 1972 for temperature simulations at four different depths (0.0 m, 2.5 m, 7.5 m, and 30.0 m). The percentage of runs that remains valid at any day, given an error tolerance, is plotted against time (days). The error criterion is the coefficient of variation around the deterministic value.

ACKNOWLEDGMENTS

This research was supported by the Ministère de l'Environment et du Cadre de Vie (convention 78-127) and the CNRS (ATP Analyse des Systèmes 37-02). The authors are indebted to P. Matherat (Computer Centre, Ecole Normale Supérieure) who designed the graphic display terminal, and to J.M. Frailong (LIMSI, CNRS) and J.P. Vila (Laboratoire de Biométrie, Jouy-en-Josas, INRA) for many fruitful discussions. The Lake Nantua field data were collected by the staff of the Station d'Hydrobiologie Lacustre (INRA, Thonon-les-Bains), whose help is gratefully acknowledged.

REFERENCES

Bella, A.D. (1970). Simulating the effect of sinking and vertical mixing on algal population dynamics. Journal of the Water Pollution Control Federation, 42(5) part 2: 140–152.

Bengtsson, L. (1978). Wind induced circulation in lakes. Nordic Hydrology, 9: 75–94.

Bowden, K.F. and Hamilton, P. (1975). Some experiments with a numerical model of circulation and mixing in a tidal estuary. Estuarine and Coastal Marine Science, 3: 281–301.

Chahuneau, F. and des Clers, S. (in preparation). Use of routine meteorological data for simulation of thermal seasonal evolution of temperate lakes.

Ford, D.E. and Thornton, K.W. (1979). Time and length scales for the 1-dimensional assumption and its relation to ecological models. Water Resources Research, 15(1): 113–120.

Lehman, R.S. (1977). Computer Simulation and Modelling. An Introduction. Lawrence Erlbaum Associates, Hillsdale, New Jersey.

Leonard, B.P., Vachtsevanos, G.J., and Abood, K.A. (1978). Unsteady-state two-dimensional salinity intrusion model for an estuary. In C.A. Brebbia (Editor), Applied Numerical Modelling. Pentech Press, Plymouth, UK, pp. 113–123.

Lerman, A. (1979). Geochemical Processes. Water and Sediment Environments. Wiley, New York.

Li, Y.-H. (1973). Vertical eddy diffusion coefficient in Lake Zürich. Schweizerische Zeitschrift für Hydrologie, 35(1): 1–7.

Marquardt, D.W. (1963). An algorithm for least-squares estimation of non-linear parameters. Journal of the Society for Industrial and Applied Mathematics, 11: 431–441.

Meeter, D.A. (1968). Program GAUSHAUS. Numerical Analysis Laboratory, University of Wisconsin at Madison, Madison, Wisconsin.

Munk, W.H and Anderson, E.R. (1948). Notes on a theory of the thermocline. Journal of Marine Research, 7(3): 276–295.

Newbold, J.D. and Ligget, J.A. (1974). Oxygen depletion model for Cayuga Lake. Journal of the Environmental Engineering Division, American Society of Civil Engineers, 100(1): 41–59.

O'Neill, R.V. and Gardner, R.H. (1979). Sources of uncertainty in ecological models. In B.P. Zeigler, M.S. Elzas, G.J. Klir, and T.I. Ören (Editors), Methodology in Systems Modelling and Simulation. North-Holland, Amsterdam, pp. 447–463.

Orlob, G.T. and Selna, L.S. (1970). Temperature variations in deep reservoirs. Journal of the Hydraulics Division, American Society of Civil Engineers, 96(2): 391–410.

Remson, I., Hornberger, G.H., and Holz, F.J. (1971). Numerical Methods in Subsurface Hydrology. Wiley–Interscience, New York.

Ryan, P.J. and Harleman, D.R.F. (1971). Prediction of the annual cycle of temperature changes in a stratified lake or reservoir: mathematical model and user's manual. MIT Report 137. MIT Press, Cambridge, Massachusetts.

Svensson, U. (1978). Examination of the summer stratification. Nordic Hydrology, 9(2): 105–120.

Tetra Tech. (1978). Rates, constants and kinetics formulations in surface water quality modeling. Report TC-3689, EPA/600/3-78/105. US Environmental Protection Agency, Duluth, Minnesota.

Tiwari, J.L., Hobbie, J.E., Reed, J.P., Stanley, D.W., and Miller, M.C. (1978). Some stochastic differential equation models of an aquatic ecosystem. Ecological Modelling, 4: 3–27.

Tucker, W.A. and Green, A.W. (1977). A time-dependent model of the lake-averaged vertical temperature distribution of lakes. Limnology and Oceanography, 22(4): 687–699.

Vila, J.P. (1980). Modélisation non-linéaire. Le problème de l'identification. Considérations numériques et statistiques. Internal Report. Laboratoire de Biométrie, INRA–CNRS, 78350 Jouy-en-Josas, France.

Walters, R.A., Carey, G.F., and Winter, D.K. (1978). Temperature computation for temperate lakes. Applied Mathematical Modelling, 2: 41–48.

MULTIDIMENSIONAL SCALING APPROACH TO CLUSTERING MULTIVARIATE DATA FOR WATER-QUALITY MODELING

Saburo Ikeda*

Department of Applied Mathematics and Physics, Faculty of Engineering, Kyoto University, 606 Kyoto (Japan)

Hidekiyo Itakura**

Department of Electrical Engineering, Faculty of Engineering, Kyoto University, 606 Kyoto (Japan)

1 INTRODUCTION

This paper is concerned with a statistical treatment of multivariate water-quality data to help regulatory and operational personnel engaged in monitoring, control, and managing problems of water pollution and eutrophication to obtain a comprehensive view of water quality in their own areas. Because of the variety of parameters observed as water-quality data, and the complexity and uncertainty involved in pollution and eutrophication mechanisms in aquatic environments, it is necessary to develop a methodology that is able to identify common and differing aspects of water quality in data from various sources. In particular, in order to build a water-quality model of the compartment type which is better able to identify regional characteristics, it is necessary to have an integrated index of regional water quality which makes it possible to divide a designated area into several compartments.

In this paper a statistical method is presented in which expert knowledge or experience can be utilized together with extracted regional statistics based on an observed data set obtained at various points in the area of interest. The method consists of aggregating the data available on many measures of water quality, for example, transparency, chemical oxygen demand (COD), nutrients, chlorophyll-a, etc., obtained over many years, into

---

* Present address: Institute of Socio-Economic Planning, University of Tsukuba, Sakura, 305 Ibaraki, Japan.
** Present address: Department of Electrical Engineering, Chiba Institute of Technology, Narashino, 275 Chiba, Japan.

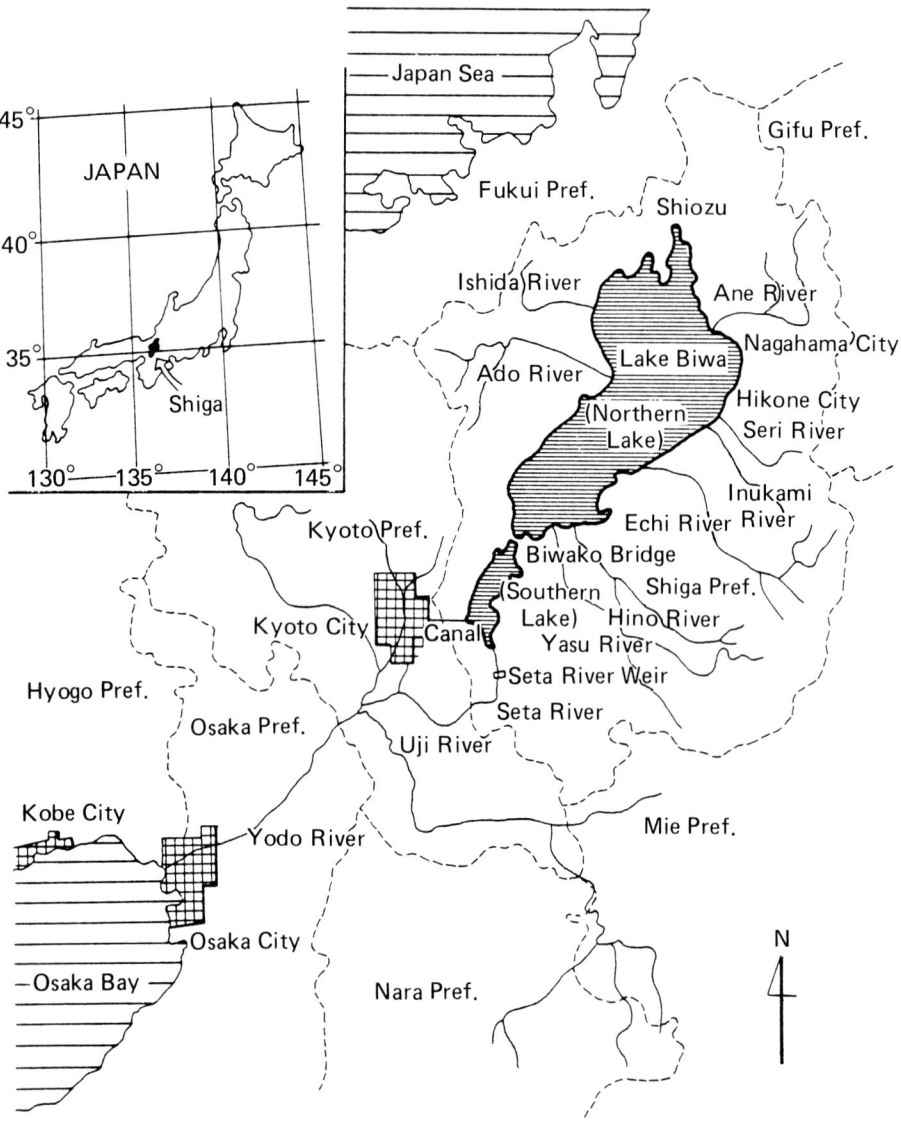

FIGURE 1 Lake Biwa and the Yodo River region.

a smaller number of indices which define criteria to represent the differences in water quality at various sampling points; then a visible representation of those differences in two-dimensional space is obtained by means of multidimensional scaling (MDS) (Shepard et al., 1972; Itakura et al., 1979). This visible representation gives model-builders information from which subjective groupings may be made, in contrast to the rigid grouping obtained by using the formal procedure of sampling points in a spatial segmentation for water-quality modeling under the specific situation of uncertainty involved when using multivariate water-quality data.

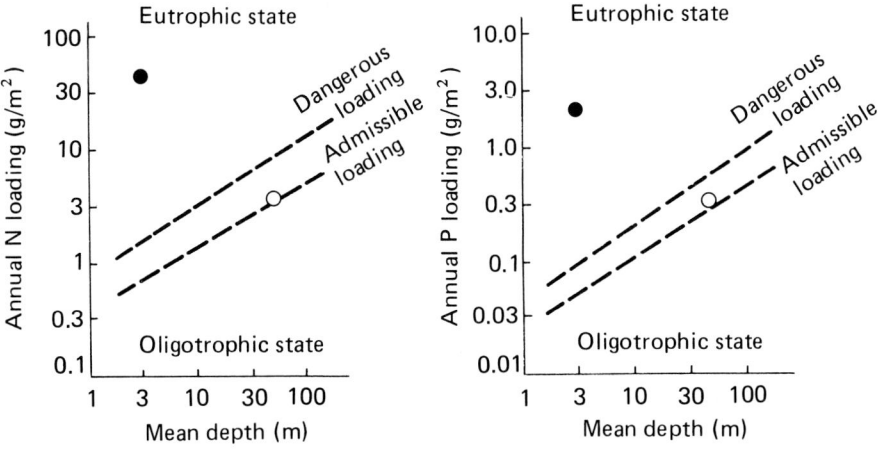

FIGURE 2 Loading of nitrogen (N) and phosphorus (P) in Lake Biwa (Lake Biwa Office, Ministry of Construction, 1974). The symbols ○ and ● refer to the Northern and Southern Lakes, respectively.

The method is applied to the water-quality assessment of Lake Biwa, the largest lake in Japan, which supplies drinking water to twelve million people, and industrial water to the economic center of western Japan, as shown in Figure 1. Recent socioeconomic features of the lake basin have made the water quality worse, owing to an increase in both the demand for water and the discharge of various pollutants into the lake. Much research and data collection work has been done on the water quality of Lake Biwa (Japan Society of Civil Engineering, 1970-1977; Ikeda and Adachi, 1978). Information gathered by those efforts shows that the lake is now in a hazardous situation, being tranformed from an oligotrophic to a eutrophic state (see Figure 2) (Lake Biwa Office, Ministry of Construction, 1974). The main purpose of the present study is to gain more knowledge about the eutrophication phenomena in various parts of the lake by making use of recent monitoring data on water quality.

2 WATER-QUALITY DATA FOR LAKE BIWA

Lake Biwa is 680 km² in area and 27.56 km³ in volume. Several million years ago the lake was formed by a disastrophism and it is thought to be as old as lakes such as Baikal and Tanganyika. The lake is composed of two parts: the northern part, the main lacustrine, is called the Northern Lake and is in an oligotrophic state; the southern part, a smaller and shallower sublacustrine, is called the Southern Lake and is in a eutrophic state. More than 100 rivers flow into the lake, but it has only one outlet, situated at the end of the Southern Lake. The general characteristics of Lake Biwa are given in Table 1.

The regulatory office conducts regular monthly monitoring of a number of water-quality parameters at 23 points in the lake; 12 of these are situated in the Northern Lake and 11 in the Southern Lake, as shown in Figure 3. A set of data on water quality which includes biological data, on, e.g., chlorophyll-a, has been collected from 1975 to date

TABLE 1 Characteristics of Lake Biwa (Lake Biwa Office, Ministry of Construction, 1974).

Normal water level	84.371 m above sea level
Total area of lake	680 km^2
Average depth	41.2 m
Maximum depth	103.6 m
Volume of water	27.5 km^3
Average annual precipitation	1900 mm
Average annual runoff	53×10^8 m^3
Average monthly air temperature	3.1–26.2°C
Number of tributaries	121
Retention time	5.2 years
Ratio of catchment area to lake surface	5.7
Population around the lake	95×10^4
Land usage in catchment area (%)	
Forests, etc.	54.8
Lake Biwa	17.7
Rice paddies	17.4
Other	10.1

TABLE 2 Measured parameters of water quality.

Parameter	Unit	Symbol
Transparency	m	TRANSP
pH		PH
Dissolved oxygen	mg l^{-1}	DO
Chemical oxygen demand	mg l^{-1}	COD
Suspended solids	mg l^{-1}	SS
Ammonia nitrogen	mg l^{-1}	NH4
Nitrite nitrogen	mg l^{-1}	NO2
Nitrate nitrogen	mg l^{-1}	NO3
Organic nitrogen	mg l^{-1}	ORGN
Total nitrogen	mg l^{-1}	TN
Orthophosphorus	mg l^{-1}	PO4
Total phosphorus	mg l^{-1}	TP
Chlorophyll-a	µg l^{-1}	CHLORA
Water temperature	°C	WATTMP
Water level	cm (± standard)	WATLEV
Solar radiation	cal cm^{-2}	SOLAR
Wind velocity	m s^{-1}	WNDVEL

(Lake Biwa Office, Ministry of Construction, 1974–1979). Thus for the present analysis a set of data exists for 17 parameters (listed in Table 2) for every month from April 1975 to March 1979. Among these are data on physical parameters for the lake which might affect water quality, such as water temperature, water level, solar radiation, and wind velocity.

Figure 4 shows the difference in annual averaged values of typical water-quality parameters such as *COD, SS*, NO$_3$ (*NO3*), and chlorophyll-a (*CHLORA*) between the

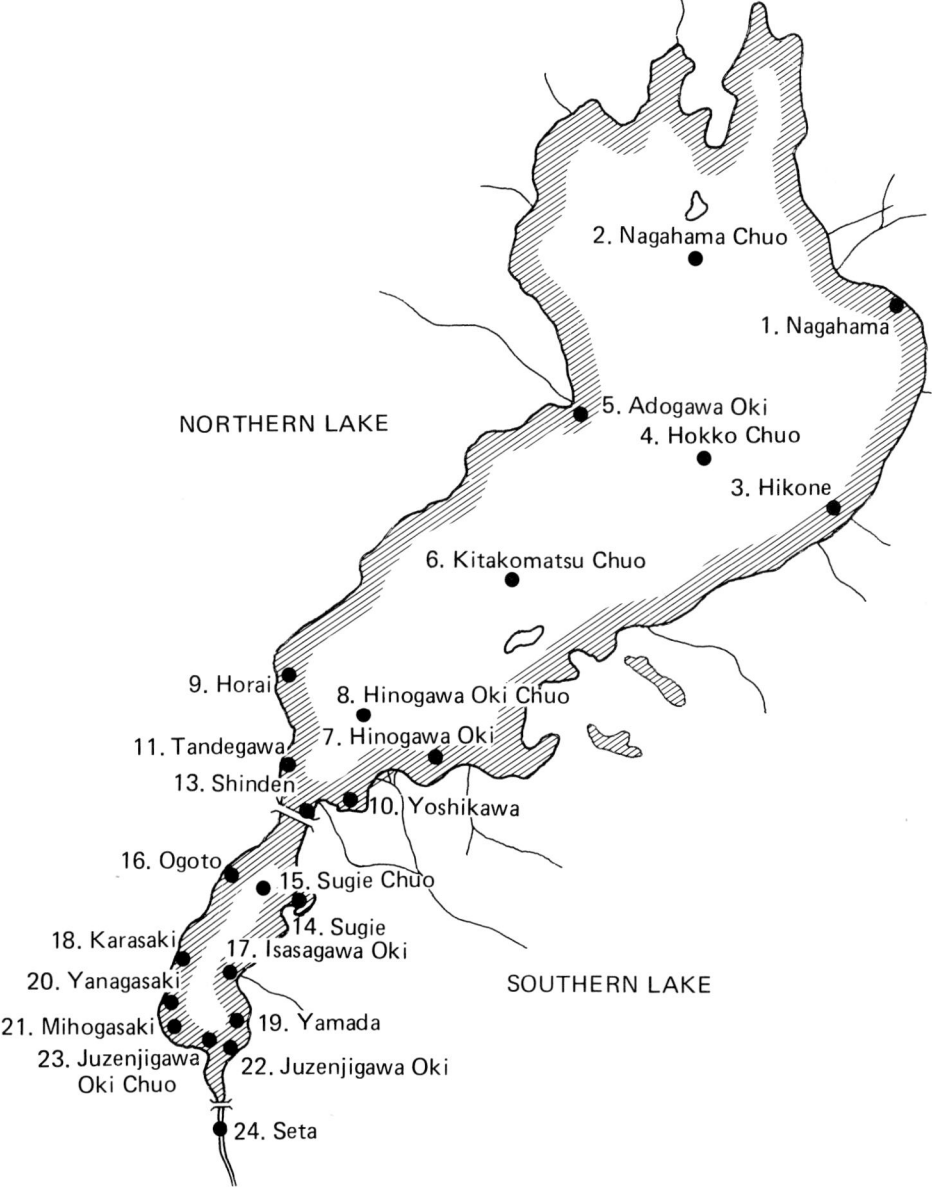

FIGURE 3 The observation points.

Northern and Southern Lakes over a four-year period. Apparently, the lake has different pollution levels in the northern and southern parts. Furthermore, since it has been pointed out that a clear distinction exists between the dominant species of phytoplankton and zooplankton in the two parts of the lake, a separate analysis should be done for each part.

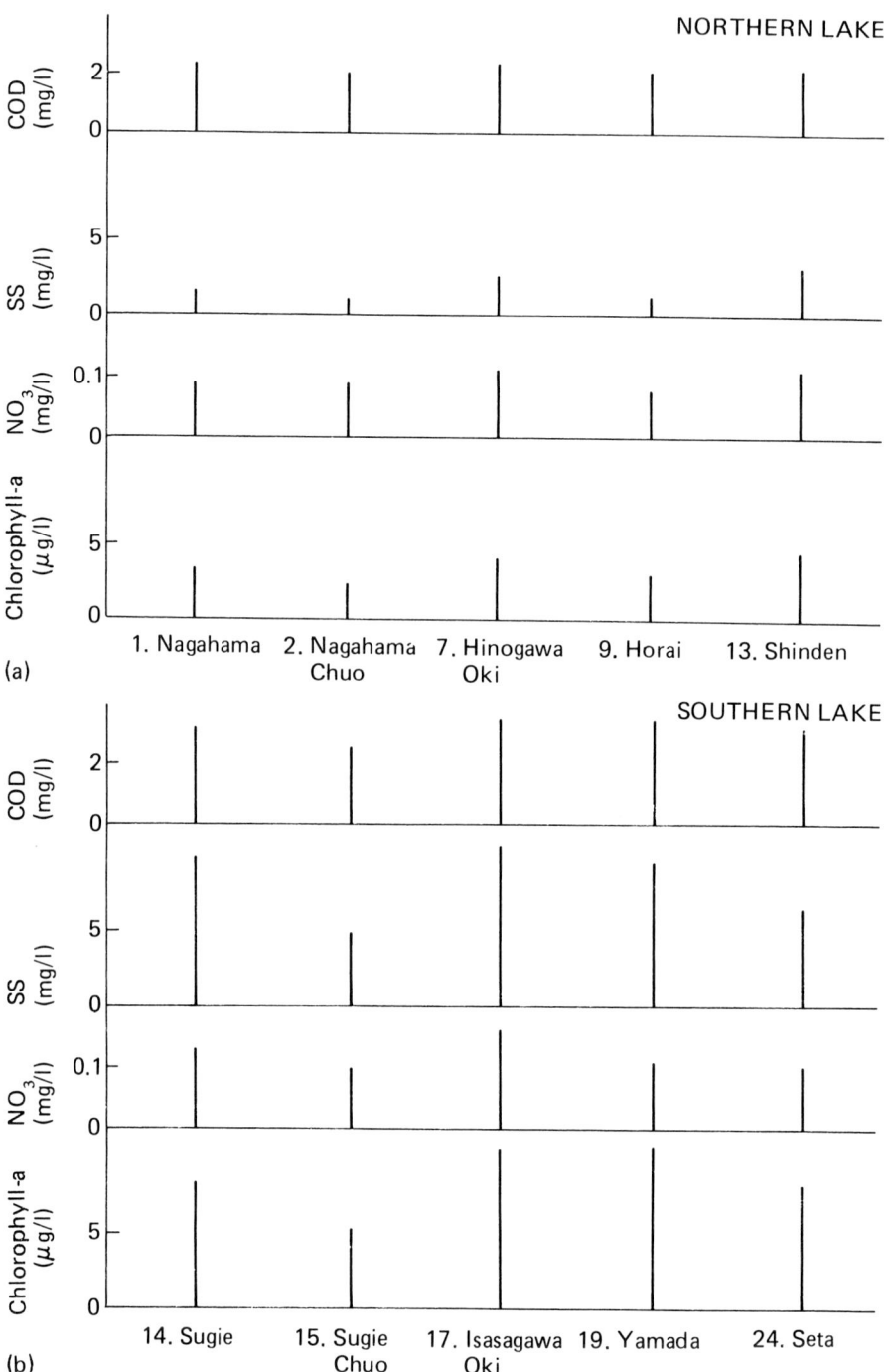

FIGURE 4 Annual averaged water quality from 1975 to 1979: (a) in the Northern Lake; (b) in the Southern Lake.

MDS Approach to clustering multivariate data 211

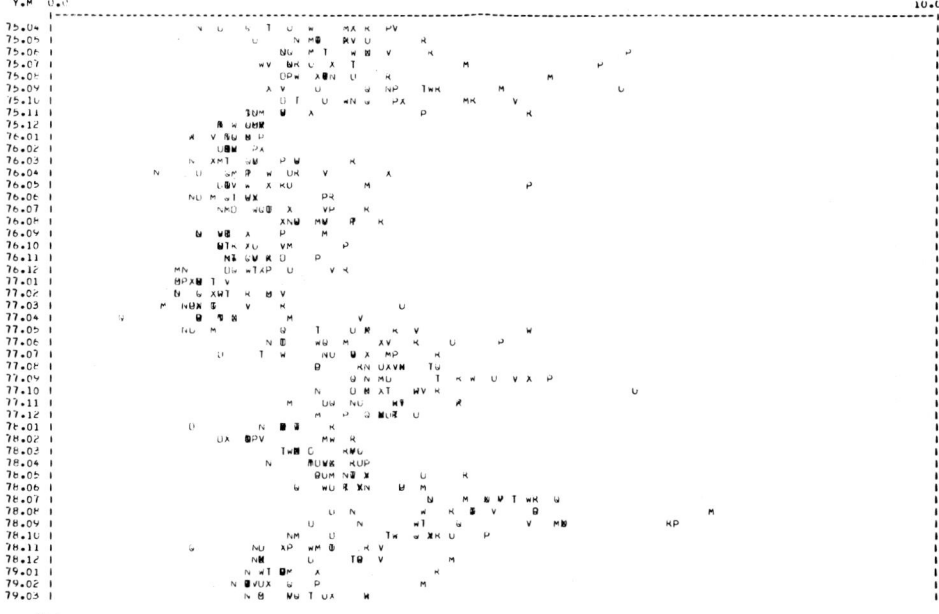

FIGURE 5 Variation of *COD* with time: (a) in the Northern Lake; (b) in the Southern Lake. The capital letters A, B, C, etc., in the diagrams refer to the following observation points whose names are given in Figure 3: A, 1; B, 2; C, 3; D, 4; E, 5; F, 6; G, 7; H, 8; I, 9; J, 10; K, 11; L, 13; M, 14; N, 15; O, 16; P, 17; Q, 18; R, 19; T, 20; U, 21; V, 22; W, 23; X, 24.

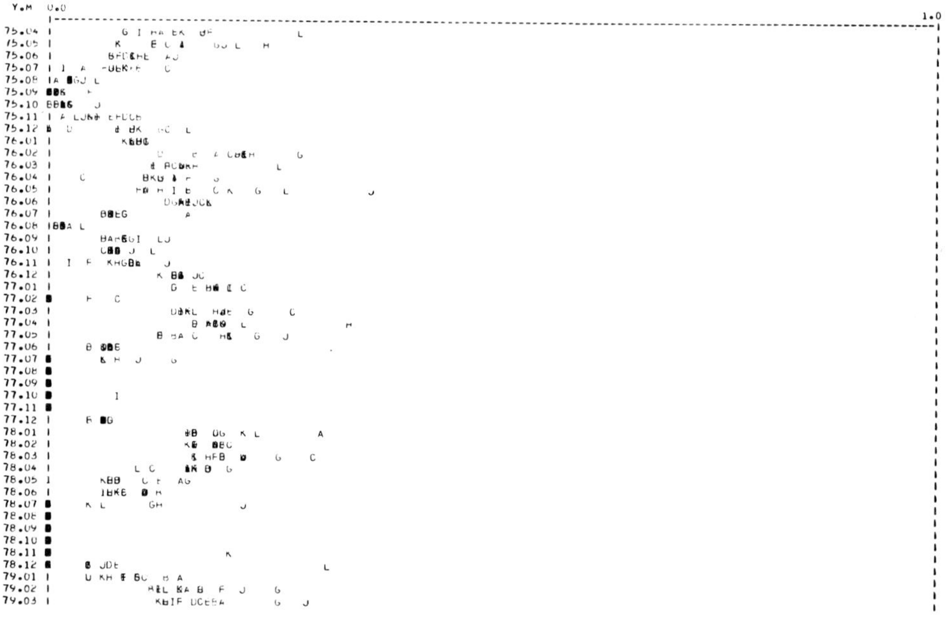

(a)

(b)

FIGURE 6 Variation of *NO3* with time: (a) in the Northern Lake; (b) in the Southern Lake. The capital letters used are the same as those in Figure 5.

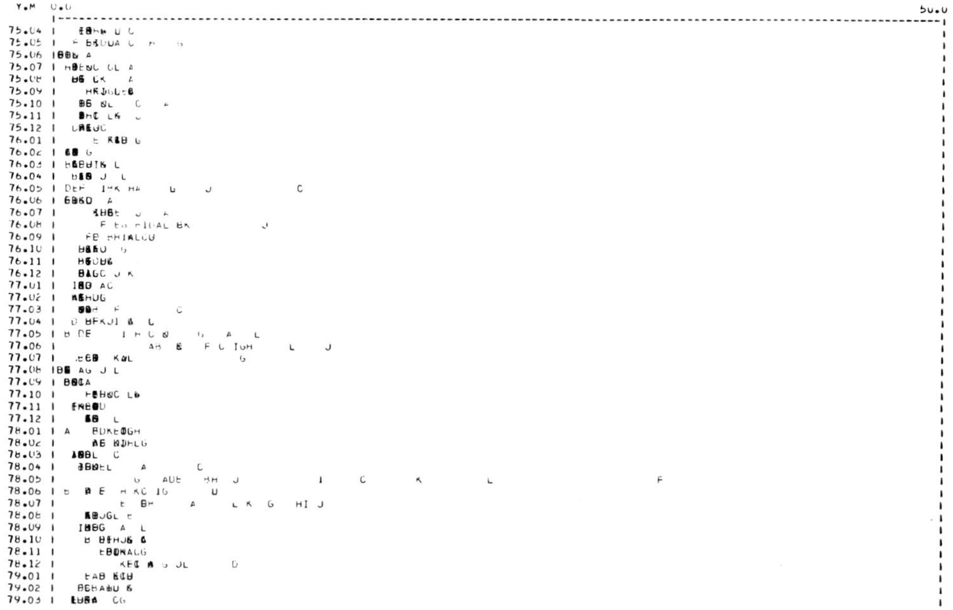

MDS Approach to clustering multivariate data 213

(a)

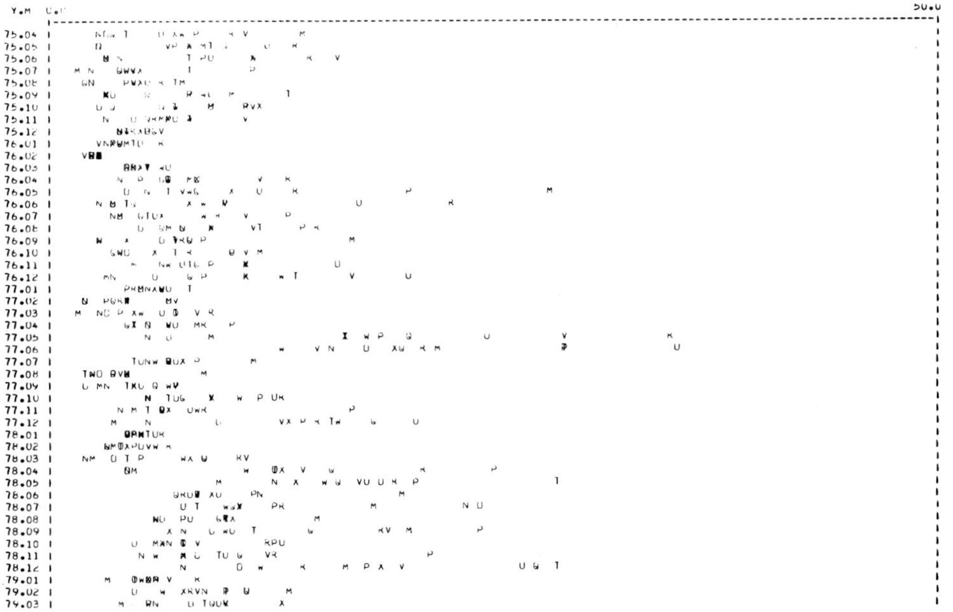

(b)

FIGURE 7 Variation of *CHLORA* with time: (a) in the Northern Lake; (b) in the Southern Lake. The capital letters used are the same as those in Figure 5.

Figures 5–7 show the monthly variations of three of the parameters of Figure 4, which have been measured at the 23 points indicated in Figure 3. It can be seen that the value of each parameter differs from point to point in the lake. However, note that some common interrelationships exist between certain parameters in these figures. Motivated by these observations, an attempt was made to use the MDS statistical technique, outlined in the next section, to extract some quantitative information from the analysis.

3   MDS APPROACH TO DATA ANALYSIS CONCERNING WATER QUALITY

This section outlines the analytical techniques of multidimensional scaling (MDS) and principal component analysis (PCA). The approach is tailored for the analysis of a set of data associated with the water quality of Lake Biwa. The principal aim is to group the observation points such that the water quality at points within a group exhibits a common aggregate index.

The simplest way of grouping the observation points is to use cluster analysis. Once a measure representing the difference in water quality between observation points has been defined, the analysis gives unique grouping through a formal procedure. But unique grouping is not always efficient because it is too rigid and account cannot always be taken of *a priori* knowledge or information about water quality in the area of interest. To incorporate such expert knowledge and experience into the clustering technique, it is necessary to integrate the data on many parameters over several years, define a measure that represents the difference in water quality, and make a visual representation of these differences.

First, the PCA technique, which aims to derive an integrated index of water quality, will be discussed. The technique is generally used for aggregating statistical data on a small number of hypothetical variables. This aggregation is done through linear combinations of the original variables.

Consider a set of data containing n variables $x_1, x_2, \ldots, x_n$, and also consider the linear combination

$$z_1 = a_{11}x_1 + a_{12}x_2 + \ldots + a_{1n}x_n \tag{1}$$

Now assume that the data on all variables have been normalized with a zero mean and a variance of unity. The values of unknown coefficients a_{1j} are determined such that the resulting value of z_1 has maximum variance, under the restriction

$$a_{11}^2 + a_{12}^2 + \ldots + a_{1n}^2 = 1 \tag{2}$$

The procedure used to obtain a_{1j} consists of reducing the calculation to an eigenvalue problem on a correlation-coefficient matrix of x_j. The coefficients a_{1j} obtained weight the parameters x_j and contain information about each variable x_j ($j = 1, 2, \ldots, n$). Although the variable z_1 aggregates most of the information contained in the data on $x_1, x_2, \ldots, x_n$, the rest of the information will be missed.

Now consider a second linear combination

$$z_2 = a_{21}x_1 + a_{22}x_2 + \ldots + a_{2n}x_n \tag{3}$$

The values of a_{2j} are determined such that the correlation of z_2 to z_1 is a minimum and the variance of z_2 is a maximum under the restriction

$$a_{21}^2 + a_{22}^2 + \ldots + a_{2n}^2 = 1 \tag{4}$$

The quantity z_1 is called the first principal component and z_2 the second principal component. The third, fourth, etc., components are designated in a similar way. The ratio of the quantity of information contained in each component to that contained in all the variables is called the component contribution, and this is evaluated by the eigenvalue associated with the corresponding component. For instance, in the present case of Lake Biwa, the three components up to z_3 contain 65–75% of all information in the original data from parameters x_1 to x_n. Once the a_{ij} values are determined, substituting the normalized real data x_j into eqns. (1) and (3) yields the value of z_i. Let z_{imp} be the value of z_i at observation point p in the lake and in month m; this is designated the principal-component (PC) value.

Next, based on the PC value of z_i, the parameter d_{pq} that represents the difference in water quality between points p and q is defined by

$$d_{pq} = \left[\sum_m \sum_i (z_{imp} - z_{imq})^2\right]^{1/2} \tag{5}$$

Several ways exist of making this summation over i, the PC number, on only one of the components or over a number of components. If the values x_j are exactly the same at two points, then $d_{pq} = 0$ for these points; if they are very different, then d_{pq} is large. Thus, d_{pq} is considered to represent totally the differences between the water quality at the different points. A table could be drawn up in which the values of d_{pq} were listed, and this could provide a summary of the differences, but the numerical values in such a table would be of doubtful worth. Therefore, an attempt was made to introduce a technique that would give a visual representation of d_{pq}. This is the MDS technique, which provides a simple graphic display of d_{pq}.

The MDS technique is useful for the analysis of a variety of data from the social and behavioral sciences. Its purpose is to extract the structure that is hidden within a set of data and to represent this structure in the form of a more accessible geometrical picture. The objects studied are represented by points in an r-dimensional space, in such a way that the significant features of the data are revealed in the geometrical relations between the points.

Although there are a variety of MDS procedures, corresponding to the varying features of the data to be analyzed, the basic framework is as follows. Let $\tilde{\xi}_p$ be the coordinate-value vector corresponding to observation point p in r-dimensional space. To obtain $\tilde{\xi}_p$, the least-squares criterion

$$J = \sum_{p,q} (d_{pq} - \hat{d}_{pq})^2, \qquad \hat{d}_{pq} = \|\tilde{\xi}_p - \tilde{\xi}_q\| \tag{6}$$

is introduced. The vector ξ_p is determined by minimizing J. There are $N \times r$ parameters to be estimated in eqn. (6), where N is the number of sampling points (in practice, some of the parameters need not be estimated; for example the first point may be placed at the origin, the second point on one of the axes, etc.). This estimation problem can be solved by a conventional unconstrained nonlinear optimization method, for instance, that of Hooke and Jeeves (1961). The quantity $\hat{d}_{pq}$ represents a Euclidean distance between points p and q distributed in r-dimensional space. Plotting the ξ_p values obtained in the r-space gives a geometrical picture representing the differences in water quality between the observation points. Unless the dimension r is very high and/or the d_{pq} values exactly satisfy the so-called triangle law, the minimum value of J is not zero in most cases. In addition, a very high-dimensional space is unsuitable for visual representation and so a two- or three-dimensional space is often used. Thus the aim is to incorporate the original d_{pq} values into the $\hat{d}_{pq}$ values in a lower-dimensional space. This naturally results in nonzero J, but the results give more useful information than would a table containing only numerical values of d_{pq}.

4 ANALYSIS AND DISCUSSION

Among the seventeen parameters of water quality listed in Table 2, the following variables were selected to construct the integrated index to describe the physical nature of water quality: *TRANSP, PH, DO, COD, SS, NO3, TN, TP,* and *CHLORA*. The correlation-coefficient matrix of these variables for the Northern Lake is given in Table 3, and the corresponding matrix for the Southern Lake in Table 4.

TABLE 3 Correlation-coefficient matrices of variables for the Northern Lake.

	TRANSP	PH	DO	COD	SS	NO3	TN	TP	CHLORA
TRANSP	1.0								
PH	−0.267	1.0							
DO	−0.028	−0.160	1.0						
COD	−0.287	0.600	−0.230	1.0					
SS	−0.606	0.221	0.079	0.366	1.0				
NO3	−0.013	−0.434	0.543	−0.464	0.085	1.0			
TN	−0.260	−0.149	0.272	−0.192	0.258	0.558	1.0		
TP	−0.343	0.004	0.077	0.056	0.405	0.202	0.301	1.0	
CHLORA	−0.388	0.399	0.159	0.394	0.398	−0.013	0.168	0.222	1.0

Taking the correlation matrices of Tables 3 and 4 into account, a PCA was undertaken for a series of cases with different combinations of these nine water-quality variables. Four examples from these cases are shown in Table 5; they were selected simply because of the (relatively) good figures for the first three principal components in terms of component contribution in the PCA.

For convenience of explanation of the proposed procedure, case 3 will be discussed, since it has well-suited PC coefficients a_{ij} for both the physical meaning of the

TABLE 4 Correlation-coefficient matrices of variables for the Southern Lake.

	TRANSP	PH	DO	COD	SS	NO3	TN	TP	CHLORA
TRANSP	1.0								
PH	−0.015	1.0							
DO	0.089	−0.231	1.0						
COD	−0.395	0.344	−0.398	1.0					
SS	−0.586	−0.041	0.009	0.330	1.0				
NO3	−0.049	−0.463	0.401	−0.313	0.074	1.0			
TN	−0.267	−0.163	0.053	0.169	0.267	0.627	1.0		
TP	−0.327	−0.063	−0.004	0.228	0.231	0.267	0.383	1.0	
CHLORA	−0.309	0.328	−0.082	0.417	0.192	−0.152	0.170	0.120	1.0

TABLE 5 Combinations of water-quality variables chosen for PCA.

Variable	Case 1	Case 2	Case 3	Case 4
TRANSP	X	X	X	X
PH	X	X	X	X
DO	X	X		
COD	X	X	X	X
SS	X	X	X	X
NO3	X		X	
TN	X	X	X	X
TP	X	X	X	X
CHLORA	X	X	X	X

TABLE 6 Cumulative values of the contributions of the first three components (%).

	Case 1	Case 2	Case 3	Case 4
Northern Lake	69.2	69.0	71.8	72.8
Southern Lake	65.0	64.2	70.2	70.0

water-quality index and for the total information contained in the original data. Cumulative values of the contributions of the first three components are given in Table 6. Furthermore, the fewer factors a particular PCA has, the higher the percentage contribution that can be obtained by summing over the first three components, because the relative proportion of information involved becomes larger than for other cases which have more factors. Hence a tradeoff must be made between the number of factors in the PCA and the cumulative value of the contributions of the first three components.

The PC coefficients a_{ij} for case 3 are given in Table 7. If we examine the italicized values in Table 7, it seems that the first PC tends to describe a quantity associated with organic substances, while the second PC tends to describe a quantity associated with inorganic substances. This is because coefficients such as *SS, COD, PH,* and *CHLORA* in the first PC, and *NO3, TN,* and *TP* in the second PC, have much larger positive values

TABLE 7 Principal component (PC) coefficients a_{ij} of case 3.

	Northern Lake			Southern Lake		
	1st PC	2nd PC	3rd PC	1st PC	2nd PC	3rd PC
TRANSP	−0.265	−0.184	0.183	−0.307	0.039	0.394
PH	0.233	−0.332	0.277	0.038	−0.508	0.377
COD	0.254	−0.325	0.019	0.249	−0.352	0.071
SS	0.272	0.212	−0.242	0.277	0.010	−0.551
NO3	−0.074	0.559	0.278	0.078	0.599	0.220
TN	0.059	0.525	0.382	0.239	0.369	0.393
TP	0.156	0.339	−0.578	0.232	0.185	0.220
CHLORA	0.255	0.043	0.523	0.209	−0.295	0.377
Contribution (%)	34.5	27.6	9.7	31.7	26.5	12.0

than the others. Figures 8 and 9 illustrate the distributions of monthly variations in the first and second PC values for each observation point. The two figures (compare, for instance, (a) and (b) in Figures 8 and 9) display different seasonal behavior in the variation of these two PCs.

By using the calculated results from the MDS procedure described in the previous section, a graphic display of observation points is obtained, based on the following four different versions of metric distances d_{pq}: (a) 1st PC, (b) 2nd PC, (c) 1st + 2nd + 3rd PCs, and (d) all components. Figures 10 and 11 show these four types of graphic display obtained using the MDS approach, where the numbers 1–23 correspond to the locations in the two parts of the lake (see Figure 3).

Taking account of the geographical location of the observation points and past trends in water quality, it is reasonable to choose the following clusters of points from Figures 10(a) and 11(a):

Northern Lake: (1,3), (2,4,5), (7,10), (8,9,11), (6), (13) (7)

Southern Lake: (15,16,18), (21,22,24), (20,23), (14), (17), (19) (8)

Clustering patterns (7) and (8) agree with those in Figures 10(d) and 11(d) except for point 5 in the Northern Lake and point 21 in the Southern Lake, although patterns (7) and (8) are much more consistent with existing knowledge about the water quality of Lake Biwa. This means that it is not necessary to use all the original information and only three components are needed to derive the fundamental characteristics of water quality in the area concerned in terms of filtered information from the noisy multivariate data. Furthermore, compared with the clustering of points encircled in parts (b) and (c) of Figure 11, it is clear that the relative location of clusters (21, 22,24) and (15,16,18) is reversed for point 17, which has the worst water quality in the entire area. The other points are in almost the same positions. This fact indicates that the two clusters have different characteristics with respect to the worst point, 17, depending on whether organic or inorganic water pollution is considered, as has

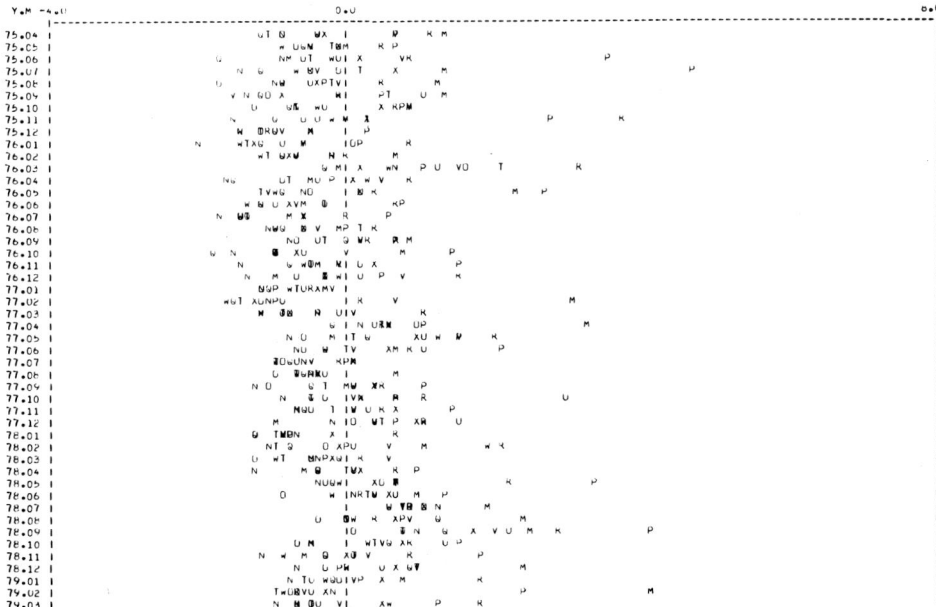

(a)

(b)

FIGURE 8 Variation of the values of the 1st PC with time: (a) in the Northern Lake; (b) in the Southern Lake. The capital letters used are the same as those in Figure 5.

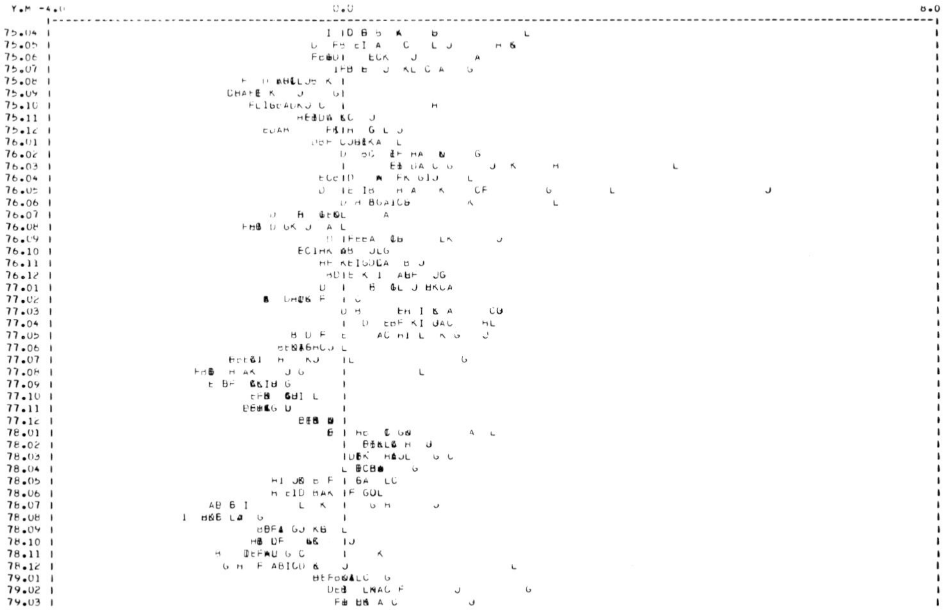

(a)

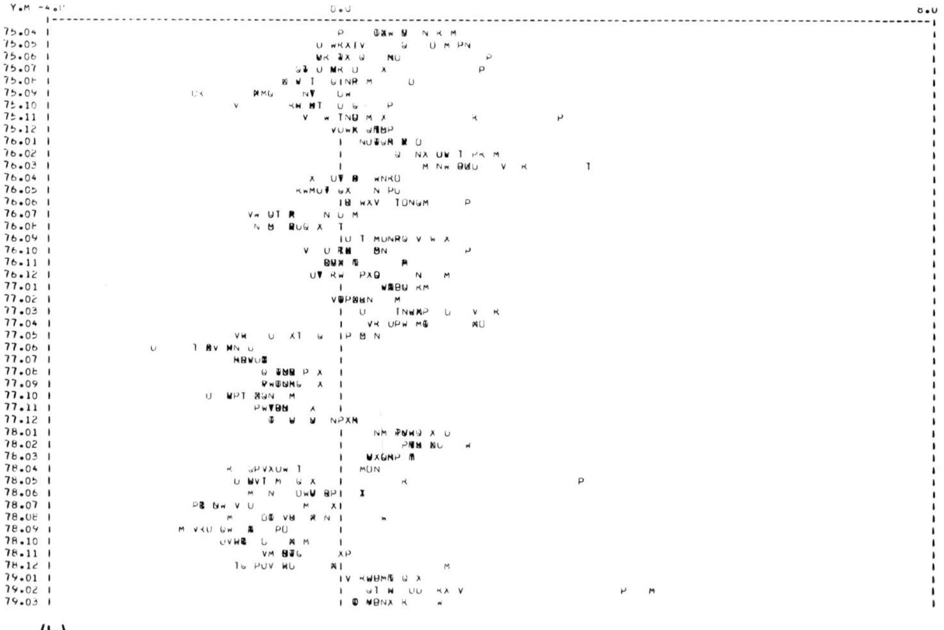

(b)

FIGURE 9 Variation of the values of the 2nd PC with time: (a) in the Northern Lake; (b) in the Southern Lake. The capital letters used are the same as those in Figure 5.

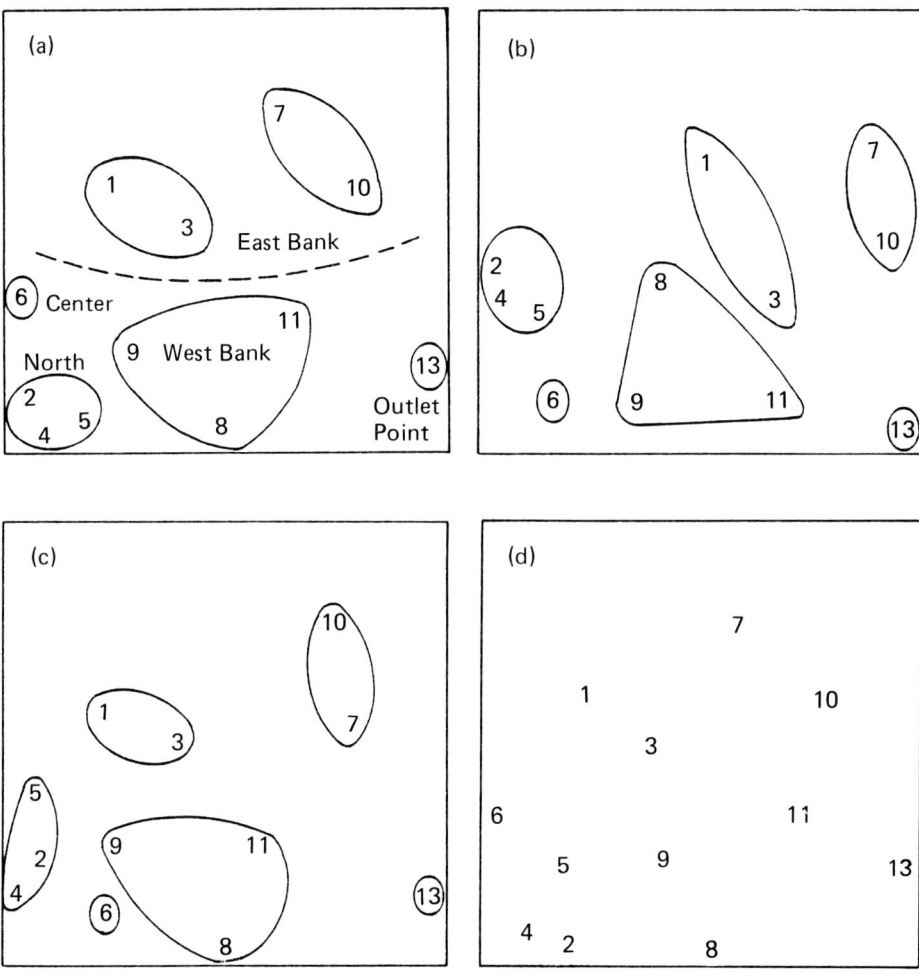

FIGURE 10 Configuration of the observed points in the Northern Lake, with distance defined using: (a) the first three PCs; (b) the 1st PC; (c) the 2nd PC; (d) all the components.

already been described for the difference of PC coefficients. That is, the southeast section of the Southern Lake has a greater similarity to the most polluted point, 17, than does the northeast section in terms of organic pollution. A similar observation can be made for the Northern Lake, comparing parts (a) and (b) in Figure 10, as regards the relative position of the northeast section (1,3) with respect to effluent point 13.

Thus, various interpretations can be made from these figures depending on our knowledge and experience of the area concerned. We consider that the more general participation of specialists in this analysis could yield even more useful and practical information for producing a comprehensive survey of water quality and, particularly, for the further elaboration of water-quality modeling.

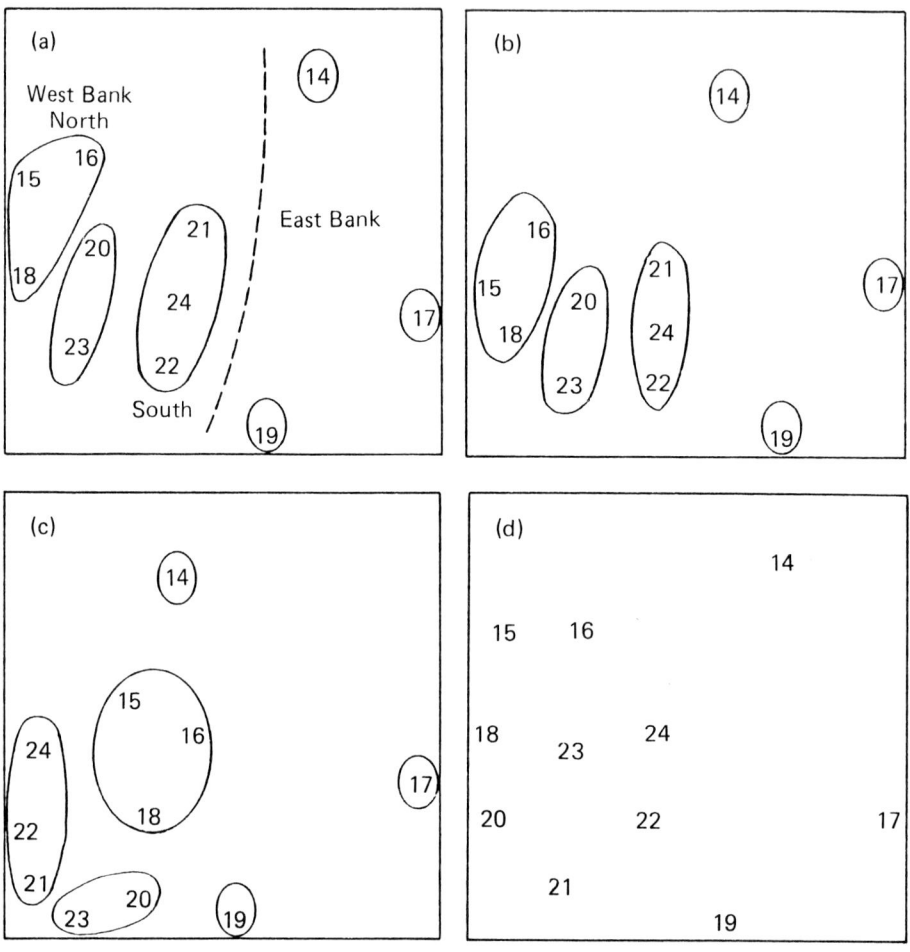

FIGURE 11 Configuration of the observation points in the Southern Lake, with distance defined using: (a) the first three PCs; (b) the 1st PC; (c) the 2nd PC; (d) all the components.

5 CONCLUSIONS

An MDS approach to clustering multivariate water-quality data has been presented. This yields a graphic display of regional characteristics with respect to an integrated index of water quality. In this paper, only the physical and chemical aspects of water quality have been discussed, in terms of similar monthly variation patterns in the data concerned. In the next step of the work, the biological aspects of water quality will be examined using other types of measure, for example, coefficients of regression analysis for a particular biological element such as the biomass of different types of algae. This approach might extract effective information from the original data by applying expert knowledge and experience to the analysis of aggregate measures plotted in two-dimensional space.

REFERENCES

Hooke, R. and Jeeves, T.A. (1961). Direct search solution of numerical and statistical problems. Journal of the Association for Computing Machinery, 8:212-229.

Ikeda, S. and Adachi, N. (1978). A dynamic water quality model of Lake Biwa – a simulation study of the lake eutrophication. Ecological Modelling, 4:151-172.

Itakura, H., Nishikawa, Y., and Yamauchi, T. (1979). A class of individual differences multidimensional scaling with weighted inner-product model and its applications to social survey data analysis. Paper presented at the International Symposium on Policy Analysis and Information Systems, 1st, Durham, North Carolina, June.

Japan Society of Civil Engineering (1970-1977). Reports on Future Water Quality of Lake Biwa. JSCE, Tokyo.

Lake Biwa Office, Ministry of Construction (1974). Eutrophication in Lake Biwa. (In Japanese.)

Lake Biwa Office, Ministry of Construction (1974-1979). Annual Reports on Regular Water Quality Monitoring.

Shepard, R.N., Romney, A.K., and Nerlove, S.B. (Editors) (1972). Multidimensional Scaling: Theory and Applications in the Behavioral Sciences. Seminar Press, New York.

NONLINEAR STEADY-STATE MODELING OF RIVER QUALITY BY A REVISED GROUP METHOD OF DATA HANDLING

Hiroyuki Tamura and Tadashi Kondo*

Department of Precision Engineering, Osaka University, 2-1 Yamada-oka, Suita, Osaka 565 (Japan)

1 INTRODUCTION

In river-quality systems there are many complex phenomena at work, such as biochemical reactions, thermal behavior, sedimentation, and photosynthetic oxygen production; therefore the structure of any physical model that considers the influence of these phenomena is necessarily very complex (Rinaldi et al., 1976, 1979). Parameter estimation procedures for physical models (Rinaldi et al., 1976) that have been used for predicting pollution levels of river quality are also very complicated.

The Group Method of Data Handling (GMDH) (Ivakhnenko, 1970, 1971) is a useful technique of data analysis for identifying these complex nonlinear systems through statistical analysis of input–output data, especially when only few data are available. The basic GMDH and its modifications (Duffy and Franklin, 1975; Ikeda et al., 1976; Tamura and Kondo, 1978, 1980) have many advantages, probably the most remarkable being that they automatically select the structure (degree of nonlinearity) of the model without using *a priori* information on relationships among the input–output variables. Therefore, if the system is predominantly nonlinear and its mechanistic structure is not known explicitly, GMDH can be a useful technique for modeling and identification. However, using a conventional GMDH it is difficult to identify a physically meaningful structure among the input–output variables because the partial polynomials, in which the intermediate variables are used as the input variables in each selection layer, have been estimated and accumulated in the multilayered structure.

In this paper, a nonlinear steady-state river-quality system is identified using a revised GMDH (Kondo and Tamura, 1979), which generates optimal intermediate polynomials instead of partial polynomials in each selection layer. The optimal intermediate

---

* Present address: Toshiba Corporation, Fuchu, Tokyo 183, Japan.

polynomials express the direct relationships among the input–output variables, and they are generated so as to minimize the Akaike information criterion (AIC) (Akaike, 1972, 1973, 1974) evaluated by using all the data. Therefore, even if the internally descriptive (mechanistic) model is not known explicitly, the physically meaningful structure can be identified by using this revised GMDH when the characteristics of the system are well incorporated in the measured data. By using various measures of river quality such as biochemical oxygen demand (BOD) and dissolved oxygen (DO) levels, for the example of the Bormida River, Italy, two kinds of nonlinear models of steady-state river quality are constructed, and the structures and prediction accuracies are compared with those of Rinaldi's linear physical (mechanistic) model.

2 MODELING THE STEADY-STATE RIVER QUALITY

BOD and DO levels have been widely accepted as the most important indexes of organic river quality. The dynamic behavior of these levels can be described using a generalized Streeter–Phelps model (Rinaldi et al., 1979)

$$\partial b/\partial t + V \partial b/\partial l = -(k_1(T) + (k_3(V)/A))b \tag{1a}$$

$$\partial c/\partial t + V \partial c/\partial l = -k_1(T)b + (k_2(T,Q)/H(Q))(c_s(T) - c) + k_4/A \tag{1b}$$

where

b = BOD (mg l^{-1}),
c = DO (mg l^{-1}),
c_s = saturation level of DO (mg l^{-1}),
k_1 = deoxygenation rate (day^{-1}),
k_2 = reoxygenation rate (m day^{-1}),
k_3 = suspended BOD sedimentation rate (m^2 day^{-1}),
k_4 = photosynthetic oxygen production rate ((mg l^{-1})(m^2 day^{-1})),
t = time (day),
l = distance (km),
T = water temperature (°C),
A = cross-sectional area (m^2),
Q = flow rate (10^3 m^3 day^{-1}),
V = (Q/A) = average stream velocity (km day^{-1}), and
H = mean river depth (m).

Here, for simplicity, it is assumed that the cross-sectional area A does not vary along the river and that the velocity V is constant over space and time. Then, the steady-state BOD and DO levels satisfy the differential equations

$$db/dl = -K_1(T,Q)b \tag{2a}$$

$$dc/dl = -K_2(T,Q)b + K_3(T,Q)(c_s - c) + K_4(Q) \tag{2b}$$

where the functions K_h ($h = 1, 2, 3, 4$) depend upon the two independent variables Q and T, i.e.

$$K_1(T, Q) = k_1(T)/V(Q) + k_3(V(Q))/Q \tag{3a}$$

$$K_2(T, Q) = k_1(T)/V(Q) \tag{3b}$$

$$K_3(T, Q) = k_2(T, Q)/(H(Q)V(Q)) \tag{3c}$$

$$K_4(Q) = k_4/Q \tag{3d}$$

The solutions to eqns. (2) are

$$b(l, K_1, b_0) = b_0 e^{-K_1 l} \tag{4a}$$

$$c(l, K_1, K_2, K_3, K_4, b_0, c_0) = c_s + K_4/K_3 - [c_s + (K_4/K_3) - c_0] e^{-K_3 l}$$
$$+ [K_2 b_0/(K_1 - K_3)] [e^{-K_1 l} - e^{-K_3 l}] \tag{4b}$$

where b_0 and c_0 are the BOD and DO levels, respectively, near the discharge point, and it is assumed that there is no discharge inside the subject range.

Data are measured for n different steady states. The ith steady state is characterized by the flow rate Q^i and the temperature T^i. The BOD and DO levels are measured at r points along the river as shown in Figure 1. We will assume that the following measured data are available:

$$(b_0^i, c_0^i) \quad (i = 1, 2, \ldots, n) \tag{5a}$$

$$(b_j^i, c_j^i) \quad (i = 1, 2, \ldots, n; \ j = 1, 2, \ldots, r) \tag{5b}$$

where j denotes the jth measuring point along the river.

2.1 Parameter Estimation of the Physical Model (Rinaldi et al., 1976)

Here, the estimation method for the parameters contained in eqns. (4a) and (4b) is introduced briefly. The structures of the functions K_h contained in eqns. (4a) and (4b) are assumed to be

$$K_h = K_h(\underline{\theta}_h, T, Q) \quad (h = 1, 2, 3, 4)$$

where the $\underline{\theta}_h$ denote the parameters contained in K_h. By using the measured data (5a) and (5b), the parameters $\underline{\theta}_h$ are estimated so as to minimize the criterion

$$J = \sum_{i=1}^{n} J^i \tag{6a}$$

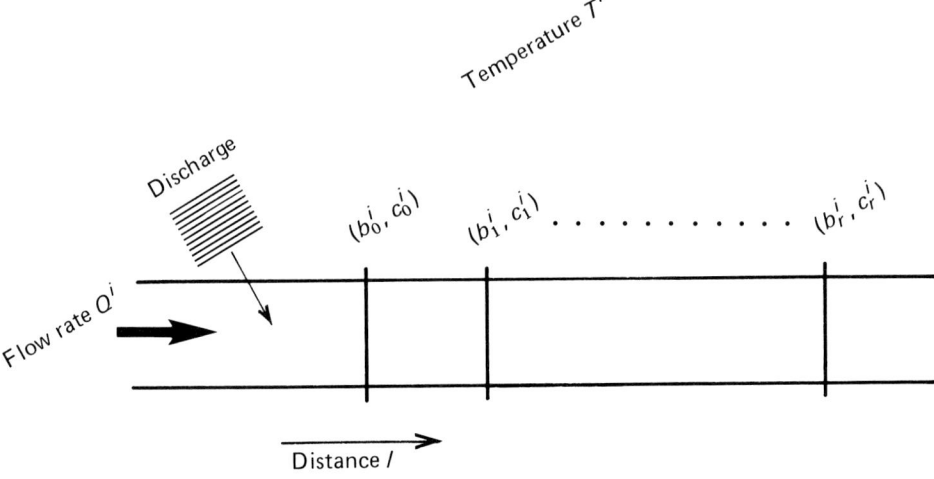

FIGURE 1 The variables measured in a river. Superscript i denotes the ith steady state in the river.

where

$$J^i = \sum_{j=1}^{r} [\lambda \epsilon_b^{ji} + (1-\lambda)\epsilon_c^{ji}] \quad (0 \leq \lambda \leq 1) \tag{6b}$$

$$\epsilon_b^{ji} = [b(l_j, K_1^i, b_0^i) - b_j^i]^2 \tag{6c}$$

$$\epsilon_c^{ji} = [c(l_j, K_1^i, K_2^i, K_3^i, K_4^i, b_0^i, c_0^i) - c_j^i]^2 \tag{6d}$$

and ϵ_b^{ji} is a square error between the measured value of the BOD level of the ith steady state at the jth point and the estimated value from eqn. (4a). ϵ_c^{ji} is a square error for the DO level, and λ is a weight for the BOD level. It is very difficult to estimate the parameters $\underline{\theta}_h$ directly so as to minimize J in eqn. (6a) because the dimension of $\underline{\theta}_h$ is very high; therefore, the following procedure is used for this estimation. Firstly, by using the data measured in each steady state, functions K_h^i ($h = 1, 2, 3, 4; i = 1, 2, \ldots, n$) are estimated so as to minimize J^i ($i = 1, 2, \ldots, n$). Then, by using the estimated values of K_h^i, the parameters $\underline{\theta}_h$ are estimated so as to minimize

$$J' = \sum_{i=1}^{n} \sum_{h=1}^{4} (K_h(\underline{\theta}_h, T^i, Q^i) - K_h^i)^2 \tag{7}$$

A more precise description of this procedure can be found in Rinaldi et al. (1976).

2.2 Modeling the Steady-State System using the Revised GMDH

Here, the steady-state model of river quality is constructed using the revised GMDH algorithm. In this algorithm, optimal intermediate polynomials, which express the direct relationships between the input and output variables, are generated automatically in each selection layer so as to minimize the AIC, and the final model is obtained from the optimal intermediate polynomial remaining in the final layer. Using the revised GMDH algorithm, the following two steady-state models are constructed.

2.2.1 Steady-State Model I

A steady-state model in the form of eqns. (2) is constructed. Two variables, $b(j+1)$ and $c(j+1)$, are used as output variables and five variables, $b(j)$, $c(j)$, Q^{-1}, $Q^{-0.5}$, and T, are used as input variables. Here it is assumed that the measuring points for the BOD and DO levels are equally spaced along the river. The steady-state model to be identified by the revised GMDH is

$$b(j+1) = f_1(b(j), c(j), Q^{-1}, Q^{-0.5}, T) \tag{8a}$$

$$c(j+1) = f_2(b(j), c(j), Q^{-1}, Q^{-0.5}, T) \tag{8b}$$

Equations (8) can be transformed to

$$(b(j+1) - b(j))/\Delta l = (1/\Delta l)\{f_1(b(j), c(j), Q^{-1}, Q^{-0.5}, T) - b(j)\} \tag{9a}$$

$$(c(j+1) - c(j))/\Delta l = (1/\Delta l)\{f_2(b(j), c(j), Q^{-1}, Q^{-0.5}, T) - c(j)\} \tag{9b}$$

If the left-hand sides of eqns. (9a) and (9b) are accepted as approximations for db/dl and dc/dl, respectively, a steady-state model in the form of eqns. (2) can be obtained.

2.2.2 Steady-State Model II

A steady-state model in the form of eqns. (4) is constructed. Two variables, $b(l)$ and $c(l)$, are used as output variables and seven variables, b_0, c_0, l, l^{-1}, $Q^{0.5}$, $Q^{-0.5}$, and T, are used as input variables. In this case, no physical interpretation of the model constructed by the revised GMDH is possible, because eqns. (4) cannot be described as physically meaningful polynomials in terms of these input variables. That is, the revised GMDH model obtained is a nonphysical model. The steady-state model to be identified by the revised GMDH is

$$b(l) = g_1(b_0, c_0, l, l^{-1}, Q^{0.5}, Q^{-0.5}, T) \tag{10a}$$

$$c(l) = g_2(b_0, c_0, l, l^{-1}, Q^{0.5}, Q^{-0.5}, T) \tag{10b}$$

For constructing this model, the measuring points for BOD and DO levels need not necessarily be equally spaced along the river.

3 THE REVISED GMDH

From the many kinds of mathematical models available, such as polynomials, Bayes formulas, trigonometrical functions, etc., the Kolmogorov–Gabor polynomial

$$\phi = a_0 + \sum_i a_i x_i + \sum_i \sum_j a_{ij} x_i x_j + \sum_i \sum_j \sum_k a_{ijk} x_i x_j x_k + \ldots \tag{11}$$

is widely used in the GMDH as a complete description of the system model. If a conventional multiple regression analysis is followed, it is necessary to estimate the enormous number of parameters in eqn. (11) simultaneously, which is impossible from both the statistical and the computational points of view. Equation (11) can be constructed by combining so-called partial polynomials

$$y_k = b_0 + b_1 x_i + b_2 x_j + b_3 x_i x_j + b_4 x_i^2 + b_5 x_j^2 \tag{12}$$

of two variables in multilayers, where the y_k values are called the intermediate variables. On going to the second layer, the intermediate variables $y_1, y_2, \ldots, y_L$ are regarded as the input variables of the second layer. That is, the partial polynomials generated in the second layer are of the form

$$z_k = b_0' + b_1' y_i + b_2' y_j + b_3' y_i y_j + b_4' y_i^2 + b_5' y_j^2 \tag{12'}$$

In the basic GMDH originated by Ivakhnenko the available data were divided into two sets: the training data and the checking data. The training data were used for estimating the parameters in the partial polynomials, and the checking data were used for selecting intermediate variables. Much research was done by Ivakhnenko's group on the best method of dividing the data into these two data sets (Ivakhnenko et al., 1979).

In the revised GMDH used in this paper, this artificial differentiation between training and checking data is eliminated. Furthermore, instead of partial polynomials (eqns. (12) and (12')), the intermediate polynomials are used. These intermediate polynomials are constructed from the direct relationships among the original input/output variables (while, as seen from eqn. (12'), the partial polynomials were constructed from the relationships among the intermediate variables and the output variables), and they are generated so as to minimize the AIC evaluated by using all the available data. A detailed discussion of the mathematical form for the intermediate polynomials can be found in Kondo and Tamura (1979). By using this revised GMDH a physically meaningful structure can be identified when the characteristics of the system are well incorporated in the data, even if the internally descriptive (mechanistic) model is not known explicitly.

Figure 2(a) shows the block diagram of a conventional GMDH, while Figure 2(b) shows the revised GMDH used in this paper. A detailed discussion on the algorithm of the revised GMDH can be found in Kondo and Tamura (1979).

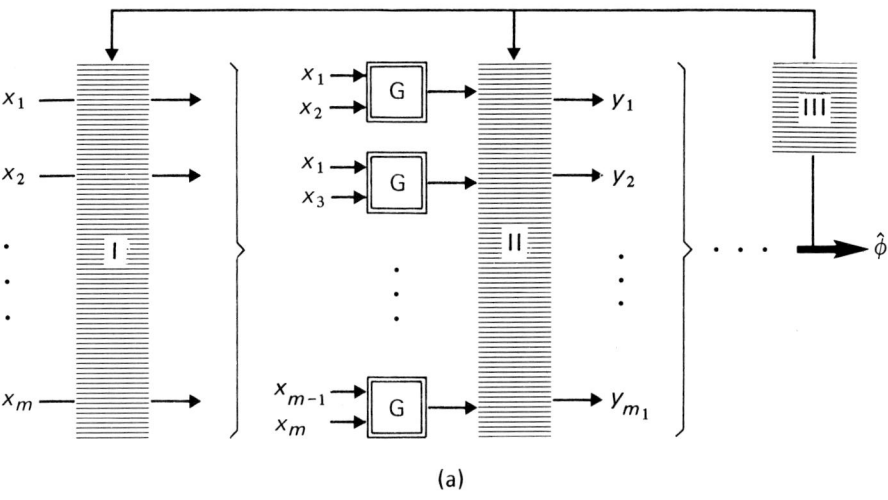

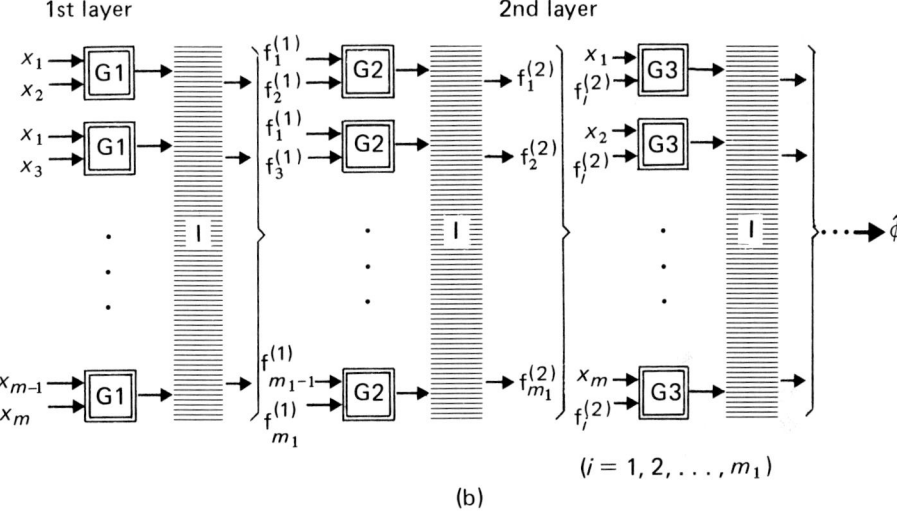

FIGURE 2 Block diagrams of: (a) basic GMDH (I, division of original data into two sets; II, self-selection of the intermediate variables; III, optimization of the threshold; G, generator of the partial polynomials); and (b) revised GMDH (I, self-selection of the optimal intermediate polynomials; G1, G2, G3, generators of the optimal intermediate polynomials).

4 MODELING THE STEADY-STATE WATER QUALITY OF THE BORMIDA RIVER

The steady-state model of the Bormida River shown in Figure 3 is constructed by applying the revised GMDH algorithm to the data shown in Table 1; the predicted results obtained using the revised GMDH model are compared with those obtained from the physical model estimated by Rinaldi et al.

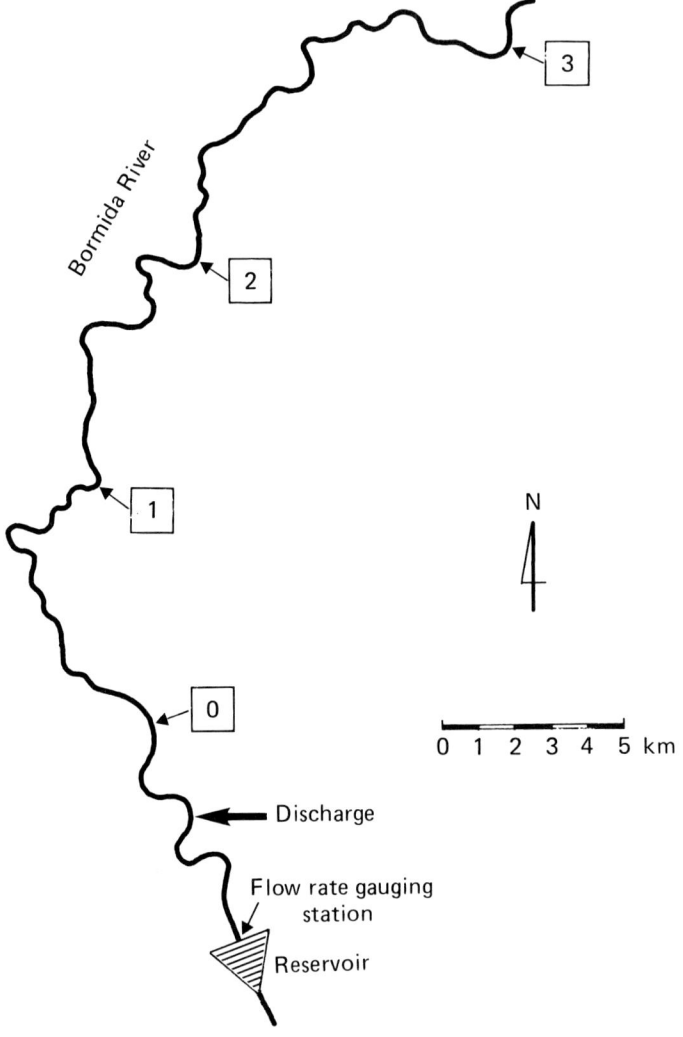

FIGURE 3   The Bormida River and locations of measurements stations (Rinaldi et al., 1976).

The data measured in the Bormida River (Rinaldi et al., 1976) are used; four variables, namely BOD level b, DO level c, flow rate Q, and temperature T, are measured, and these are shown in Table 1. Data for BOD and DO levels are daily average values measured at six points located at intervals of about 10–15 km along the river. Here, the data obtained for the fourth point are not the measured values but values obtained by linear interpolation. Temperature data are average values obtained at six points but the measurement time is different for each steady state, and therefore it is difficult to give a significant interpretation. The effect of temperature variation was, in fact, simply neglected. Fifteen steady states are measured ($n = 15$). From these, data from thirteen

TABLE 1 The data used for modeling and model validation (Rinaldi et al., 1976)[a].

Steady state	Station number and distance (km)						Flow rate (10^3 m^3 day^{-1})	Water temperature (°C)	
	0 4.20	1 14.00	2 25.00	3 40.00	(4) (54.00)	5 68.00		Average	Range
1	200.0	118.0	64.0	38.0	24.0	10.0	55	17.5	4.2
	0.0	4.5	5.5	6.5	7.8	9.0			
2	120.0	92.0	72.0	58.0	41.0	24.0	60	9.0	5.1
	3.0	5.5	9.0	9.5	9.5	9.5			
3	162.0	126.0	110.0	66.0	53.0	40.0	125	0.5	5.0
	1.0	3.0	5.0	6.5	8.5	10.5			
4	105.0	84.0	70.0	44.0	41.0	38.0	100	19.0	3.0
	2.0	5.0	5.5	6.0	6.8	7.5			
5	125.0	78.0	46.0	18.0	16.0	14.0	75	18.0	3.2
	1.5	3.5	4.5	5.5	6.3	7.0			
6	125.0	86.0	70.0	46.0	33.0	20.0	80	17.0	3.3
	2.0	5.0	6.0	6.0	6.3	6.5			
7	68.0	56.0	50.0	34.0	29.0	24.0	225	5.0	2.5
	2.0	6.0	7.0	9.5	10.8	12.0			
8	145.0	72.0	68.0	30.0	23.0	16.0	100	25.0	3.7
	0.0	1.2	2.2	3.6	4.7	5.8			
9	200.0	104.0	98.0	60.0	59.0	58.0	55	10.0	8.9
	0.0	4.0	6.0	6.0	6.5	7.0			
10	90.0	70.0	68.0	58.0	40.0	22.0	200	1.8	3.5
	4.0	4.0	8.0	9.0	9.0	9.0			
11	80.0	60.0	50.0	36.0	30.0	24.0	250	3.5	2.4
	6.0	8.0	10.0	10.5	10.8	11.0			
12	135.0	100.0	85.0	62.0	56.0	50.0	125	11.8	2.4
	0.5	4.0	5.0	6.0	7.0	8.0			
13	70.0	60.0	44.0	46.0	34.0	22.0	200	16.0	2.5
	3.0	6.0	7.0	7.5	7.8	8.0			
14	85.0	70.0	55.0	40.0	30.0	20.0	200	11.5	5.5
	3.0	6.0	7.0	9.0	9.3	9.5			
15	80.0	40.0	30.0	20.0	16.0	12.0	150	16.0	6.0
	2.5	5.0	7.0	8.5	8.8	9.0			

[a] The first row for each steady state refers to BOD (mg l^{-1}) and the second to DO (mg l^{-1}) observations.

4.1 Results of Parameter Estimation of the Physical Model (Rinaldi et al., 1976)

Parameters of the physical model are estimated using the procedure described in Section 2.1. The data of the steady states 1–13 are used for modeling. The structures of the functions K_h ($h = 1, 2, 3, 4$) are assumed to be

$$K_h(\underline{\theta}_h, Q) = \theta_{h1} Q^{\theta_{h2}} \tag{13}$$

where

$$\underline{\theta}_h = (\theta_{h1}, \theta_{h2})$$

Functions K_h^i ($h = 1, 2, 3, 4$; $i = 1, 2, \ldots, 13$) are estimated so as to minimize J^i ($i = 1, 2, \ldots, 13$) in eqn. (6b) and as a result

$$K_1 \simeq K_2, \quad K_4 \simeq 0 \tag{14}$$

is obtained. This result shows that the BOD and DO levels in the Bormida River can be described by a Streeter–Phelps model. Then parameters $\underline{\theta}_1$ and $\underline{\theta}_3$ are estimated so as to minimize J' in eqn. (7) and the results

$$db/dl = -0.2 Q^{-0.43} b \tag{15a}$$

$$dc/dl = -0.2 Q^{-0.43} b + 16.4 Q^{-0.8}(c_s - c) \tag{15b}$$

are obtained.

4.2 Results of Modeling Using the Revised GMDH

4.2.1 Steady-State Model I

Four variables, $b(j)$, $c(j)$, Q^{-1}, and $Q^{-0.5}$, are used as input variables. BOD models identified by the revised GMDH will be considered first and these are shown in Table 2. Model 4 is identified using all the data from the 15 steady states. It can be seen that the structure of the model varies slightly according to the measured data used for modeling. In the revised GMDH, the structure of the model is determined by using only the measured data, and therefore the dependence of the structure of the model on the statistical characteristics of the measured data cannot be avoided. However, if sufficient data can be used, the dependence can be reduced. Model 3

$$b(j+1) = -4.22 + 0.920 b(j) + 0.000037 b(j)^2 - 0.0133 Q^{-0.5} b(j)^2 \tag{16}$$

TABLE 2 Structure of the BOD model I

Model number	Prediction points	Constant	b	b^2	$bQ^{-0.5}$	$b^2Q^{-0.5}$
1	4, 5	−5.84	0.960	−0.00040	–	−0.011
2	9, 10	−2.38	1.027	−0.00070	−2.06	–
3	14, 15	−4.22	0.920	0.00004	–	−0.013
4	0	−3.82	0.900	0.00008	–	−0.013

is identified using the measured data for steady states 1–13. This model can be transformed to

$$(b(j+1) - b(j))/\Delta l = (1/\Delta l)\{-4.22 - 0.080b(j) + 0.000037b(j)^2$$

$$- 0.0133Q^{-0.5}b(j)^2\} \tag{17}$$

Since $\Delta l \simeq 10$ km, eqn. (17) can be approximately reduced to

$$db/dl = -0.422 - 0.0080b + 0.0000037b^2 - 0.00133Q^{-0.5}b^2 \tag{18}$$

From this model it is found that the second-order terms of the BOD level are contained in eqn. (18), and the structure of the model is a little more complex than physical model (2a). In order to verify the effectiveness of eqn. (16), the prediction errors for the steady states 14 and 15 of eqn. (16) are compared with those of the physical model (2a). In eqn. (16), the BOD concentration $b(1)$ is predicted using the measured data b_0, and the BOD levels $b(j+1)$ for $j = 1-4$ are obtained using the predicted values for $j = 0-3$. Predicted results for steady states 14 and 15 are shown in Figure 4. It can be seen that the prediction accuracy obtained from the revised GMDH model (16) is identical with that obtained from physical model (2a).

DO models identified by the revised GMDH will now be considered (see Table 3). Model 4 is identified by using all the data from the 15 steady states. From Table 3, it can be seen that the structure of the model varies remarkably according to the measured data used for modeling. In particular, the terms concerned with flow rate Q are very varied. The reason for this is that the number of different measurement data for the flow rate are very few compared with the number of terms contained in the model. In other words more data for different flow rates are needed before more precise information can be extracted from the data concerned with input variable Q. Model 3

$$c(j+1) = 6.72 + 0.431c(j) - 0.000203b(j)^2 + 0.00222Q^{-0.5}b(j)^2$$

$$- 46.1Q^{-0.5} + 3.91Q^{-0.5}c(j) \tag{19}$$

is identified using the measured data for steady states 1–13. This model can be transformed to

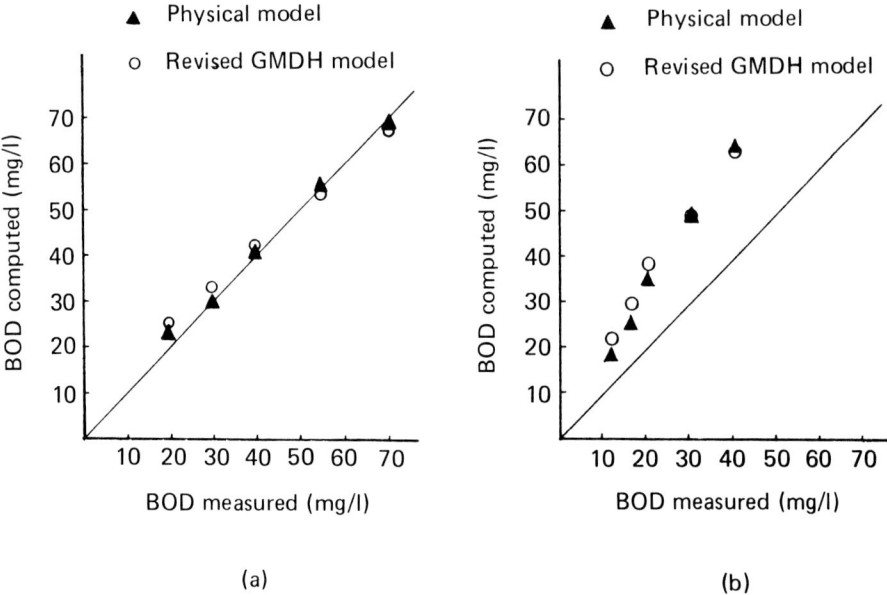

FIGURE 4 Measured and computed values of BOD from model I-3: (a) steady state 14; (b) steady state 15.

$$(c(j+1) - c(j))/\Delta l = (1/\Delta l)\{6.72 - 0.569c(j) - 0.000203b(j)^2$$
$$+ 0.00222Q^{-0.5}b(j)^2 - 46.1Q^{-0.5} + 3.91Q^{-0.5}c(j)\} \quad (20)$$

Once again using $\Delta l \simeq 10$ km, eqn. (20) can be approximately reduced to

$$dc/dl = 0.672 - 0.0569c - 0.0000203b^2 + 0.000222Q^{-0.5}b^2$$
$$- 4.61Q^{-0.5} + 0.391Q^{-0.5}c \quad (21)$$

It can be seen that the second-order terms b^2 and $Q^{-0.5}b^2$ are contained in both the BOD model (18) and the DO model (21). The terms $Q^{-0.5}$ and $Q^{-0.5}c$ are similar to the terms $Q^{-0.8}$ and $Q^{-0.8}c$, respectively, contained in the physical model (2b). In order to verify the effectiveness of eqn. (19), the prediction errors for steady states 14 and 15 of eqn. (19) are compared with those of the physical model (2b). In eqn. (19), the DO level $c(1)$ is predicted using the measured data b_0 and c_0, and the DO levels $c(j+1)$ for $j = 1-4$ are obtained using the predicted values for $j = 0-3$. Predicted results for steady states 14 and 15 are shown in Figure 5. From Figure 5(a), it can be seen that the revised GMDH model (19) gives much better prediction accuracy for steady state 14 than does the physical model (2b). From these prediction results, it can be seen that the steady-state model I identified by the revised GMDH algorithm is fairly reliable as a prediction model. Furthermore, the structure of steady-state model I is a little more complex than that

TABLE 3 Structure of the DO model I.

Model Number	Prediction points	Constant	c	b^2	$b^2 Q^{-0.5}$	$b^2 Q^{-1.0}$	$Q^{-0.5}$	$Q^{-1.0}$	$cQ^{-0.5}$	$cQ^{-1.0}$
1	4, 5	2.39	0.895	0.00003	—	—	—	—	—	—
2	9, 10	7.75	0.993	−0.00020	0.0024	—	−54.2	—	−1.19	78.6
3	14, 15	6.72	0.431	−0.00020	0.0022	—	−46.1	—	−10.4	—
4	0	10.3	0.553	−0.00008	—	0.0080	−118	382	3.91	18.3

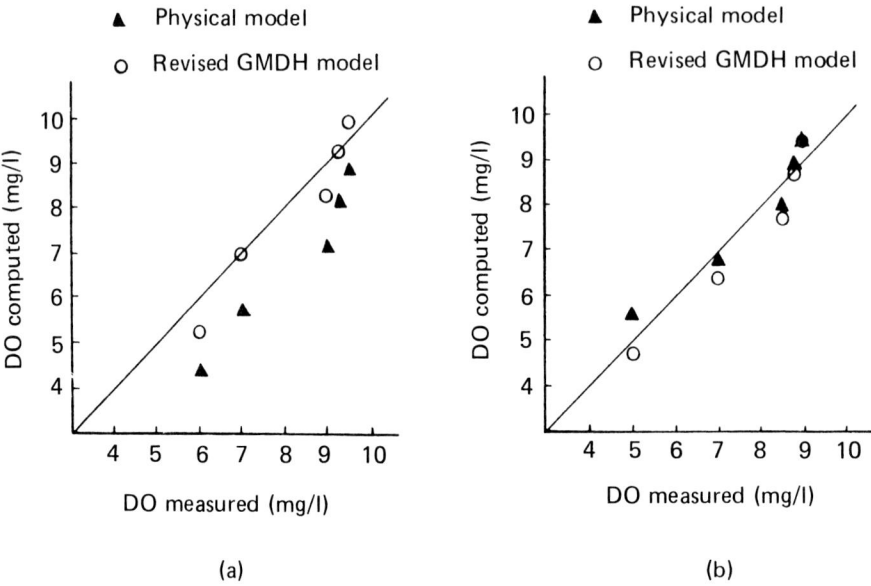

FIGURE 5 Measured and computed values of DO from model I-3: (a) steady state 14; (b) steady state 15.

of the physical model but the two are very similar. This shows that statistical analysis of the input and output data by the revised GMDH algorithm using intermediate polynomials gives important information about the physical structure of the system.

4.2.2 Steady-State Model II

Six variables, b_0, c_0, l, l^{-1}, $Q^{0.5}$, and $Q^{-0.5}$, are used as input variables. The BOD model identified by the revised GMDH will be considered first. By using the measured data of steady states 1–13, the BOD model is identified as

$$b(l) = 28.9 - 0.0268 b_0 l - 0.0217 l^2 + 1.523 l + 0.000261 b_0 l^2 + 10.2 b_0 Q^{-0.5} l^{-1}$$

$$+ 0.0004 b_0^2 Q^{0.5} + 0.871 b_0 c_0 Q^{-0.5} - 0.000042 b_0^2 c_0 Q^{0.5} \qquad (22)$$

It can be seen that the structure of eqn. (22) is more complex than that of the steady-state model I (16). In order to verify the effectiveness of eqn. (22), the prediction errors for steady states 14 and 15 of eqn. (22) are compared with those from the physical model (4a). Predicted results for steady states 14 and 15 are shown in Figure 6. It can be seen that the revised GMDH model (22) has the same prediction accuracy as the physical model (4a).

The DO model identified by the revised GMDH will now be considered. By using the measured data for steady states 1–13, the DO model is identified as

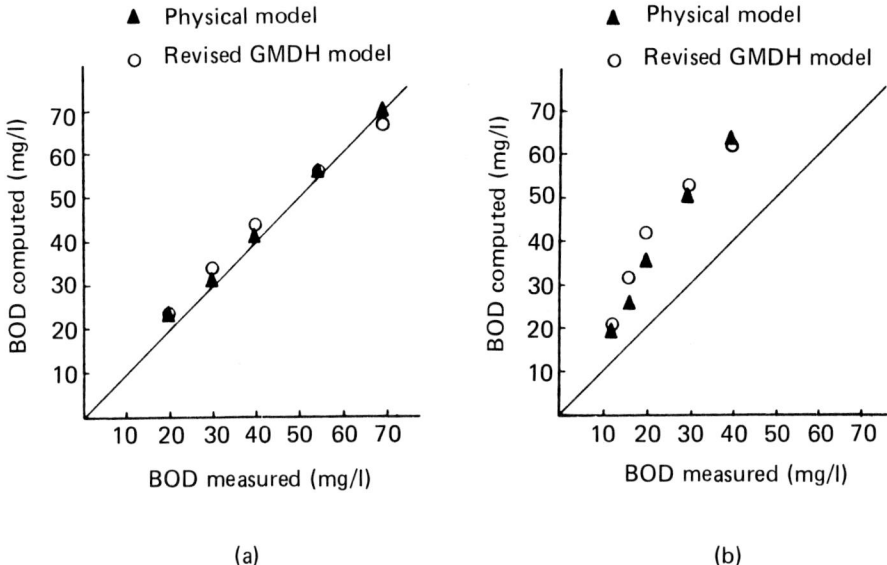

FIGURE 6 Measured and computed values of BOD from model II: (a) steady state 14; (b) steady state 15.

$$c(l) = -34.8 + 1.74Q^{0.5} - 11.6l^{-1} - 0.00104l^2 + 189Q^{-0.5} + 9.26c_0 Q^{-0.5}$$
$$+ 0.0106 Q^{0.5} l - 0.000436 b_0 c_0 l + 0.000004 b_0 l^2 + 0.000003 b_0^2 c_0 l \quad (23)$$

It can be seen that the structure of eqn. (23) is once again more complex than that of the steady-state model I (19). In order to verify the effectiveness of eqn. (23), the prediction errors for steady states 14 and 15 of eqn. (23) are compared with those of the physical model (4b). Predicted results for steady states 14 and 15 are shown in Figure 7. From Figure 7(b), we see that the revised GMDH model (23) gives a worse prediction accuracy for steady state 15 than does the physical model (4b). The reason for this is that the structure of the system for the DO level is very complex and cannot just be described as a polynomial approximation of the six input variables used here; in other words, more suitable input variables are needed for this model.

5 CONCLUSIONS

In this paper, two kinds of steady-state river-quality models are constructed by applying the revised GMDH algorithm to the measured data for the Bormida River. On comparing the revised GMDH model with the physical model identified by Rinaldi et al. the following results are obtained.

For the steady-state model I identified by the revised GMDH, the second-order terms of the BOD level are contained in both the BOD and DO models. It is interesting to see that the remaining terms are quite similar to those in the physical (mechanistic)

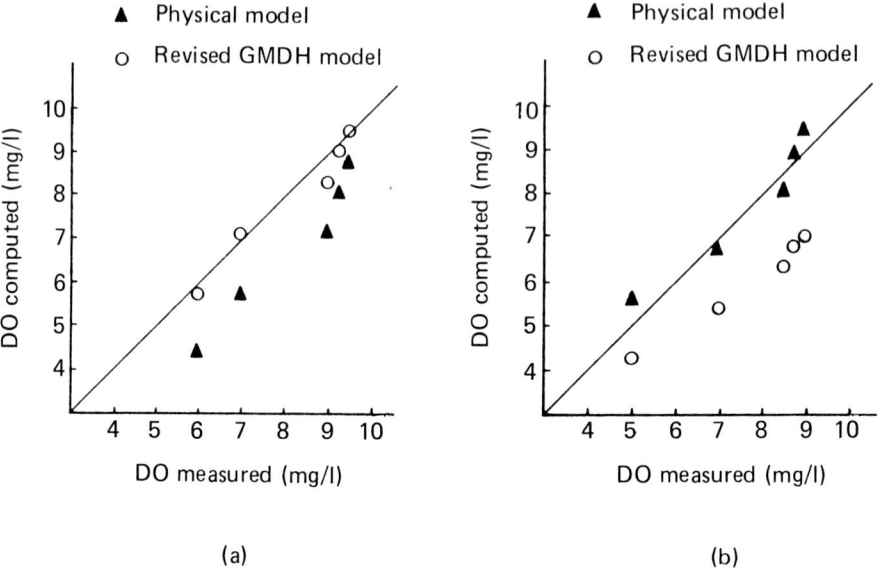

FIGURE 7 Measured and computed values of DO from model II: (a) steady state 14; (b) steady state 15.

model. This implies that the revised GMDH model using intermediate polynomials gives important information about the physical structure of the system.

Steady-state model I gives the same prediction accuracy as the physical model for the BOD level but better prediction accuracy than the physical model for the DO level.

In the revised GMDH models for the DO concentration, steady-state model I gives better prediction accuracy than does steady-state model II. This is because the structure of the system for DO concentration is very complex and cannot be described as a polynomial approximation of only the six input variables used here. More suitable input variables are needed for this case.

The structure of the revised GMDH model is heavily dependent on the statistical properties of the data used for modeling because the structure (degree of nonlinearity) of the model is determined by using input–output data only. In the case of the Bormida River example, the flow rate terms in the revised GMDH model are particularly dependent on the data because of the lack of information about different flow rates in the available data.

Finally, in the physical model the computation for estimating the parameters is quite complex, but in the revised GMDH model this is not so.

ACKNOWLEDGMENTS

The authors are greatly indebted to Dr. M.B. Beck of IIASA for his valuable discussions and comments, and also for the opportunity to present this paper at the IIASA

Task Force Meeting. This research was supported in part by the Ministry of Education, Japan, for the science research program on Environmental Science under Grant No. 403066.

REFERENCES

Akaike, H. (1972). Automatic data structure search by the maximum likelihood method. In Computers in Biomedicine, Supplement to the Proceedings of the Hawaii International Conference on System Sciences, 5th, pp. 99–101.

Akaike, H. (1973). Information theory and an extension of the maximum likelihood principle. Proceedings of the International Symposium on Information Theory, 2nd. Akademiai Kiado, Budapest, pp. 267–281.

Akaike, H. (1974). A new look at statistical model identification. IEEE Transactions on Automatic Control, AC19(6): 716–723.

Duffy, J.J. and Franklin, M.A. (1975). A learning identification algorithm and its application to an environmental system. IEEE Transactions on Systems, Man, and Cybernetics, SMC5(2): 226–240.

Ikeda, S., Ochiai, M., and Sawaragi, Y. (1976). Sequential GMDH algorithm and its application to river flow prediction. IEEE Transactions on Systems, Man, and Cybernetics, SMC6(7): 473–479.

Ivakhnenko, A.G. (1970). Heuristic self-organization in problems of engineering cybernetics. Automatica, 6(2): 207–219.

Ivakhnenko, A.G. (1971). Polynomial theory of complex systems. IEEE Transactions on Systems, Man, and Cybernetics, SMC1(4): 364–378.

Ivakhnenko, A.G., Krotov, G.I., and Visotsky, V.N. (1979). Identification of the mathematical model of a complex system by the self-organization method. In E. Halfon (Editor), Theoretical Systems Ecology. Academic Press, New York, pp. 326–352.

Kondo, T. and Tamura, H. (1979). Revised GMDH for generating optimal intermediate polynomials using the AIC. Transactions of the Society of Instrument and Control Engineers of Japan, 15(4):466–471.

Rinaldi, S., Romano, P., and Soncini-Sessa, R. (1976). Parameter estimation of a Streeter–Phelps type water pollution model. In the Proceedings of the IFAC Symposium on Identification and System Parameter Estimation, 4th, Tbilisi.

Rinaldi, S., Soncina-Sessa, R., Stehfest, H., and Tamura, H. (1979). Modeling and Control of River Quality. McGraw-Hill, New York.

Tamura, H. and Kondo, T. (1978). Nonlinear modeling for short-term prediction of air pollution concentration by a revised GMDH. Proceedings of the International Conference on Cybernetics and Society. IEEE Systems, Man, and Cybernetics Society, Tokyo and Kyoto, pp. 596–601.

Tamura, H. and Kondo, T. (1980). Heuristics free GMDH for generating optimal partial polynomials with application to air pollution prediction. International Journal of Systems Science, 11(9): 1095–1111.

Part Three

Uncertainty, Forecasting, and Control

PARAMETER UNCERTAINTY AND MODEL PREDICTIONS: A REVIEW OF MONTE CARLO RESULTS

R.H. Gardner and R.V. O'Neill
Environmental Sciences Division, Oak Ridge National Laboratory, Oak Ridge, Tennessee 37830 (USA)

1 INTRODUCTION

Uncertainty in ecological models (O'Neill and Gardner, 1979) is due to a number of factors. The total error associated with model predictions can only be assessed by a validation process (Caswell, 1976; Mankin et al., 1977) which tests the model against independent data (Shaeffer, 1979). However, such validation experiments are often infeasible, and modeling research has focused on individual factors that contribute to total error. These factors include assumptions in model construction (Harrison, 1978; Cale and Odell, 1979; O'Neill and Rust, 1979), measurement errors (O'Neill, 1973; Argentesi and Olivi, 1976), and errors in formulating ecosystem processes (O'Neill, 1979a).

Of the factors contributing to total error, parameter variability has received the greatest emphasis. Many studies have taken advantage of the availability of analytical methods for estimating the variance on model output (see, for example, Argentesi and Olivi, 1976; Beck et al., 1979; DiToro and van Straten, 1979; Lettenmaier and Richey, 1979). In the series of studies reviewed in this paper, we approached the study of parameter variability by Monte Carlo analysis, i.e., repeated simulations of the model with randomly selected parameter values. At the beginning of each simulation (or at intervals during the simulation), parameter values are chosen from specific frequency distributions. This process is continued for a number of iterations sufficient to converge on an estimate of the frequency distribution of the output variables.

Our goal in these studies was not merely to establish error bounds around model predictions, but to explore the general properties of error propagation in models. In our opinion, the Monte Carlo approach is uniquely suited to this exploration because the technique is not limited to any specific set of assumptions, the sources of model error must be explicitly considered, and the method can be quickly implemented, allowing the comparison of many different models.

In all of our studies we have assumed that parameters are measured in independent laboratory or field experiments. This assumption is appropriate for models that synthesize

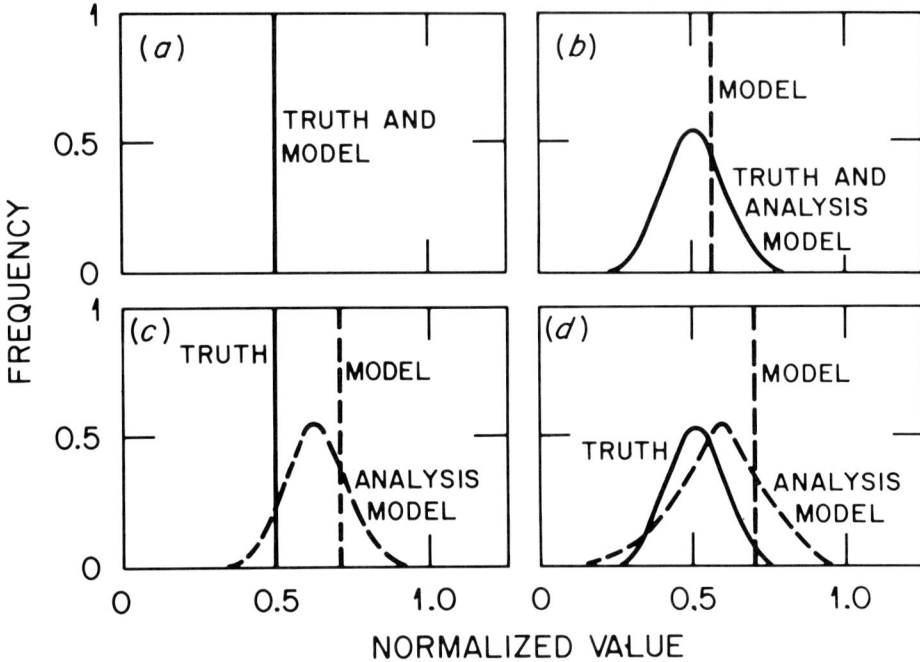

FIGURE 1 Comparison of frequency distributions of the real system ("truth") with the deterministic model solution ("model") and the deterministic model with variable parameters ("analysis model"). Four situations are illustrated: (a) the real system is deterministic, the model is deterministic and unbiased — no analysis is necessary; (b) the real world is variable, the model is deterministic, the analysis model is coincident with "truth"; (c) the real system is deterministic, the model is deterministic and biased, the analysis model is variable and less biased; (d) the real world is distributed, the model is deterministic and biased, the analysis model is more variable than the real world and less biased than the deterministic model.

individual physiological studies (see, for example, Park et al., 1974). However, the assumption is inappropriate when all parameters of a model are simultaneously fitted to sequential measurements of the state variables, either by nonlinear least-squares (see, for example, Halfon, 1975) or extended Kalman filter (see, for example, Beck, 1979) methods. In this paper we will review our recent error analysis studies, with emphasis on the counter-intuitive results produced by the Monte Carlo approach.

2 TWO SOURCES OF PARAMETER VARIABILITY

Differences in the assumed sources of error lead to differences in formulating the results. Figure 1 illustrates the interplay between two sources of error: natural variability in the ecosystem and error in parameter estimation. Figure 1(a) shows the simple case with no natural variability and no measurement error; in this case every parameter of the model is known exactly. In this case, the "true" behavior of the system can be represented by a frequency distribution which is a single vertical line of height 1.0. Assuming that there are no errors in model construction (as we will assume throughout this paper), the model predicts the distribution perfectly and there is no need for error analysis. This situation, of course, never actually occurs but will serve as a reference for more realistic cases.

In Figure 1(b), the ecosystem has natural variability but there is no measurement error. In this case, we must represent ecosystem behavior by a distribution representing the statistical population of behaviors of which the ecosystem is capable. This is an important point to keep in mind: natural variability in the ecosystem implies a population of possible behaviors. The Monte Carlo implementation of the model explicitly accounts for this variability and exactly predicts the distribution correctly if all sources of natural variability are known exactly. The deterministic model, in contrast, still predicts a single value and, even without error in measuring the mean of each parameter, this prediction is incorrect. This bias or shift in the predicted value results from attempting to represent the variable system by a deterministic model. The error results from the fact that

$$E[f(A)] \neq f[E(A)] \tag{1}$$

whenever $f(A)$ is a nonlinear function. In other words, the expected value of a function, $E[f(\ )]$, with a set, A, of randomly varying parameters is not necessarily equal to the value of the function using the expected value of each of the parameters. The two are equal only when $f(A)$ is simply the sum or product of the A terms or when the function is linear (O'Neill, 1979b), but these cases do not appear particularly relevant for ecological models. Even a system of linear differential equations produces an $f(A)$ which is an exponential function.

Figure 1(c) depicts the situation in which the ecosystem has no natural variability but where parameter values determined in independent experiments are measured with error. In this case, the deterministic model shows a shift in the predicted value because the parameters are incorrect. The mean value from the Monte Carlo implementation of the model is also incorrect, but it is possible to make a probabilistic statement about the magnitude of the error, because the distribution of model outputs is produced.

Figure 1(d) shows the most common situation, in which there is both natural variability and also uncertainty in model parameters, each of which has been measured independently. The deterministic model is incorrect because incorrect parameters are used and bias is introduced by the model (eqn. 1). The expected value of the Monte Carlo iterations is also incorrect due to measurement errors. However, the shift or bias in the expected value will be less than in the deterministic model, and an estimate of the uncertainty associated with the prediction is possible.

3   SENSITIVITY ANALYSIS AS AN APPROXIMATION TO ERROR ANALYSIS

In many studies, the contribution of error on individual parameters to overall prediction uncertainty is estimated by sensitivity analysis (see, for example, van Straten and de Boer, 1979). This approach evaluates the partial derivative of some model output (typically, the value of a state variable) with respect to each of the parameters. However, this method only approximates the contribution of each parameter because of three implicit assumptions:

(1) The expected behavior of the model is equal to the behavior of the model using the expected parameter values (Argentesi and Olivi, 1976).
(2) The contribution to total error can be approximated by examining the contribution due to each parameter separately.
(3) Small perturbations in the parameters approximate errors resulting from large uncertainties (van Straten and de Boer, 1979), i.e., higher-order effects are absent.

Since sensitivity analysis is the most commonly applied method for this type of analysis, it is important that we examine the extent to which these assumptions are valid.

It should be clear from our discussion of Figure 1 that a deterministic model of a variable system always yields a biased prediction of the expected behavior. Thus, the first assumption of sensitivity analysis is ordinarily violated. However, the magnitude of the bias can vary significantly. In one study (Gardner et al., 1980b) comparing six phytoplankton–zooplankton models, all of the models were calibrated to a single hypothetical data base. As a result, bias was small, ranging from 1% to about 10% of the total uncertainty in the model prediction. In contrast, a study of a marsh hydrology model (Gardner et al., 1980a) showed that the deterministic prediction can be in error by an order of magnitude due to bias! In the majority of the applications examined, however, the bias has been small (approximately 10%). It seems reasonable to conclude, therefore, that violation of the first assumption of sensitivity analysis will not lead to serious problems. If the first-order approximation of Hahn and Shapiro (1968) is used, the assumption will be violated only in unusual cases.

Our past studies do not provide a direct test of the second assumption that each parameter contributes independently to total error. We can, however, approach the question by comparing partial and simple correlation coefficients calculated between individual parameters and total model variability. When all parameters are varied simultaneously, the partial correlation coefficient indicates the direct contribution of that parameter to the variance of the predicted value. If there are no interaction terms between parameters, the partial correlation coefficient will approximate the individual sensitivity coefficient. The simple correlation coefficient represents the direct relationship between a parameter and predicted values when all parameters are varied simultaneously. Comparison between simple and partial correlations is a test of the second assumption if we assume that no higher-order interaction terms are present. If the partial is not equal to the simple correlation, this indicates that the variance of the other parameters has altered the relationships between parameters and predictions. We use the correlation coefficient for this analysis because, when the coefficient is squared, it expresses the fraction of the prediction uncertainty that is accounted for by variability in the parameter.

A comparison of partial and simple correlation coefficients is possible for our analysis of the marsh hydrology model (Gardner et al., 1980a; Huff and Young, 1980). For most parameters, the partial and simple coefficients are similar and the second assumption appears valid. Where they diverge, it is usually only for a portion of the simulated annual cycle. The few exceptions that were found among the 14 parameters had unusually high variances. For example, the greatest divergence occurs for a parameter with a coefficient of variation of 48%. Our analysis indicates that the sensitivity coefficients are conservative; that is, the effect of simultaneously considering all parameters is a decrease in the correlation between an individual parameter and total model error. A similar conclusion was reached by DiToro and van Straten (1979).

The third assumption is that the uncertainty in model output can be characterized by examining small variations in the parameters; that is, large variations, more characteristic of ecological measurements, will not significantly alter parameter sensitivities. We can address this assumption directly, based on unpublished analyses of the marsh hydrology model. In separate Monte Carlo simulations, we assumed all parameters to have, firstly, a coefficient of variation of 1% and, secondly, variations characteristic of real field measurements. By comparing the partial correlation coefficients generated by these two runs, we can examine how larger variations alter sensitivity patterns. If the model were linear, the coefficients would be identical for the two runs. Therefore, differences should indicate the importance of nonlinearities when variances on parameters are large.

In general, the partial correlation coefficients are similar between the two sets of simulations, and the assumption appears valid for this model. The exceptions, however, were dramatic. Figure 2 shows the partial correlation coefficient between the field capacity of the soil (*FC*) and water level for the 1% case (dashed line) and the case in which all parameters were varied at realistic levels (solid line). Allowing all parameters to take on large variations obviously has an important effect on the correlation of model error to variability in this parameter (*FC*). In the 1% case, a significant fraction of the variance in water level is explained by variance in *FC*, particularly early in the year (days 0–40), during the summer (days 160–240), and following rainfall events (sharp peaks in the graph). With high variances on parameters, nonlinear responses of the model and, especially, nonlinear interaction terms between *FC* and other parameters, cause very little of the variance in water level to be explained by variations in *FC* alone. In general, when the coefficients diverge (Figure 2), sensitivity analysis indicates that the model is more sensitive to a parameter than would be indicated by a Monte Carlo analysis. In other words, it would overestimate the reduction in prediction uncertainty resulting from better measurements of a parameter.

In an analysis of a phytoplankton–zooplankton model (O'Neill et al., 1980; Gardner et al., 1980b), increasing the variance associated with individual parameters from 4% to 10% of their means results in complex changes in the predicted values. The deterministic system is characterized by an increase in populations in the spring, a decline during the summer months, a fall recovery, and a winter decline. Prediction error (sum of squared deviation from the deterministic solution) increases in the spring and fall when the populations are increasing rapidly (Figure 3). Because all populations are declining during the warm summer months, the errors at this point are at a minimum. Seasonality in prediction uncertainty has also been noted by DiToro and van Straten (1979).

Increasing the variability of the parameters from 4% to 10% changes the magnitude of variation and the pattern of variability throughout the year (Figure 3). Herbivore

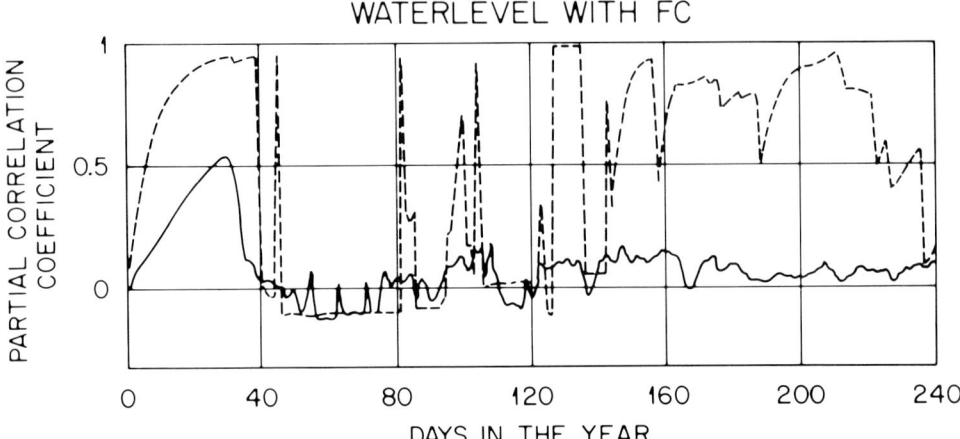

FIGURE 2 The partial correlation coefficient between *FC* (the estimated capacity for water of soil) and the predicted water level in a marsh system through 240 days of the year. The solid line is based on 500 iterations with parameters varied at realistic levels; the broken line is based on 500 iterations with parameters varied at 1%.

errors increased approximately ten-fold and carnivore errors approximately five-fold. Increase in parameter variability resulted in a smoothing and broadening of the peaks of variability, and, because the relationship between populations is nonlinear, the pattern of response of the fall carnivore populations is most affected. It should be noted that this analysis ignored covariances between parameters. There is ample evidence (DiToro and van Straten, 1979; O'Neill et al., 1980) that correlations decrease prediction uncertainty.

At the present stage of understanding, the assumptions of sensitivity analysis do not appear to cause serious problems. This is a consoling result because most analyses will be limited to this approach. However, some important exceptions to the rule were observed. The marsh study (Gardner et al., 1980a) showed that sensitivity analysis (1% variation) could lead to erroneous decisions. Based on the sensitivity analysis, one would be led to believe that model uncertainty could be significantly reduced by increasing the accuracy of a small subset of parameters. The error analysis (i.e., realistic variances) revealed that increased accuracy and precision in measuring this subset of parameters would have little practical effect on model error. Sensitivity analysis has many practical and theoretical uses for model examination, but caution must be exercised because the assumptions can be violated under some circumstances.

3.1 Establishing the Domain of Applicability of an Analysis

Unless consideration is given to the different sources of parameter variability (Figure 1), considerable ambiguity can be introduced into the interpretation of results. The inferences which can be drawn from any particular error analysis study are dependent on the selection of nominal values for the parameters and the definition of their

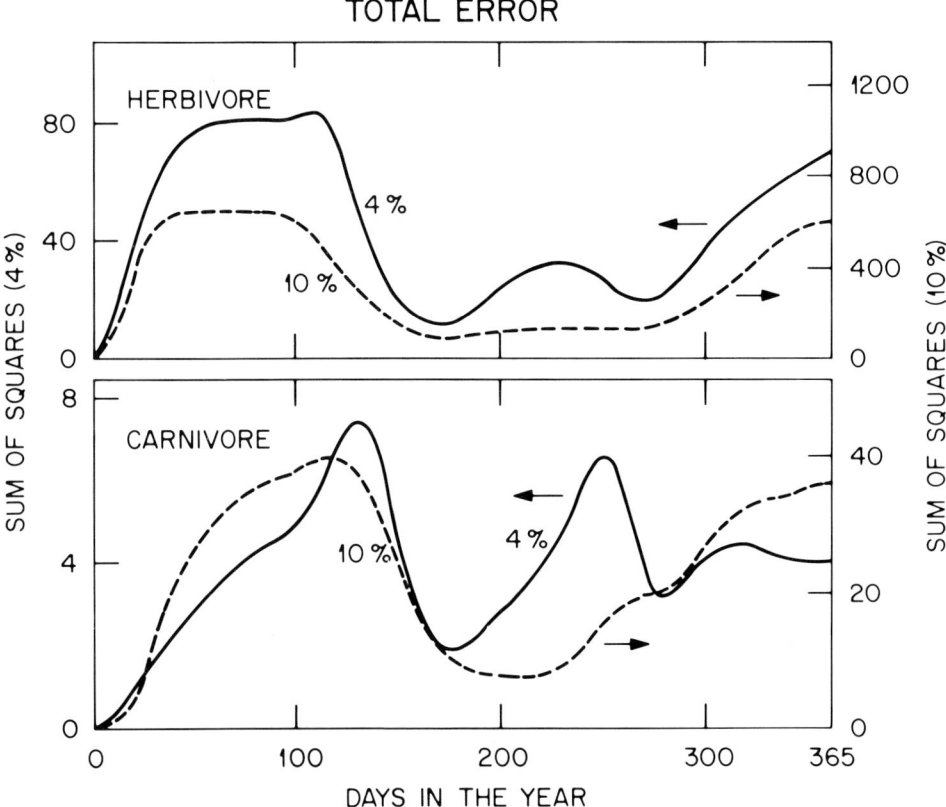

FIGURE 3 The daily total error of a coupled herbivore–carnivore system when parameters are varied at either 4% or 10% of their mean value. The total sum of squares is calculated as the daily mean of the sum of squares deviation from the deterministic system.

statistical properties (e.g., distributions, variances, maxima and minima, etc.). Any assumptions implicit in the choice of these statistical properties will strongly influence the inferences drawn from the analysis. If variances are measured for a specific system in a specific year, conclusions should not be drawn about other systems or future years. If study objectives call for inferences about a particular system, variances characteristic of an entire class of systems should not be used. The problems associated with the formulation of a particular case and inferences which can be drawn are illustrated by two recent studies.

In the analysis of the marsh hydrology model (Gardner et al., 1980a), the statistical distributions were assumed to be normal, with variances and extreme values estimated *a priori* from available information on marsh systems. One parameter, W, set the upper limit above which overland flow occurs and excess water drains rapidly into an adjacent lake. When this value is allowed to vary from iteration to iteration it determines directly the maximum water level of the marsh and, hence, the actual water level and storage when the soils are saturated. The initial investigation set the mean of W at 15.2, the variance at 48% of the mean, and the minimum and maximum at 0.0 and 30.0, respectively. The resulting relationship between W and predicted water level in the spring (March, April, and May) was quite strong, with W the second most significant contributor to water-level variability (R^2 being 44–77%).

This result had very little practical meaning for any single marsh. For an individual marsh system the variability of W would be smaller and result largely from spatial heterogeneity and measurement error. The uncertainty of the nominal value of W is obviously much less "within" a marsh than is that of the "between marsh" value used in the simulation. When the Monte Carlo experiment was repeated with the variance of W reduced from 48% to 1%, the statistical relationship between W and water level during saturated periods disappeared.

The second example concerns the frequency distribution of predicted dose to the thyroid of infants from a chronic release of radioactive iodine-131 (O'Neill et al., 1981). The model includes a Gaussian-plume atmospheric dispersion, movement of the radioisotope through the food-chain into milk, and the subsequent dose (resulting from ingestion) to the thyroid gland of infants. The extreme value for a particular parameter, B, which describes the transfer of radioisotope from the soil to pasture forage, proved to be troublesome. The parameter B is dependent on the nature of the soil and is quite variable between sites. In addition, there are few measured values of B. Variability of B at a local site is restricted to the variability of the local soils within the area and yet we can only characterize the universal variability of B across all soils. This point was not apparent when we began our investigation but we soon realized that allowing the maximum value of B to change by orders of magnitude from 1 to 1000 resulted in a change in the correlation of B with dose from 0.06 to 0.32 and a shift in the mean and maximum predicted dosages as well (Table 1). Like W in the marsh model, B has a small effect on the predicted value when variances characteristic of one site are used, but when either the region is large or knowledge of the parameter is insufficient, then the results must be cast in a different light.

Meaningful insights must come from a meaningful definition of parameter values and distributions. The approach must be applied uniformly across all parameters of the model, otherwise meaningful relationships between parameters and predictions will be obscured. It is most helpful to define first the level of resolution of the model, then define the statistical properties of parameters within that framework.

3.2 Frequency Distributions of Parameters

Any real system (Figure 1c) contains natural variability, and system behavior is most realistically represented as a frequency distribution of potential behaviors. The distribution of system behaviors is the result of the mathematical characteristics of the model and the distributions of model parameters. The factors used to select a parameter distribution should include the probabilistic properties of specific processes in the model, the empirical distribution of available data, and any information on the expected distribution of system behaviors. In our experience this information is seldom available, even for a portion of the parameters of the model. Approximations must be made based on the best available information (Morgan et al., 1978).

TABLE 1 Relationship between the maximum value of B and the predicted dose to thyroid of infants from chronic release of radioactive iodine-131[a].

B_{max}	Predicted dose (rem year^{-1})			r[b]
	Mean	95 percentile	Maximum	
1	0.87	2.9	9.4	0.06
10	0.88	2.9	9.6	0.07
100	1.03	3.4	11.5	0.14
1000	2.47	9.0	53.6	0.32

[a] Values of B were generated from a triangular distribution with minimum value equal to 0.0, expected value of 0.2, and maximum value as indicated in the table. Each row summarizes the results of 1000 Monte Carlo iterations. The expected value from the deterministic system is 0.72 rem year^{-1}.
[b] r is the simple correlation coefficient between B and the predicted dose.

Under such circumstances we concur with Tiwari and Hobbie's (1976) recommendation that the triangular distribution be selected. The few parameters necessary for this distribution (mode, maximum, and minimum) can usually be inferred from the physical process under investigation. Tiwari and Hobbie point out that the choice of any other distribution involves additional assumptions. The triangular distribution is the least biased assumption under these conditions. In addition, under many circumstances, the results generated by the triangular distribution resemble results using more complex distributions.

For purely analytical studies (for example, no statements about real confidence limits are expected), we prefer normal distributions because covariances can be specified with relative ease and the symmetrical distribution of parameters aids in interpreting the often skewed frequency distributions of predicted values since, in this case, the skewed distribution must be due to the mathematics of the model. The effects of altering the distributions of parameters by changes in the variance, by specifying covariance terms, or by selecting another distribution, can alter the frequency distribution of predicted results. For instance, in a multiplicative chain model (prediction is calculated as the simple product of a number of coefficients and variables), prediction errors can be calculated analytically (Shaeffer, 1979) if parameters are lognormally distributed. The choice of a lognormal distribution has been justified for a number of reasons, including the fact that extreme values are more likely and predicted frequency distributions will be conservative. However, the shape of the frequency distribution of predictions is largely determined by the mathematics of the model rather than by the assumed distribution of model parameters. The mean and variance of the output distribution are affected by the choice of the particular parameter distribution.

The effect of altering the parameter distributions is illustrated by holding the parameter means and variances constant, but changing the frequency distributions from lognormal to normal distributions for the radiation dose model (Shaeffer, 1979; O'Neill et al., 1981). When parameters are lognormally distributed, the mean, the 95 percentile, and the maximum value of dose from 1000 iterations are 0.86, 2.9, and 9.5 rem year^{-1}, respectively. When the parameters are normally distributed, the mean value is 1.4, the 95 percentile 4.5, and the maximum 109.5 rem year^{-1} (calculations based on an arbitrary release of 1 Curie of iodine-131 per year).

Examination of lognormal distributions with different variances (Hahn and Shapiro, 1968) shows that as variance increases, the peak of the distribution shifts farther to the left, i.e., more values lie below the mean. The complementary normal distribution is symmetrical, and a proportionately larger fraction of the values lies above the mean. The practical result is that the normal distribution predicts a slightly higher mean dose and a much higher extreme dose. For this example, the choice of distribution affects the mean, variance, and extreme value, and a choice of lognormal distribution for conservative results is a poor one. However, no matter what parameter distribution is used, the predictions appear to be lognormally distributed.

Another factor which affects the frequency distribution of predicted values is the method chosen to simulate the problem. We recently studied density-independent Leslie models of striped bass populations which predict abundances in each of 15 age classes for 40 years, based on age-specific fecundities (F_i) and survival (P_i) parameters.

Leslie matrix models (Leslie, 1945, 1948) tend to predict population behavior that approaches infinity or zero through time. The probability of choosing random parameter values for the matrix that will result in a stable age distribution is very small. One way of forcing the model to produce a stable result is to calculate the survival from eggs to young-of-the-year, P_0, based on the remaining parameters (Van Winkle et al., 1978). If parameters are chosen randomly only at the beginning of a Monte Carlo run, if P_0 is calculated to ensure a stable age distribution, and if the parameters remain unchanged for the 40 years of the simulation, the predicted mean population size is equal to the deterministic solution of the model and the coefficient of variation is 5.9%. If parameter values are chosen each year of the simulation (and P_0 recalculated each year), the mean is 35% greater than the deterministic solution and the coefficient of variation is 19%. The upward shift of the mean results from the continual adjustment of P_0 to reach a new, stable age distribution based on current parameter and age distribution.

If we make P_0 a random variable, like the other parameters of the model, a stable age distribution is no longer guaranteed. Now the predicted mean population size is 140% greater than the deterministic solution and the coefficient of variation is 254%! The shift to a higher mean value results from exponential increase in the population for some parameter combinations. This exponential increase can result in very large populations for a few iterations, resulting in an increased mean.

The fish population model also reflects an important advantage of the Monte Carlo approach. The simulation in which all parameters, including P_0, are assumed to vary randomly, predicts an increased mean population size. Using an analytical approach which reflects only the shift in the mean, this result indicates that the fish population would be larger than the deterministic prediction. The Monte Carlo simulation produces a complete frequency distribution of predictions. It is clear from this distribution (Figure 4) that the most likely result is a reduction in population size. In this case, the mode of the distribution is of greater significance because the mean is strongly influenced by a few large numbers. The frequency distribution reveals that most populations will lie well below the mean value. In this respect, dealing only with the means and variance is deceptive, and the frequency distribution of predicted behaviors is needed to arrive at a reasonable understanding of the result.

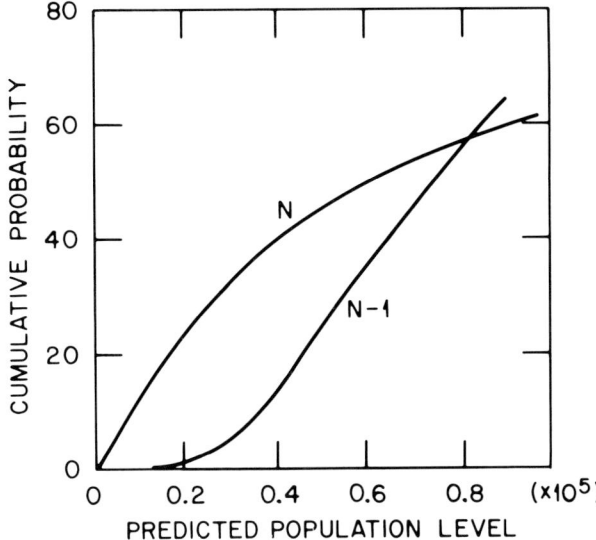

FIGURE 4 The cumulative probability of predicted levels of a Leslie model of striped bass populations when all parameters of the model are varied at 10% (N case) or when P_0 (the probability of survival from eggs to young-of-the-year) is calculated for each iteration ($N-1$ case). Each line represents the results obtained from 1000 iterations.

4 CONCLUSIONS

Much of this paper has been concerned with showing that prediction error is a complex phenomenon that requires careful analysis in order to avoid confusion in interpreting results. Figure 1 shows that natural variability in an ecosystem results in a bias in any deterministic model of the system. Changing the variances on parameters or changing the frequency distributions will affect conclusions drawn from the analysis. Particular attention must be paid to any implicit assumptions involved in the selection of statistical properties of parameters. If parameter distributions are representative of an entire class of ecosystems, results will not be characteristic of prediction error applied to a specific site.

In some cases, the purposes of the study will not be satisfied by simply stating a confidence interval around model predictions. In the fish population study, the entire frequency distribution of predictions was required to recognize that it was the mode and not the mean value that was of greatest interest. In our analysis of the radiation dose model, the probability of predicting an extremely high dose is of greater potential importance than predicting the mean.

It is clear that the current state of information about prediction uncertainty for ecosystem models is inadequate and in a state of rapid change. It would appear unwise at present to advocate any single technique for error estimation to the exclusion of other possible approaches. The Monte Carlo approach has distinct advantages during the present exploratory stages because it is not limited to any particular set of assumptions about the nature of errors or their magnitude. The Monte Carlo approach may not always be an efficient method for estimating error bounds on a prediction, but it may well be the most effective approach for exploring the mechanisms involved in propagating uncertainty and the factors involved in minimizing and controlling these uncertainties.

ACKNOWLEDGMENTS

This research was sponsored by the National Science Foundation under interagency agreement 40-700-78 with the US Department of Energy under contract W-7405-eng-26 with Union Carbide Corporation.

REFERENCES

Argentesi, F. and Olivi, L. (1976). Statistical sensitivity analyses of a simulation model for the biomass–nutrient dynamics in aquatic ecosystems. Proceedings of the 4th Summer Computer Simulation Conference. Simulation Councils, La Jolla, California, pp. 389–393.
Beck, M.B. (1979) Model structure identification from experimental data. In E. Halfon (Editor), Theoretical Systems Ecology. Academic Press, New York, pp. 259–289.
Beck, M.B., Halfon, E., and van Straten, G. (1979). The Propagation of Errors and Uncertainty in Forecasting Water Quality; Part I: Method. WP-79-100. International Institute for Applied Systems Analysis, Laxenburg, Austria.
Cale, W.G. and Odell, P.L. (1979). Concerning aggregation in ecosystem modeling. In E. Halfon (Editor), Theoretical Systems Ecology. Academic Press, New York, pp. 55–77.
Caswell, H. (1976). The validation problem. In B.C. Patten (Editor), Systems Analysis and Simulation in Ecology, Volume IV. Academic Press, New York, pp. 313–325.
DiToro, D.M. and van Straten, G. (1979). Uncertainty in the Parameters and Predictions of Phytoplankton Models. WP-79-27. International Institute for Applied Systems Analysis, Laxenburg, Austria.
Gardner, R.H., Huff, D.D., O'Neill, R.V., Mankin, J.B., Carney, J., and Jones, J. (1980a). Application of error analysis to a marsh hydrology model. Water Resources Research, 16(4):659–664.
Gardner, R.H., O'Neill, R.V., Mankin, J.B., and Kumar, D. (1980b). Comparative error analysis of six predator–prey models. Ecology, 61(2):323–332.
Hahn, G.J. and Shapiro, S.S. (1968). Statistical Models in Engineering. Wiley, New York.
Halfon, E. (1975). The systems identification problem and development of ecosystem models. Simulation, 38: 149–152.
Harrison, G.W. (1978). Compartmental lumping in mineral cycling models. In D.C. Adriano and I.L. Brisban (Editors), Environmental Chemistry and Cycling Processes. CONF-760429. National Technical Information Service, Springfield, Virginia, pp. 121–137.
Huff, D.D. and Young, H.L. (1980). The effect of a marsh on runoff: I. A water-budget model. Journal of Environmental Quality, 9:633–640.
Leslie, P.H. (1945). On the uses of matrices in certain population mathematics. Biometrika, 33: 183–212.
Leslie, P.H. (1948). Some further notes on the use of matrices in population mathematics. Biometrika, 35: 213–245.
Lettenmaier, D.P. and Richey, J.E. (1979). Use of first order analysis in estimating mass balance errors and planning sampling activities. In E. Halfon (Editor), Theoretical Systems Ecology. Academic Press, New York, pp. 79–104.
Mankin, J.B., O'Neill, R.V., Shugart, H.H., and Rust, B.W. (1977). The importance of validation in ecosystem analysis. In G.S. Innis (Editor), New Directions in the Analysis of Ecological Systems. Simulation Councils, La Jolla, California, pp. 63–71.

Morgan, M.G., Morris, S.C., Meier, A.K., and Shenk, D.L. (1978). A probabilistic methodology for estimating air pollution health effects from coal-fired power plants. Energy Systems Policy, 2: 287–310.

O'Neill, R.V. (1973). Error analysis of ecological models. In D.J. Nelson (Editor), Radionuclides in Ecosystems. CONF-710501. National Technical Information Service, Springfield, Virginia, pp. 898–908.

O'Neill, R.V. (1979a). Natural variability as a source of error in model predictions. In G.S. Innis and R.V. O'Neill (Editors), Systems Analysis of Ecosystems. International Cooperative Publishing House, Fairland, Maryland, pp. 23–32.

O'Neill, R.V. (1979b). Transmutations across hierarchical levels. In G.S. Innis and R.V. O'Neill (Editors), Systems Analysis of Ecosystems. International Cooperative Publishing House, Fairland, Maryland, pp. 59–78.

O'Neill, R.V. and Gardner, R.H. (1979). Sources of uncertainty in ecological models. In B.P. Zeigler, M.S. Elzas, G.J. Klir, and T.I. Oren (Editors), Methodology in Systems Modelling and Simulation. North-Holland Publishing Company, Amsterdam, pp. 447–483.

O'Neill, R.V. and Rust, B. (1979). Aggregation error in ecological models. Ecological Modelling, 7: 91–105.

O'Neill, R.V., Gardner, R.H., and Mankin, J.B. (1980). Analysis of parameter error in a nonlinear model. Ecological Modelling, 8: 297–311.

O'Neill, R.V., Gardner, R.H., Hoffman, F.O., and Swartz, G. (1981). Effects of parameter uncertainty on estimating radiological dose to man – a Monte Carlo approach. Health Physics, 40:760–764.

Park, R.A., O'Neill, R.V., Bloomfield, J.A., Shugart, H.H., Booth, R.S., Goldstein, R.A., Mankin, J.B., Koonce, J.F., Scavia, D., Adams, M.S., Clesceri, L.S., Colon, E.M., Dettmann, E.H., Hoopes, J., Huff, D.D., Katz, S., Kitchell, J.F., Kohberger, R.C., LaRow, E.J., McNaught, D.C., Petersen, J., Titus, J., Weiler, P.R., Wilkinson, J.W., and Zahorcak, C.S. (1974). A generalized model for simulating lake ecosystems. Simulation, 23:33–50.

Shaeffer, D.L. (1979). A model evaluation methodology applicable to environmental assessment models. ORNL-5507. Oak Ridge National Laboratory, Oak Ridge, Tennessee.

Tiwari, J.L. and Hobbie, J.E. (1976). Random differential equations as models of ecosystems: Monte Carlo simulation approach. Mathematical Bioscience, 28: 25–44.

van Straten, G. and de Boer, B. (1979). Sensitivity to uncertainty in a phytoplankton–oxygen model for lowland streams. WP-79-28. International Institute for Applied Systems Analysis, Laxenburg, Austria.

Van Winkle, W., DeAngelis, D.L., and Blum, S.R. (1978). A density-dependent function for fishing mortality rate and a method for determining elements of a Leslie matrix with density-dependent parameters. Transactions of the American Fisheries Society, 107(3): 395–401.

A MONTE CARLO APPROACH TO ESTIMATION AND PREDICTION

Kurt Fedra
International Institute for Applied Systems Analysis, Laxenburg (Austria)

1 INTRODUCTION

The model representation of complex environmental systems requires numerous simplifications; frequently, arbitrary choices of how to formally represent the relationships between causes and effects have to be made, since these relationships are neither obvious nor easy to detect. Environmental systems *in toto* do not easily yield to the classical scientific tool of planned experimentation. Consequently, the analyst has to utilize whatever bits of information may be available, which as a rule are very few and not strictly appropriate in terms of the problems addressed. *A priori* knowledge about the structure and function of any ecosystem is generally poor, and reliable quantitative information on the governing processes and their rates and interrelationships insufficient. Consequently, building and testing models and finally applying them for predictive purposes often consists of a more or less formalized trial-and-error iterative process of estimation, testing, and improvement. The following discussion proposes an approach for formalizing this process of model building, calibration, and application; it emphasizes the interdependences of the individual steps in the process. The approach proposed is based on the recognition of uncertainty as an inevitable element in modeling, and uses straightforward Monte Carlo techniques to cope with this uncertainty.

2 SOURCES OF UNCERTAINTY

2.1 System Variability

Ecosystems are diverse, complex (see, for example, Pielou, 1975), and mostly large-scale systems. The number of component elements is usually extremely high, and the relationships among these elements are complex. They are characterized by a rich behavioral repertoire, are variable in time and highly structured in space (see, for example, Steele, 1978), they are driven by (generally unpredictably) fluctuating external conditions, and exhibit complex feedback and control mechanisms (see, for example,

Conrad, 1976; Straskraba, 1976, 1979) such as adaptation and self-organization. Most functional relationships in such systems are nonlinear and time-variable, and even the boundaries of the system must in many cases be defined quite arbitrarily. When attempting formal description and representation, numerous sources of uncertainty can be identified (see, for example, Beck et al., 1979; DiToro and van Straten, 1979; O'Neill and Gardner, 1979; O'Neill and Rust, 1979; Reckhow, 1979; Fedra et al., 1981). Summarizing, ecosystems seem to be just about the least desirable subjects for deterministic mathematical modeling!

2.2 Theoretical Background

All the above features are well reflected in the theoretical background of systems ecology. There is no well-established, unifying theory in systems ecology. At best, one can find a mosaic of unrelated concepts and approaches (see, for example, Halfon, 1979). Quite often, ecological theories (or rather hypotheses) are contradictory. The processes governing ecological systems are generally poorly understood, especially at a high "systems level" of organization (or rather abstraction) — the level used in systems modeling. This is due at least in part to the fact that much of the available information stems from microscale laboratory experimentation. Usually, in such physiologically oriented experiments, all but one (or a few) variables are kept constant, and the response of the system (usually an individual organism or a monoculture) to changes in one external condition is observed. Such experiments are difficult to interpret at the "ecosystem level", where nothing is constant, everything affects (almost) everything else, and the "unit" of interest is a functionally heterogeneous, diverse, adapting, multispecies, multiage and size-class, more or less arbitrarily lumped aggregate. Generally, the empirical basis or the data available are singular measurements, so that their reliability in terms of the spatial or functional macrolevel used in the model cannot be estimated. Consequently, ruling out or rejecting any hypothesis put forward is rather difficult (Fedra, 1981a), and in fact, examples of more than one contradictory hypothetical construct, each "possible" in terms of the data to be described, are known (Nihoul, 1975; Bierman et al., 1981). However, as *a priori* knowledge about a system is essential for the first steps in model-building, the lack of reliable and unambiguous knowledge adds considerable uncertainty to the problem.

2.3 Environmental Data Base

All the above features are, again, reflected in the data available on environmental systems. Not only do spatial and temporal variability make data collection under logistic constraints an art rather than a scientific procedure, but in many cases it is simply impossible to sample or measure what is described (conceptualized) in a model. Most state variables used in model descriptions are more easily represented in a flow diagram than measured, as the level of abstraction in the model representation is completely inaccessible to direct measurement. Consequently, ecological data are scarce, scattered, distorted by sampling error, and usually only exist for the "wrong" variables in any given numerical analysis. Monitoring programs, as a rule designed independently of subsequent evaluation and analysis, traditionally tend to concentrate on what other monitoring programs have included. And as only theory can tell the observer or experimenter what to measure (an only *seemingly* trivial truth ascribed to Albert Einstein), the "wrong" variables are measured. Also, different variables tend to get measured at different places and at different times. Even the most ambitious, money-consuming attempts at data collection like the IFYGL do not result in the smooth, unambiguous curves one would (probably rather naively) hope to find (compare Scavia, 1980a).

2.4 Model Uncertainty

Mathematical models designed to describe and simulate environmental systems cover a wide range of detail and complexity: they range from very simple statistical black-box models (see, for example, Vollenweider, 1969, 1975) to the "all-inclusive", multicompartment, spatially disaggregated, physical or "explanatory" model (see, for example, Park et al., 1974). But even for the most detailed and spatially disaggregated models, elements or compartments treated as being homogeneous (either in space or functionally), are very large when compared to the sampling units from the field or experiment, and are highly aggregated (see O'Neill and Rust, 1979, on the subject of aggregation errors). What models really describe are extremely simplified conceptualizations of the real-world system, which are very difficult to relate directly to the point-samples from these systems. Models and data are on two different levels of abstraction and aggregation, and therefore traditional data from a spatial or functional microlevel can hardly be used directly. Instead, from the available data one can try to derive information about the system studied at an appropriate level of abstraction, for comparison with the respective model equivalents. Ideally, the measurements should be made directly at the appropriate level, but some of the more promising techniques in environmental data collection are still in their infancy, at least as far as scientific applications are concerned (see, for example, Gjessing, 1979).

Simulation models consist of numerous, more or less arbitrary assumptions which are made about certain relationships within the system, about boundary conditions, and about the "meaning" of data in terms of the model and *vice versa*. Many authors admit that their assumptions are arbitrary, that their simplifications are gross, and that, by necessity, they ignore some more detailed (and confusing) knowledge. However, the effects of such assumptions on the reliability and usefulness of a model are rarely examined. Instead, the meta-assumption, "our assumptions will not affect the results significantly", is often made. There seems to be little doubt that such models contain a high degree of uncertainty.

3 AN ALTERNATIVE APPROACH

Considering all the above sources of uncertainty, the traditional, deterministic mathematical approach to modeling does not seem to be an appropriate tool to cope reliably with complex, environmental, real-world problems. One ought at least to explore the effects of uncertainty on the reliability and usefulness of model applications.

If one recognizes that the entities used in a simulation model and those measured in the field or in a laboratory experiment are quite different, it is obvious that they cannot be compared directly or used to estimate one from the other without taking into account these differences and the resulting uncertainty. Since the model, due to its high degree of abstraction, simulates average patterns or general features of a system (as conceptualized in the model), it is necessary to derive these patterns and estimate these features at an appropriate level of abstraction and aggregation from the measured data. Only such derived values should be compared with the terms generated using the model, in order to test and improve model performance. For a discussion of the concept of "problem-defining behavior" see Hornberger and Spear (1980) and Spear and Hornberger (1980).

If we begin the discussion with problems of model-structure identification (in a rather general and inclusive sense (cf. Beck, 1979a,b)), it should be recognized that any model structure proposed for a complex system will itself be a complex, composite hypothesis which has to be tested. Because of the very high number of interactions between the numerous elements of ecological systems, considerable conceptual simplification is needed to make the theories formulated about the structural properties and functions of the systems traceable, interpretable, and useful. Universal statements describing those properties of a system which are invariant in space and time may be called models, whether they are based on informal verbal or mental descriptions or on formalized mathematical structures. These scientific theories, or models, have to be testable; that is to say, when one puts a set of specific singular statements (the initial conditions, which, in the case of a mathematical model, also include the model parameters in a general sense (cf. Fedra et al., 1981; Fedra, 1982)) into the model, it must be possible to deduce or predict testable singular statements (observations or experimental results). Disagreement between the prediction deduced from the hypothesis or model and the available observations would then make it necessary to reject the given hypothesis, to modify and improve it, or to look for alternative hypotheses, which would then be subjected to the same procedure. This method, representing one strategy of scientific research proposed by Popper (see, for example, Popper, 1959) has, however, one major drawback when applied to complex simulation models or dynamic hypotheses describing ecological systems: namely, the so-called initial conditions used with the basic structure of the theory to deduce the testable predictions are not known exactly. Nonunique inputs, however, will produce nonunique outputs (see, for example, Tiwari et al., 1978). This certainly could be viewed as the result of two basic shortcomings, one in the measurement techniques available, and the other in the formulation of the models themselves — if the models require unknowns as inputs, they are not well formulated. The latter is certainly a generic shortcoming of ecological models, or of ecological theory in general.

The same line of argument can be followed with regard to the observations used for model-output comparison in hypothesis testing. The degree of abstraction and aggregation is quite different in the measurements and in the model conceptualization, so that the measurements can only serve as samples of the properties of the units conceptualized. As these units are generally heterogeneous (in terms of their measurable properties), and are generally characterized by a high degree of variability, that is to say, the repeatable part of the observations is only a certain range, further uncertainty has to be dealt with in the hypothesis-testing procedure. A formal concept of "disagreement" under uncertainty has yet to be formulated (Fedra, 1982). But whatever the objective for a formal approach to the analysis of a complex, dynamic environmental system may be, the testability of the models involved is an essential criterion when evaluating them as useful scientific tools. Testability, however, has to be achieved with the information available, that is to say with ranges and semiquantitative relations, if models are to be used at all. There is no scientific way of identifying the "best" model structure for a given system under uncertainty. Nonuniqueness on both sides of the testing procedure will always result in nonunique answers. All that can be done on a rigid formal basis is to rule out grossly inadequate model structures. Since initial conditions and reference behavior for the test are uncertain, they can both be specified in terms of acceptable ranges for the plausible inputs and acceptable outputs. The test is then as follows: is there at least one set of plausible input conditions which will result in an acceptable model response? The criterion is certainly one-sided and weak; however, it implies a minimum of implicit arbitrariness in a formal approach.

Given a satisfactory model structure which passes the above test of adequacy, the next step is to explore the full range and structure of the admissible initial conditions, which are largely parameters in classical terminology. Estimating appropriate parameter values is generally referred to as model calibration. To calibrate and run any given simulation model, one needs a set of numbers (the parameters, forcings and imports, and the initial conditions) to be put into the model and a set of numbers to compare with the model output. The comparison, together with a recursive tuning of the parameters (by whatever method) may be called calibration (see, for example, Lewis and Nir, 1978; Benson, 1979). To specify the above numbers, first one must understand their meaning (in terms of the real-world system or the measurements derived from this system) as conceptualized in the model. As discussed above, these numbers are average, aggregate features of the system, so that the available singular measurements can only be used as a first, rough approximation in the estimation procedure. Considering uncertainty, the data describing the system behavior (in terms of the model output) can be specified as ranges (Hornberger and Spear, 1980; Spear and Hornberger, 1980) or, given enough information about the system, probability distributions. As the model represents average and general aspects of system behavior, a data set for more than one year might be used (in the absence of obvious trends), so that the basic data set will not only contain sampling and measurement uncertainty, but also evidence of the variability of the system over time. From the empirical data base, a formal definition of the system behavior is derived in terms of ranges, for measures such as states and process rates or flows at a given point in time, as well as derived relational or integrated properties. This description of the system can be understood as a region in a hyperspace, where each measure describing the system behavior defines one dimension. Of course, the kind of measures to be used depends on what is described in the model as well as on the available information about the system.

Given this reference behavior for the estimation of model parameters, it is obvious that more than one parameter set (including the initial conditions and the coefficients used to parameterize time-variable forcings, which can also be viewed as parameters) will generate a model response within the behavior-space region taken to describe the system. Again, the concept of a parameter hyperspace can be useful. The estimation procedure now tries to identify that parameter-space region where the corresponding model response is within the defined behavior-space region. This parameter-space region, or ensemble of parameter sets with its characteristic variance–covariance structure, reflects the basic uncertainty of the (deterministic) modeling exercise. Using such an ensemble of parameter sets for predictions results in an ensemble of forecasts, where the variability of the forecast is an effect of the initial uncertainty and can be understood as an estimate of the reliability of a prediction. This approach also demonstrates the intimate coupling between "estimation" (which I prefer to use rather than the somewhat misleading term calibration, which implies some objective reference point) and prediction.

3.1 The Concept of Allowable Ranges

The two sets of numbers to be specified in testing the model structure and using the model for predictions – one describing the "expected" model behavior, the other the parameters of the model – can only be estimated roughly from the information available on the system studied, if at all. However, the raw data available will be quite different from case to case, depending on the complexity and variability of the system, and the amount and quality of the measurements. At best, time series of physical, chemical, and biological variables will be available, together with an estimate of their distribution and sample statistics within the spatial elements of the model. In addition, some independent experimental data on process rates may be available. However, generally only singular measurements are available, so that no estimation of their reliability in terms of the model's spatial and functional aggregation is possible. Also, the available measurements often do not relate exactly to the state variables of the model; for example, it may be necessary to simulate algae populations in terms of phosphorus, but if measurements are only available in terms of chlorophyll then dubious (and quite arbitrary) conversions may be needed. Also, the time intervals between measurements are often large (and traditionally constant), so that more transient dynamic features are rarely detected.

Nevertheless, the available information generally allows for the specification of ranges within which any of the observed features can reasonably be expected. Such "allowable ranges" can be formulated for the behavior-describing data – for example, the value or timing of the spring algae peak – as well as for the initial conditions, the forcings, and the parameters. Wherever such allowable ranges cannot be derived from the specific set of data available, additional *a priori* information from similar systems described in the literature, or simply ranges defined by physical laws or ecological plausibility can be used. Certainly, such ranges will not be able to describe an individual system unambiguously, but they will help to constrain the model response to realistic, or rather plausible, regions. Calibrating a model using state variables only may well lead to seemingly reasonable results (in terms of the state variables), but at the expense of unrealistic process rates (see, for example, Scavia, 1980b).

4 TESTING THE MODEL STRUCTURE: A MARINE ECOSYSTEM EXAMPLE

As an example to illustrate the first step in the approach, a data set from the southern North Sea was used. Most of the information stems from the yearly reports of

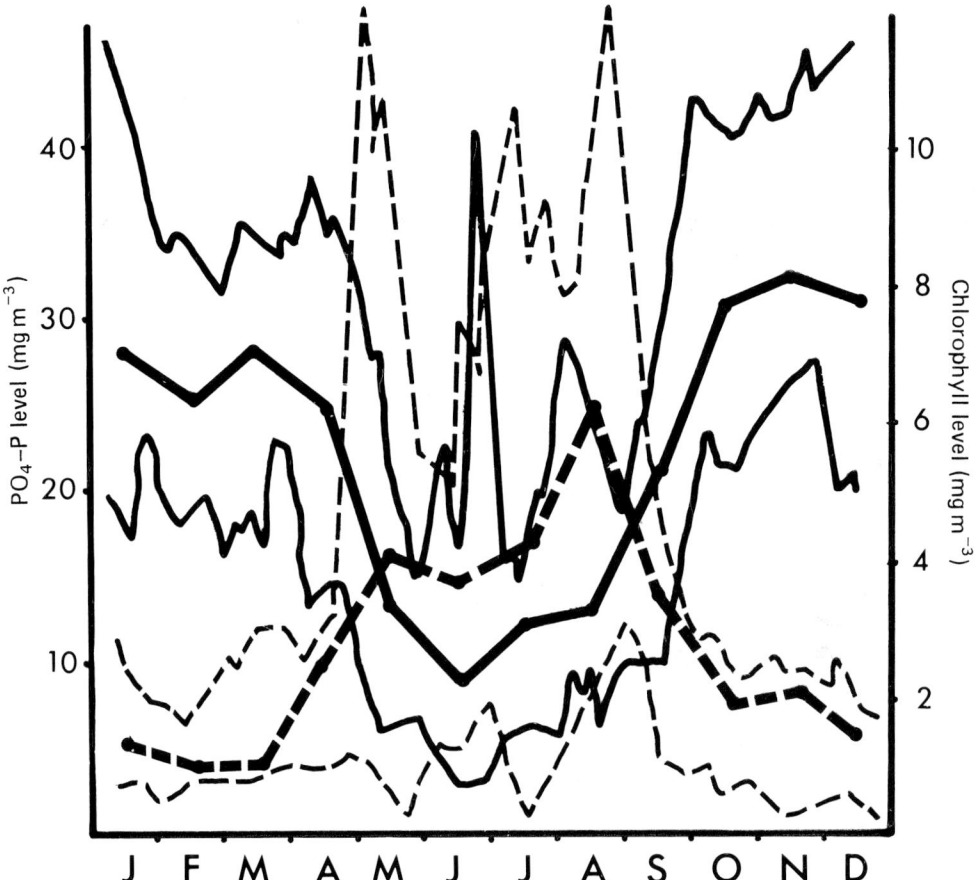

FIGURE 1 Systems behavior (German Bight, North Sea): phosphorus dynamics (PO_4-P), solid lines, from 1964 to 1979, thick line indicates monthly averages for the years 1965 to 1975; and chlorophyll dynamics (broken lines), from 1964 to 1979, thick line indicates monthly averages for the years 1965 to 1975. After unpublished data from: Weigel and Mangelsdorf; Harms; Harms and Hagmeier; Harms, Mangelsdorf, and Hagmeier; Mangelsdorf; Weigel, Hagmeier, and Treutner; and Hagmeier, Kanje, and Treutner.

the Helgoland Biological Station, and describes physiochemical as well as biological variables at the sampling station "Helgoland-Reede" for the period 1964–1979 (Biologische Anstalt Helgoland, 1964–1979, including otherwise unpublished data of Hagmeier, Hickel, Mangelsdorf, Treutner, Gassmann, and Gillbricht; Hagmeier, 1978; Lucht and Gillbricht, 1978). However, various other sources have been used for additional information (for example, Steele, 1974; Nihoul, 1975) to compile a data set typical of an arbitrary location representative of the German Bight, southern North Sea.

Figure 1 gives an example of the data used. The driving environmental variables water temperature and radiation were sufficiently smooth and well behaved for direct utilization of their long-term averages, approximated by simple sine waves. Data for nutrients (PO_4–P) and algae (measured as chlorophyll as well as in terms of carbon, recalculated from counts) showed consistent yearly patterns. However, when including the year-to-year variations (as well as the implicit sampling errors), the high variability of the observations and the difficulty in averaging over time (several years) become obvious. Although the average phytoplankton dynamics show a single but extended peak around July/August, the individual years exhibit at least two peaks in the summer, which, because of their variable timing, are averaged out when looking at the long-term mean (Figure 1). Also, the long-term mean is about one order of magnitude below the spiky peaks of the individual year's data. Little information was available on zooplankton biomass values. However, some additional information from independent experiments, mainly on primary production, was available.

Certain nonvariable and general features could be derived from the observations; these are formulated in terms of the "allowable ranges" discussed above:

— Primary producers remain below a chlorophyll level of $4\,\text{mg m}^{-3}$ during the first three months of the year; between days 120 and 270 of the calendar year there is an increase of at least twofold in biomass.
— At least two biomass peaks occur during this latter period, with a reduction of at least 25% of the first peak value between the two peaks.
— After day 270, biomass again remains below a chlorophyll level of $4\,\text{mg m}^{-3}$.
— The higher of the two peak values does not exceed a chlorophyll level of $25\,\text{mg m}^{-3}$.
— Yearly primary production falls within the range 300–$700\,\text{g carbon m}^{-2}$.
— The first biomass peak value (defined as an increase of at least twofold over initial biomass before a subsequent decline) is reached later for herbivorous consumers (zooplankton) than for phytoplankton.
— The maximum density of herbivorous consumers does not exceed $1000\,\text{mg carbon m}^{-3}$.
— The level of PO_4–P stays above $20\,\text{mg m}^{-3}$ between calendar days 1 and 90; on average, it stays below $20\,\text{mg m}^{-3}$ between days 120 and 240; after day 270 it returns to values above $20\,\text{mg m}^{-3}$. Throughout the whole year the PO_4–P level does not move outside the range 2–$50\,\text{mg m}^{-3}$.
— All state variables must be cyclically stable (with a ±20% tolerance range).

This description of the observed system features, defining a region in the behavior hyperspace of the system, should be understood as a rough and at best semiquantitative description of persistent patterns rather than a quantitative description of the system for any specific period. Certainly, more resourceful analysis of the available data and the incorporation of additional information would allow this description to be refined.

5   HYPOTHESIS GENERATION: DESIGNING ALTERNATIVE MODELS

There are several implicit assumptions hidden in the way the data are interpreted and the description derived. Ignoring short-term spatiotemporal variations (e.g., those caused by the tides) and looking instead at average features implies that we are considering a hypothetical body of water that is not absolutely fixed in space. The horizontal extension of this body of water is rather arbitrarily limited by the requirement of homogeneity within this spatial element. In the vertical, the body of water considered is defined by the extent of the measurements used, but again homogeneity has to be assumed. At the lower boundary, an "endless sink" of constant chemical properties is assumed, that is to say, one which is very large compared to the productive upper layer, and exchange between the upper layer and this sink is controlled by eddy diffusivity.

All these assumptions are more or less unrealistic if we think in terms of specific physical units in time and space. However, their precise description is not the aim of our modeling. The basic idea behind all our assumptions is that the simplified processes considered largely dominate the behavior of the conceptual system and that the processes ignored are relatively unimportant.

5.1 Hypothesis 1: Two Compartments in a Simple Physical Framework

An attempt will now be made to formulate one very simple hypothesis about the pelagic food web described in the data set above. The system is conceptualized as consisting of only two compartments, namely particulate, photosynthesizing organic matter, and mineral nutrients, which are coupled by the processes of primary production and nutrient uptake, mortality, and respiration/mineralization; one implicit assumption is that nonphotosynthesizing organic matter is in a constant proportion with the living fraction. The system is driven by light and temperature, and by turbulent mixing (eddy diffusivity). Controlling mechanisms are light limitation and nutrient limitation of primary production, self-shading of algae, and the temperature dependence of all the biological processes. Figure 2a shows a diagrammatic representation of this system.

The model description uses Monod kinetics to describe nutrient limitation of primary production, using a constant half-saturation concentration. Light limitation is described using the double time–depth integral of DiToro et al. (1971) for Steele's (1962) equation (for a discussion of the implications of this formulation see Kremer and Nixon, 1978). Mortality is described as a nonlinear, concentration-dependent function of algae biomass, and is directly coupled to remineralization, without any time lag or further control. Mixing with a "deep layer" is described as the exchange of a constant fraction of the volume of the upper layer (the top 10 m), where the PO_4–P concentration of the deep layer equals the initial (winter) concentration of the upper layer, and the algae concentration of the deep layer is zero, that is to say, algae can only be lost from the system. The rate of mixing is varied by a step function, triggered by temperature, such that the initial high (January) value is reduced to one-tenth of the initial value as soon as the surface temperature reaches three times its starting level; the mixing rate is reset to the initial high value as soon as the surface temperature drops below the trigger level.

This model requires only six parameters to be estimated, given the initial conditions and that the driving variables are "known". For each of these parameters or rate coefficients, a possible allowable range can be specified, depending on available knowledge. In the worst case, the mortality rate, for example, has to be greater than zero and smaller than one. To circumvent the problem of uncertainty in initial conditions, a set

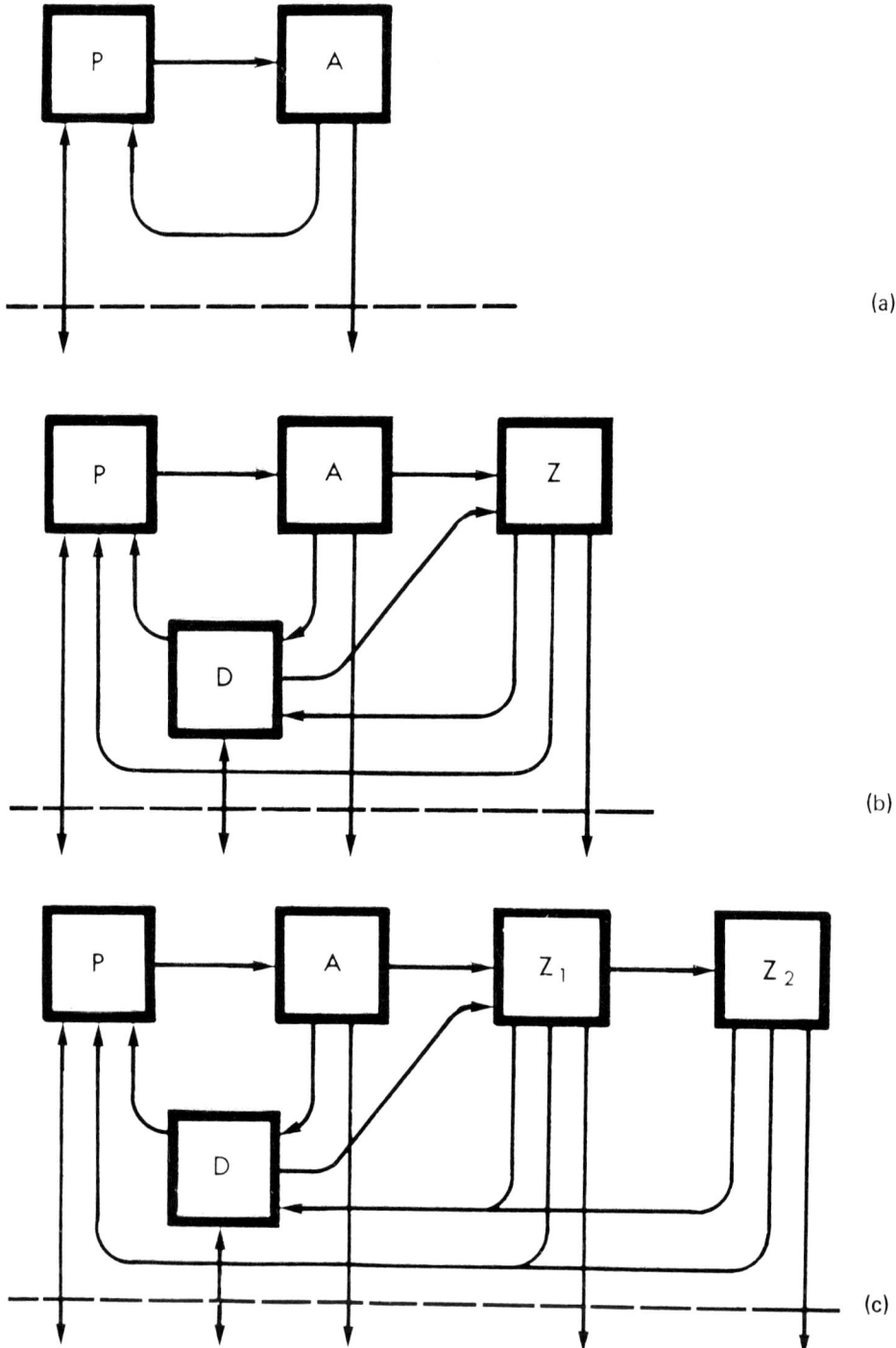

FIGURE 2 Flow diagrams for the models compared: P, phosphate; A, phytoplankton; D, detritus; Z, zooplankton; Z_1, herbivores; and Z_2, carnivores (from Fedra, 1981b).

of likely values (estimated from the available data) was taken and allowed to self-adjust by letting the model run for three years. This strategy (using the results of the third year after arbitrarily specifying the initial-state condition for year one instead of adding more dimensions to the input-search-space) was followed with all the models described below. The model is formulated in terms of phosphorus, with constant stoichiometric conversions to carbon and a time-variable carbon–chlorophyll ratio. A discussion of the description of the major biological processes can be found in Fedra (1979a).

5.1.1 Testing Hypothesis 1

To test the hypothesis formulated in model 1, the model was incorporated into a Monte Carlo framework, which randomly sampled a set of model parameters (the initial conditions) from the allowable ranges, ran the model for a period of three years – to allow the arbitrary initial values of the state variables to adjust – and finally tested for violations of the constraint conditions. This process was repeated for a sufficiently large number of trials (in fact, more than 100,000 model runs were performed with each of the model structures).

Summarizing, model 1 could fulfill all of the constraint conditions except one: it was not possible to reproduce two algae peaks during the summer period (without violating several other conditions). Figure 3 shows a sample output from model 1. Hypothesis 1 consequently had to be rejected. To construct an improved hypothesis, the distributions and correlation structure of parameters and output variables from those runs violating only condition 3 (the two algae peaks) were analyzed. For an example of the output of the analysis programs used, see Table 1 (parts A–D). The analysis clearly indicated that phytoplankton mortality is the critical process, and consequently that it deserves a more refined treatment.

5.2 Hypothesis 2: a Four-Compartment Web

As a slightly more detailed alternative to model 1, a second version was formulated which incorporates detritus and omnivorous zooplankton, to allow for a more detailed description of phytoplankton mortality. The description of primary production as well as the physical framework are essentially the same as in the first version. Model 2, however, splits the phytoplankton mortality into a natural, background mortality, which is described as concentration-dependent, and losses due to grazing. Background mortality as well as zooplankton mortality now feed into the detritus pool, which in turn (being temperature-dependent) feeds back into the nutrient pool; detritus is also consumed by zooplankton, for which, however, a certain preference for living algae is assumed. Zooplankton respiration also feeds into the nutrient pool. Figure 2b shows the flow chart for this model. Grazing was described using a simple encounter theory, but the resulting model performance was still not satisfactory: for low values of the grazing rate constant, the zooplankton did not survive phytoplankton lows in winter, and died away; for high values of the feeding rate, in contrast, phytoplankton was removed very quickly, as soon as it started to grow in the spring, with a consequent collapse of the zooplankton population itself. In between, the system was able to produce classical prey–predator oscillations which were, however, unstable in the long run. Consequently, the encounter theory

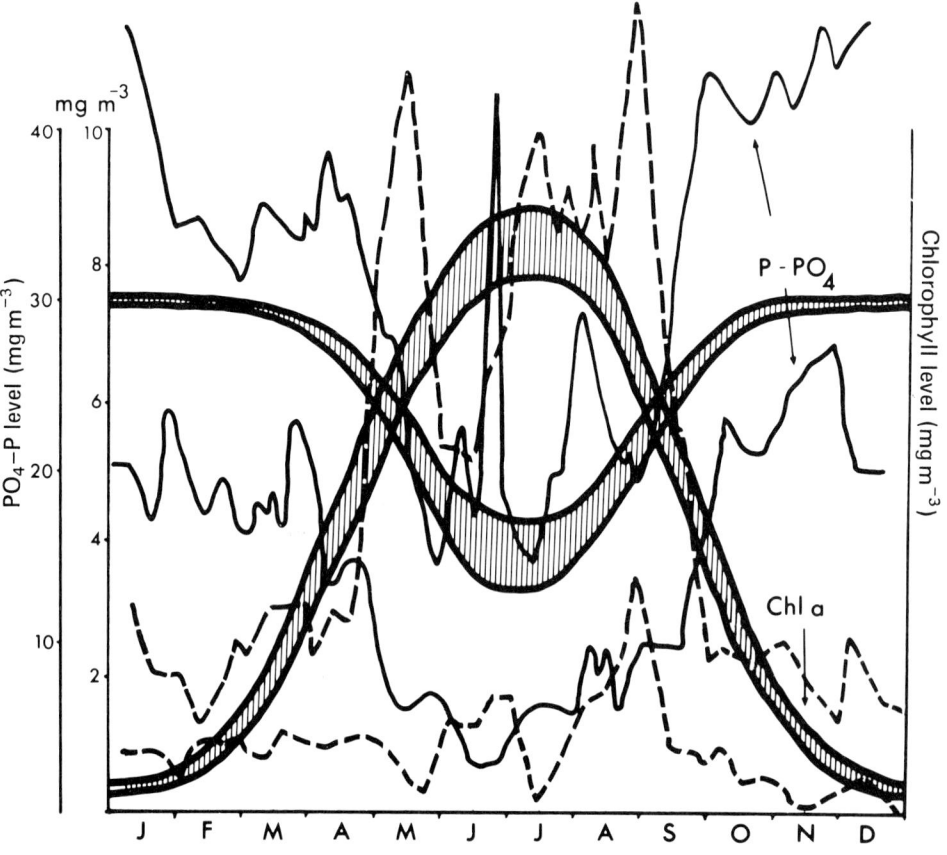

FIGURE 3 Sample output from model 1, showing an envelope for the state variables phosphate and phytoplankton for a set of runs that only violate the condition of two phytoplankton peaks. Thin lines indicate envelopes for the raw data for nutrients (phosphorus) and algae biomass (chlorophyll-a). Compare Figure 2.

was rejected and the description of grazing was reformulated based on a saturation curve (similar to Michaelis–Menten kinetics) using a temperature-dependent maximum feeding rate coefficient, with the same temperature dependence as used for respiration and remineralization.

Again this version was subjected to the simulation procedure described above, and the resulting response was analyzed and used as the basis for yet another modification, namely the introduction of another trophic level (carnivorous zooplankton), to explore its importance in controlling the herbivores (Greve, 1981). This last and most complex version (see Figure 2c) finally passed the test of adequacy, after some more refinements in the formulations of thresholds for grazing and starvation of zooplankters.

TABLE 1A Automatic parameter space analysis program: various results.

Behavior definition applied	
Minimum primary production $g\,C\,m^{-2}\,year^{-1}$	50.00
Maximum primary production $g\,C\,m^{-2}\,year^{-1}$	150.00
Time range for biomass peak value	day 60 to day 210
Upper limit for biomass peak $mg\,P\,m^{-3}$	15.00
Minimum relative increase of biomass max/min	2.00
Orthophosphate maximum in mixed period $mg\,m^{-3}$	2.50
Total phosphorus output range	
Upper limit, metric tons $year^{-1}$	8.00
Lower limit, metric tons $year^{-1}$	2.00
Maximum ratio of total P relative change	0.50
Automatic analysis results	
Number of simulation runs evaluated	10,000
Number of well-behaved runs	293
Number of nonbehavior runs	9,707 including 26 aborted runs
Constraint conditions violated by BAD class	
Primary production too low	849 cases
Primary production too high	937 cases
Biomass peak too early	4,991 cases
Biomass peak too late	1,480 cases
Biomass peak too high	4 cases
Relative biomass increase too low	0 cases
Orthophosphate level too high	7,517 cases
Phosphorus output too low	2,089 cases
Phosphorus output too high	1 case
Relative change in P content too high	2,250 cases

TABLE 1B Automatic parameter space analysis program: constraint violations %-coincidence matrix.

	Condition									
	1	2	3	4	5	6	7	8	9	10
1	100.00	0.00	74.79	3.53	0.00	0.00	54.30	55.01	0.00	26.74
2	0.00	100.00	13.02	39.27	0.43	0.00	100.00	0.00	0.11	9.39
3	12.72	2.44	100.00	0.00	0.00	0.00	65.02	31.48	0.00	36.59
4	2.03	24.86	0.00	100.00	0.27	0.00	93.11	6.22	0.07	14.26
5	0.00	100.00	0.00	100.00	100.00	0.00	100.00	0.00	0.00	50.00
6	0.00	0.00	0.00	0.00	0.00	0.00	0.00	0.00	0.00	0.00
7	6.13	12.47	43.17	18.33	0.05	0.00	100.00	10.02	0.01	13.70
8	22.36	0.00	75.20	4.40	0.00	0.00	36.05	100.00	0.00	43.03
9	0.00	100.00	0.00	100.00	0.00	0.00	100.00	0.00	100.00	0.00
10	10.09	3.91	81.16	9.38	0.09	0.00	45.78	39.96	0.00	100.00

TABLE 1C Automatic parameter space analysis program: correlation analysis and model output data.

Correlation analysis for well-behaved runs (runs evaluated = 293)

	Unit	Mean	Standard deviation	Minimum	Maximum
1 Michaelis constant	$mg\,m^{-3}$	1.0042	0.5135	0.2041	1.9930
2 Respiration/mineralization epilimnion	day^{-1}	0.1134	0.0526	0.0219	0.2000
3 Respiration/mineralization hypolimnion	day^{-1}	0.0141	0.0034	0.0100	0.0244
4 Net sedimentation velocity epilimnion	$m\,day^{-1}$	0.2571	0.1604	0.0119	0.7438
5 Net sedimentation velocity hypolimnion	$m\,day^{-1}$	1.4196	0.4241	0.3231	1.9983
6 Minimum production rate	day^{-1}	0.3770	0.0734	0.2505	0.4994
7 Maximum production rate	day^{-1}	6.3429	2.2301	1.1798	9.9996
8 Time lag of productivity maximum	day	218.4715	25.4016	180.0387	269.8400
9 Diffusion coefficient hypolimnion	$cm^2\,s^{-1}$	0.2788	0.1396	0.0200	0.4986
10 Diffusion coefficient thermocline	$cm^2\,s^{-1}$	0.1318	0.0685	0.0100	0.2498
11 Extinction coefficient	m^{-1}	0.2885	0.0562	0.2003	0.3979
12 Self-shading coefficient	$m^2\,mg^{-1}$	0.0150	0.0030	0.0101	0.0200
13 Initial thermocline depth	m	4.4748	0.8464	3.0053	5.9889
14 Final thermocline depth	m	17.6395	1.4145	15.0251	19.9728
15 Start of stratified period	day	155.3691	16.8105	120.9221	179.8988
16 End of stratified period	day	302.5700	14.8651	280.0211	329.7845
17 Thickness of thermocline	m	7.5441	1.4705	5.0062	9.9848
18 Orthophosphate input	$mg\,m^{-2}\,day^{-1}$	0.1067	0.0509	0.0116	0.1989
19 Particulate phosphorus input	$mg\,m^{-2}\,day^{-1}$	0.9294	0.3339	0.2604	1.4997
20 Hydraulic loading	$m\,day^{-1}$	0.0422	0.0050	0.0301	0.0500
21 Initial orthophosphate mixed period	$mg\,m^{-3}$	1.0648	0.4933	0.2000	1.9893
22 Initial particulate phosphorus mixed period	$mg\,m^{-3}$	3.4447	0.7131	2.5156	6.1065

Model output data

	Unit	Mean	Standard deviation	Minimum	Maximum
1 Primary production	$g\,C\,m^{-2}\,year^{-1}$	80.6641	15.9106	50.3616	128.9642
2 Orthophosphate maximum mixed period	$mg\,m^{-3}$	2.1760	0.2434	1.3256	2.4989
3 Particulate phosphorus epilimnion minimum	$mg\,m^{-3}$	0.4115	0.1971	0.0714	1.1612
4 Particulate phosphorus epilimnion maximum	$mg\,m^{-3}$	4.4593	0.8348	3.0174	8.2083
5 Day of epilimnic particulate phosphorus maximum	day	135.4863	31.8503	61.0000	185.0943
6 Total phosphorus input	$ton\,year^{-1}$	17.1331	5.6574	5.7596	27.7315
7 Net phosphorus loading	$ton\,year^{-1}$	14.6376	5.5953	3.4143	25.1839
8 Total phosphorus output	$ton\,year^{-1}$	2.4955	0.3823	2.0007	3.8662
9 Lake content total P (time = start)	ton	17.7356	3.3204	11.6538	29.3214
10 Lake content total P (time = end)	ton	12.6229	1.6847	8.4954	17.5518
11 Sedimentation total P	$ton\,year^{-1}$	19.6748	5.8205	5.8417	34.4694

TABLE 1D Automatic parameter space analysis program: data-input and response correlation matrix.

6 PARAMETER ESTIMATION AND PREDICTION

6.1 A Lake Modeling Example

The second step of the approach outlined above, parameter estimation and prediction, has been applied to a lake ecosystem, or rather to a lake ecosystem model. A real-world example and an existing data set, showing all the deficiencies of the abovementioned marine example, were used to test the practical applicability of the approach. It also gave some insight into the model used, which was, admittedly, selected quite arbitrarily. However, the model structure selected passed the above test of adequacy, so that a more detailed study of parameter ranges and relationships was feasible.

6.2 Ecosystem, Data, and Model

The lake ecosystem chosen for this study was the Attersee, a deep, oligotrophic lake in the Austrian Salzkammergut. Basic lake characteristics are compiled in Table 2, and much detailed information about the lake system can be found in the yearly reports of the Austrian Eutrophication Program, 'Projekt Salzkammergutseen' (Attersee Report, 1976, 1977; Müller, 1979; Moog, 1980), which is a national followup of the OECD alpine lake eutrophication program.

TABLE 2 Attersee: basic lake data (after Flögl, 1974).

Geographical position	47°52'N, 13°32'E
Altitude	469 m above sea level
Catchment area	463.5 km²
Lake surface area	45.6 km²
Length	19.5 km
Average width	2.4 km
Maximum depth	171 m
Average depth	84 m
Volume	3934×10^6 m³
Theoretical retention time	7–8 years
Average outflow	17.5 m³s^{-1}

For the purposes of this study, it suffices to say that the available data showed considerable uncertainty and variability. This was largely due to the limited manpower and number of observations available (most of them being singular measurements). Also, the monitoring program was designed independently of any subsequent analysis, and in addition one of the key variables, namely orthophosphate concentration, varies around the minimum level of detectability. Consequently, the method described above for deriving a formal definition of the systems behavior had to account for this uncertainty, and only comparatively broad ranges could be specified for the behavior-describing measures selected. Figure 4 shows a plot of total phosphorus data, averaged for a five-year period, for the two sampling stations on the lake. The plot gives some idea of data

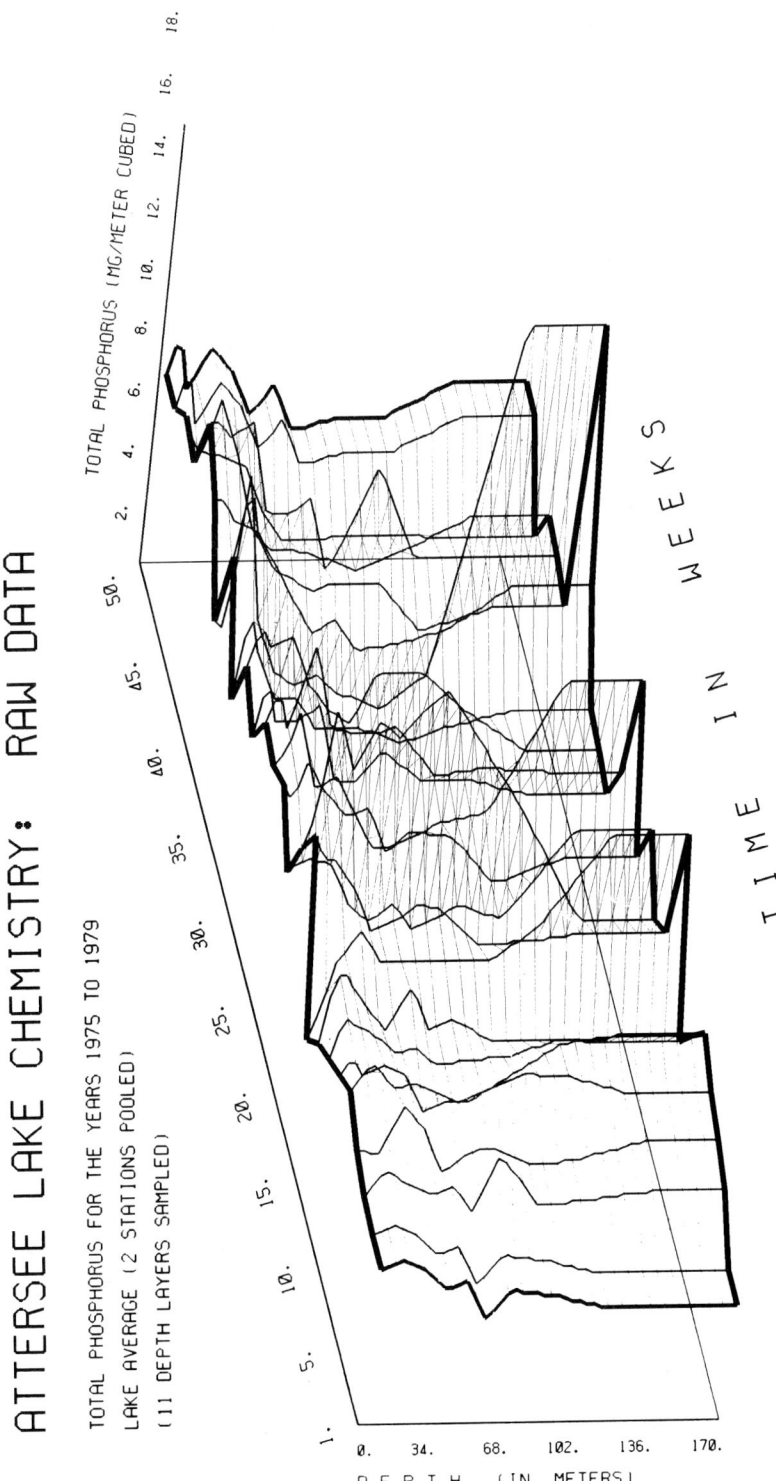

FIGURE 4 Attersee lake chemistry, raw data. Total phosphorus, lake average (2 stations), 11 depth-layers, 5 years of data pooled (raw data courtesy of F. Neuhuber).

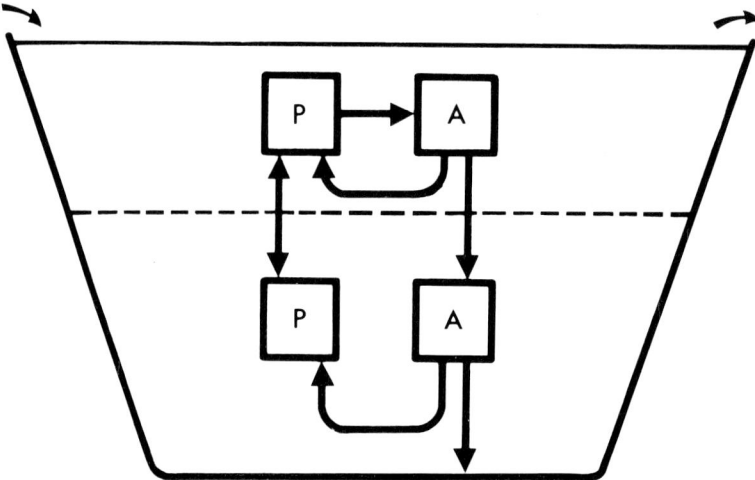

FIGURE 5 Flow diagram for the lake model: P, available phosphorus (PO_4–P); A, particulate phosphorus, representing algae biomass.

variability, and, considering the physical properties of the lake (namely the long retention time) and the almost conservative nature of total phosphorus, also gives some measure of data reliability.

The problem selected for the modeling approach was the relation of the lake's trophic state or water quality (as described by, for example, algal peak biomass, yearly primary production, or nutrient concentrations) to the nutrient imports or external loading (Fedra, 1979b). For this purpose a rather simple model of lake phosphorus dynamics (Imboden and Gächter, 1978) was used, which considers only two state variables, namely dissolved phosphorus (the available, limiting nutrient) and particulate phosphorus (algal biomass). A flow chart of the model structure is shown in Figure 5; the model structure corresponds to that in Figure 2a, although the description of depth-integrated primary production and the physical framework are different (Imboden and Gächter, 1978). The model has been applied to various lake systems with a generally satisfactory performance.

6.3 The Formal Definition of System Behavior

Depending on the problem addressed and the model selected, the behavior of the system had to be described in terms of model response, relevant to the problem, and supported by the available information. The measures selected were, in almost complete agreement with the marine example test case, yearly primary production, algae peak biomass (absolute level, as well as the timing), relative change in algae biomass during the year, nutrient concentration during the periods when the lake water is fully mixed vertically, and phosphorus export from the lake. Figure 6 shows an example of the information

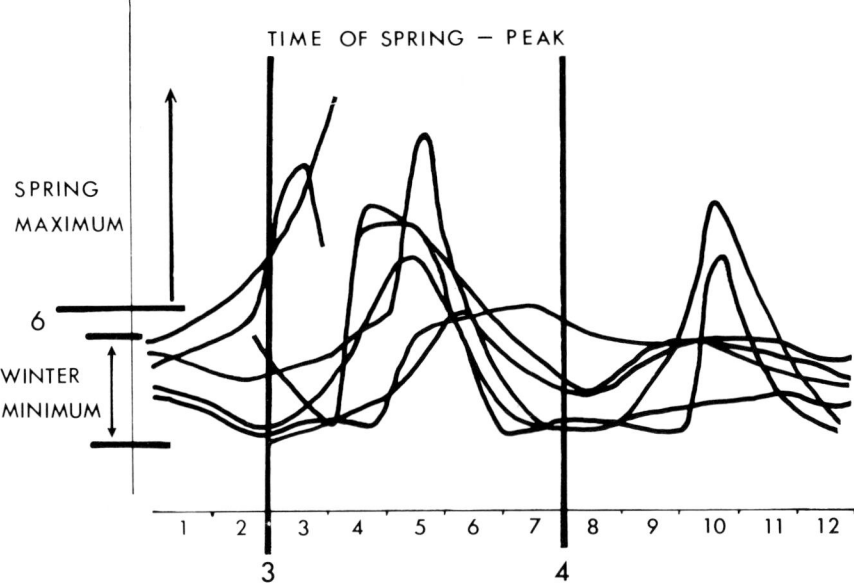

FIGURE 6 Algae dynamics (chlorophyll-a, data of several years pooled). Behavior description derived: the spring peak with a minimum extent of twice the winter minimum occurs between days 60 and 210 of the calendar year.

used to define the allowable ranges for the relative change in algae biomass, as well as the timing of the peak value. The resulting estimates (day 60 to day 210, and a minimum relative increase of 2.0) are certainly very broad. No attempt was made, however, to refine them further by more resourceful analysis of the available data, as the major objective of this study was methodological rather than problem-specific. These constraint conditions are shown in Table 1 as a part of the output of the analysis program.

Again, allowable ranges had to be specified for the parameters, the coefficients describing forcing, and the initial conditions of the model, in a manner similar to that used for the behavior-describing measures. Altogether, the chosen model required definition of 22 such ranges, after some of the time-variable forcings had been redescribed as functions of time, using auxiliary coefficients (see Fedra, 1979b; Fedra et al., 1981). A list of the parameters and inputs used is given in Table 1C. For some of the ranges specified, data to support the estimates were available, as for example in the case of the extinction coefficient, the phosphorus loadings, hydraulic loading, and depth of the thermocline. Here the ranges could be specified by a mean estimate and some observed variability around it, or at least by any estimate with an arbitrary, ample range. Whenever measurements to estimate allowable ranges were unavailable, values from the literature or simply "educated guesses" had to be used, and the additional uncertainty was reflected in the wide ranges. The subsequent estimation method, however, is not sensitive to the initial choice of the ranges, as long as they are within plausible, and physically and biologically feasible, bounds.

6.4 The Estimation Process

Given the definition of allowable ranges for all the numbers to be put into the model's computer code (parameters, coefficients describing forcing, and initial states) and the definition of allowable ranges for the model response, the estimation process involves a straightforward application of Monte Carlo methods. From the input- or parameter-space region defined, a random sample was taken (assuming rectangular, independent probability density functions for the individual ranges), and this parameter vector was then used for one simulation run. A sample output for the time–depth distribution of the two state variables "available phosphorus" and "algae biomass phosphorus" is shown in Figure 7. The resulting model behavior was compared with the predefined model response, and the parameter vector was classified according to whether it resulted in the defined response (the response being fully within the defined allowable response region) or not. This process was repeated until a sufficient number of behavior-giving parameter vectors was found. In fact, the process was repeated 10,000 times for the parameter-space definition used (after an initial 10,000 pilot runs which allowed some reduction to be made in the parameter space for the random search and established the adequacy of the model structure, as discussed above), in order to arrive at a sufficient number of behavior runs.

6.5 Model Response and Parameter-Space Structure

The 10,000 independent random combinations of 22 input (parameter) values, each used to generate a model response, resulted in a rather broad overall region for these model responses. Projections from this overall response space (a hyperspace with the axes defined by the response variables used in the behavior definition) on planes of two response or constraint variables are shown in Figure 8 to illustrate this. The figure also indicates the allowable, behavior-defining range of the constraint variables. Table 1 shows part of the output of one of the parameter-space and model-response analysis programs used.

Only a small fraction of the 10,000 runs was fully within all of the ranges; in fact, only 293 such behavior runs were found (compare Figure 7). This low "score" could be attributed to the rather broad ranges for the parameters sampled; on the other hand it has to be kept in mind that the ranges for the allowable *response* were quite liberal too. However, selecting the random samples independently, i.e., without taking into account possible correlations between them, may have been responsible for the low number of allowable responses. In fact, when analyzing the behavior-giving set of 293 input data, a marked correlation structure was found (see also Figure 8). The correlation between the parameters and the model-response variables was calculated as well, and Table 1D gives a correlation matrix for the interparameter and parameter-response correlations for the 293 behavior-giving runs.

From the results shown in Table 1, it is obvious that the allowable orthophosphate level as well as the timing of the algae spring peak are the most critical conditions to be met. Also, it seems obvious that the behavior definition is not sufficient to force the model into cyclic stability in its phosphorus budget, which seems to be an important condition for a fairly deep lake with a retention time of seven to eight years. Narrowing the admissible range for the stability criterion (relative difference in total phosphorus content between initial and final conditions) from 0.5 to 0.25, results in a lower number of behavior runs, namely 112.

Summarizing, the overall model response was found to cover a very wide range in response space, and only a very small portion of the range was satisfactory in terms of the behavior definition used. Also, the behavior-giving part of the parameter space observed shows a marked correlation structure. This correlation structure indicates not only the interdependence of the parameters, but also the possibility of arriving at one and the same response region of a model, with different parameter combinations. However, many features of the results are somewhat difficult to interpret due to the high dimensionality of the hyperspaces involved.

The analysis of model response and parameter-space structure can also be interpreted in terms of a sensitivity analysis, where again, through simultaneous variation in all the parameters, sensitivity for the whole parameter-space and associated behavior-space region covered can be studied. This is in some contrast to more-classical approaches to sensitivity analysis (see, for example, Argentesi and Olivi, 1976; van Straten and de Boer, 1979), which explore only arbitrarily selected subregions of the parameter space and response space along a very limited number of dimensions. For the behavior-giving class of parameters sets, the fourth moment, or kurtosis, of the frequency distributions of the individual parameters can be interpreted as a measure of sensitivity (this is also true for the initial conditions and the forcing-describing coefficients; as they are also estimated with a certain degree of uncertainty, an analysis of sensitivity is also meaningful). Also, the linear correlation coefficient for any paired input–output combination can be used as a measure of sensitivity. Another measure of sensitivity can be obtained from the correlation structure of the parameter space itself. Here significant correlations identify parameter combinations which strongly determine the system behavior as defined by the constraint conditions, and the sign of the correlation coefficient is an indication of whether the parameters work together in the same "direction" or against each other in opposite "directions". At the same time, the parameter-space structure thus gives some insight into the functioning of the model structure, as discussed above.

Whenever it is not possible to identify a region of increased probability for a given parameter to give the defined behavior, that is, where the final distribution is identical with the initially assumed rectangular density distribution, that parameter may be regarded as redundant. Either the available information does not allow its identification, or it just does not affect the model response; in either case a change in the model structure may be warranted. This would then lead to a repetition of the previous stage of the estimation procedure, discussed above. Also, if no behavior-giving value for a given parameter can be found within the admissible range specified, the function of this parameter in the model and the concept of the parameter and/or its real-world equivalent are not well matched, which again requires a reconsideration of the model structure. These relationships again point to the intimate coupling of the individual steps in the estimation procedure, since neither model structure nor parameter values can be estimated independently of each other.

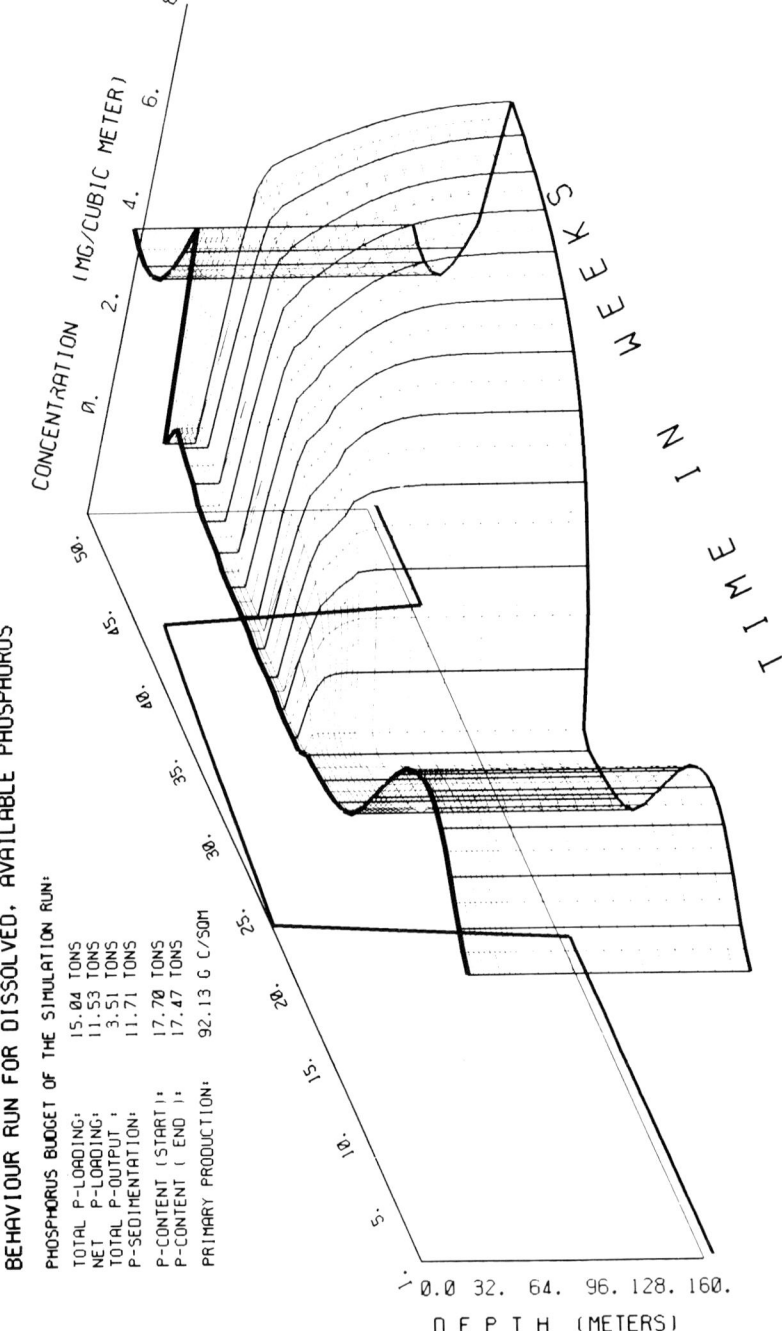

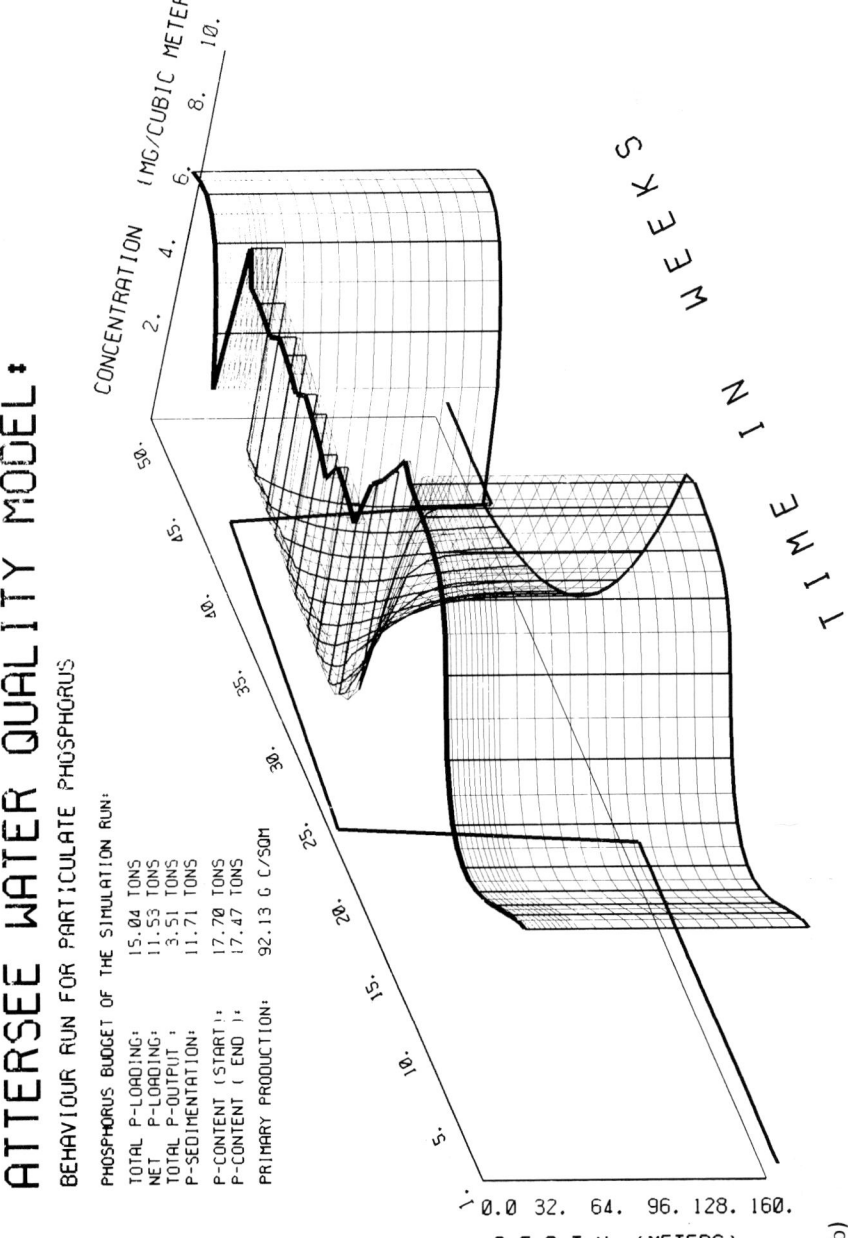

FIGURE 7 Sample output from the lake model, showing the time–depth distribution for the two state variables (a) available phosphorus and (b) algae biomass phosphorus for a yearly cycle.

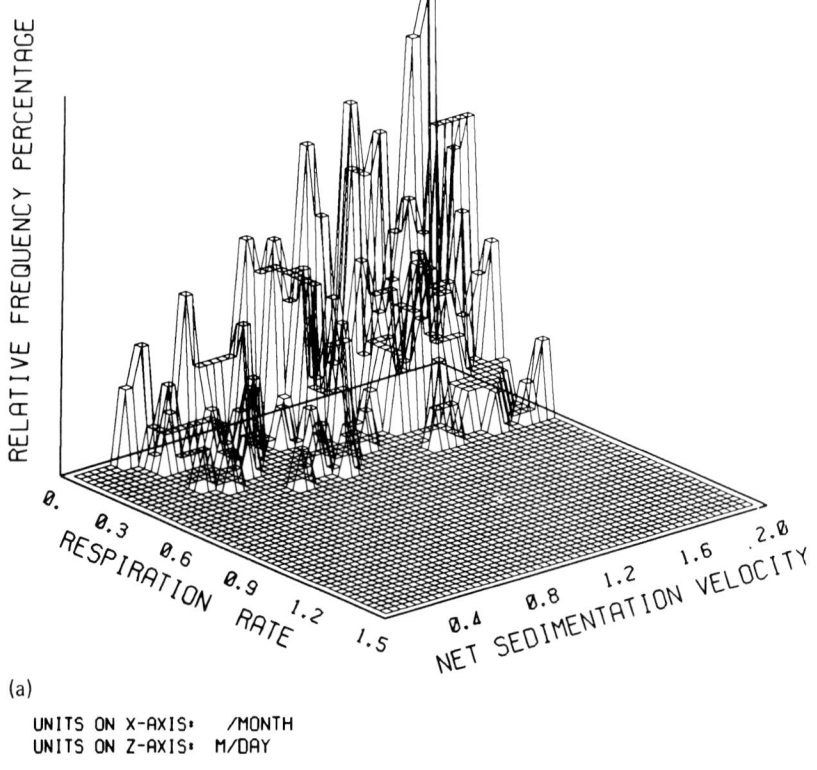

FIGURE 8 Projection from the parameter space and the model response space. (a) Frequency of parameter combinations giving rise to the defined behavior; (b) model response (10,000 runs) shown as frequency distribution over a plane of two response (behavior-constraining) variables. The allowable, behavior-defining response range is indicated.

6.6 Prediction With Ensembles

The ensemble of parameter sets or input sets identified in the estimation procedure described above, together with the model structure selected, can be viewed as the best-available (model) description of the system under study, which also represents a certain compromise between uncertainty and arbitrariness. In view of the uncertainty about the system behavior as well as the coefficients to be used, any of the data sets in this ensemble are equally good and valid descriptions of the system. This points directly to the diffuse picture of the system we are bound to have, unless this picture is arbitrarily made unambiguous.

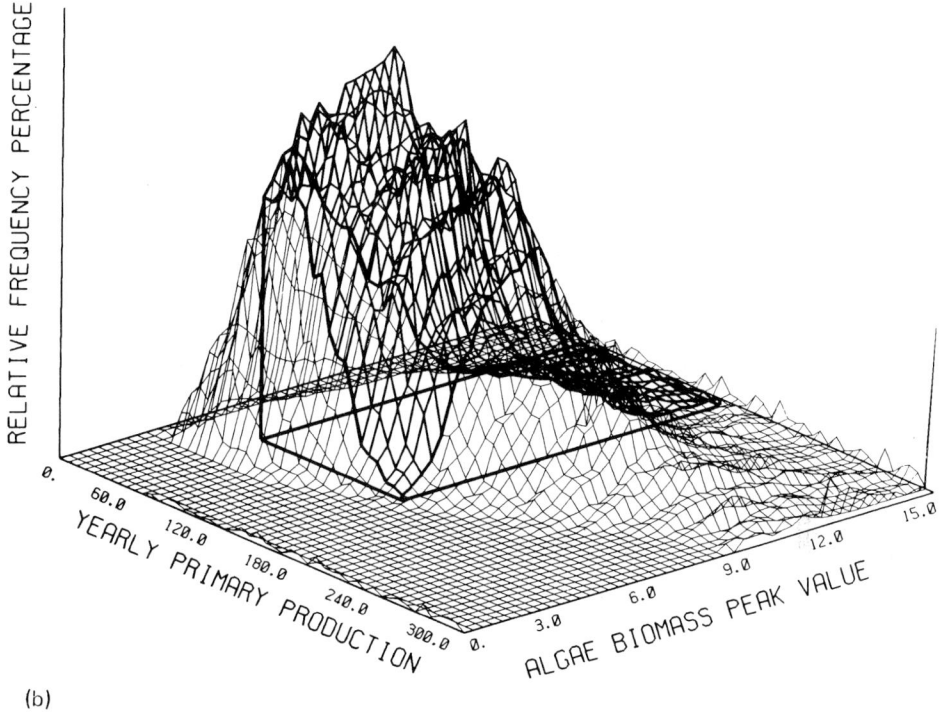

(b)

UNITS ON X-AXIS: G C/SQM AND YEAR
UNITS ON Z-AXIS: MG P/CUBIC METER

FIGURE 8 *Continued.* For details see caption opposite.

This uncertainty in the empirical basis, as well as in the theoretical understanding of the systems being modeled, should consequently be reflected in the predictions about the future behavior of these systems under changed conditions. The simplest and most straightforward way to achieve this is merely to use all the possible descriptions as a basis for a forecast, and thus project an ensemble of future behaviors. Again, such ensembles will have certain statistical properties which can be used to estimate the uncertainty (based on data error and model error) of the prediction, and its evolution over time.

In the modeling exercise described here, the loading-determining coefficients were altered to represent changes in the external conditions of the system, and for each change all (or a subset) of the behavior-giving parameter combinations were then used for runs with simulation times of up to ten years. Figure 9 shows a series of probability distributions, fitted to the model output frequencies, plotted against the changed input coefficient. Figure 10 shows a dynamic version of the model's first-year response to a pronounced relative change (twofold multiplication) in the nutrient inputs. Finally,

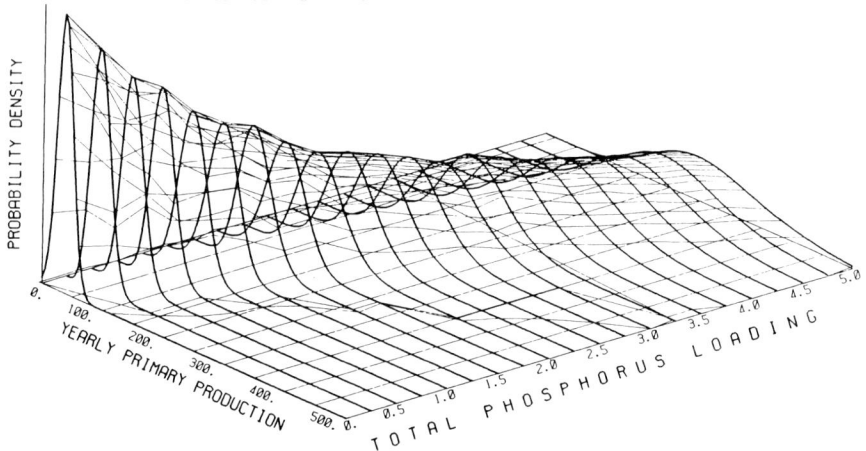

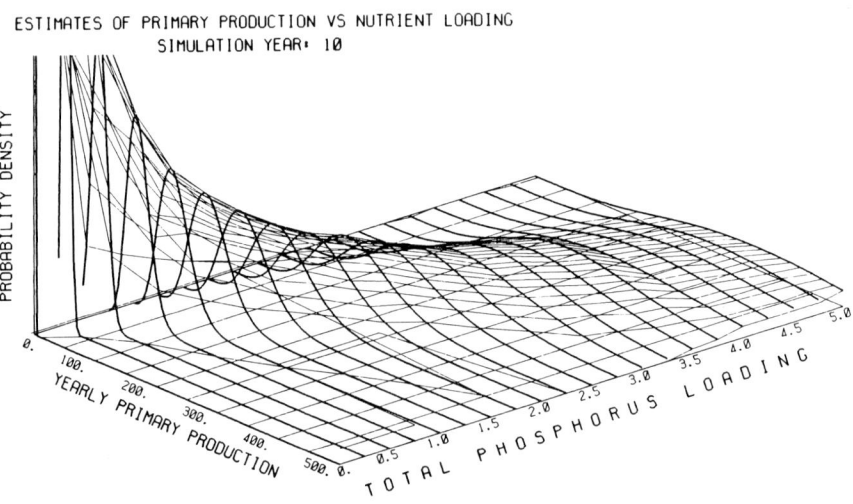

UNITS ON X-AXIS: G C/SQM AND YEAR
UNITS ON Z-AXIS: MG P/SQM AND DAY

FIGURE 9 Probability distributions for a model output variable (primary production) for various levels of an input (phosphorus-loading).

Figure 11 summarizes the general pattern observed, namely the increase of relative prediction uncertainty (e.g., measured as a coefficient of variation for any of the model outputs) with degree of change in the input conditions and with time.

Uncertainty of the prediction in terms of a coefficient of variation increases with time as well as with increasing changes in the external or input conditions. This is certainly what one would expect intuitively. The coefficients of variation reach a certain maximum in time after several years of simulation, as the model also reaches a new

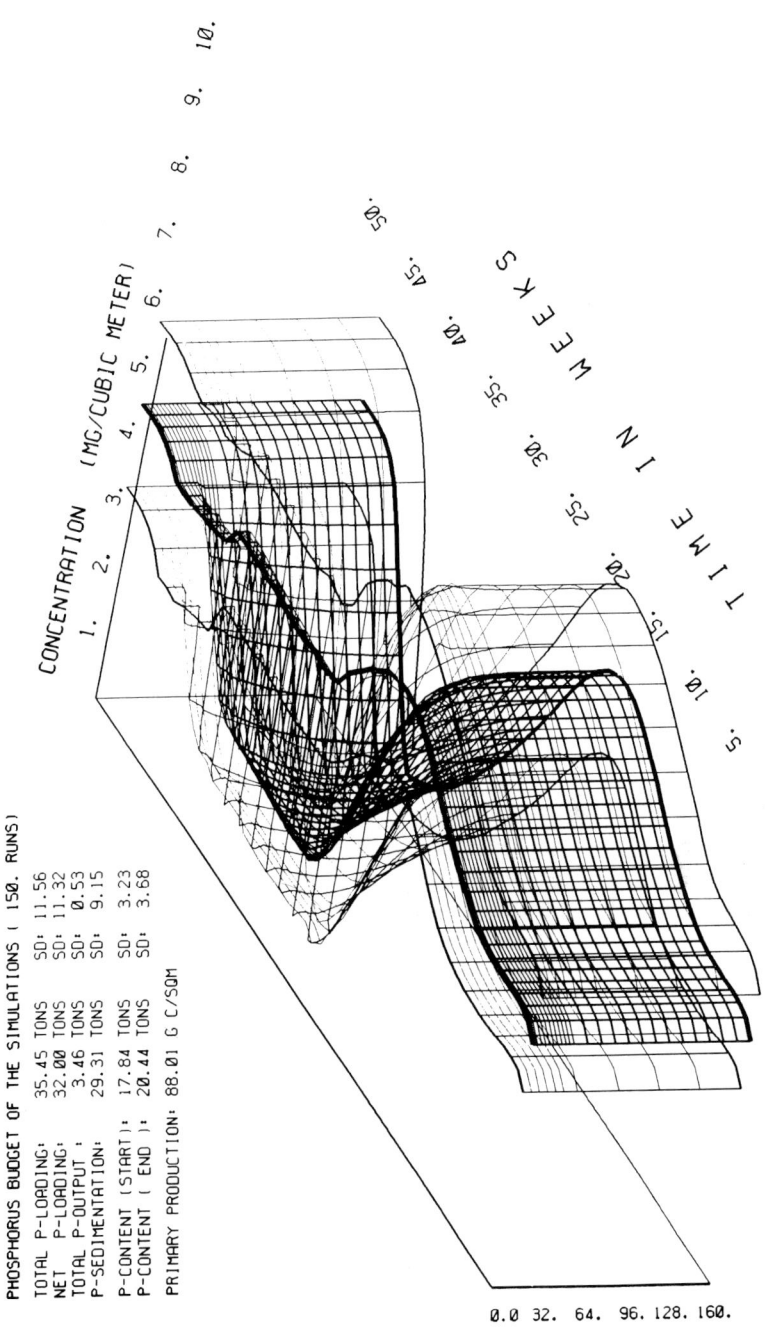

FIGURE 10 Minimum–maximum envelope and mean dynamic behavior of a model state variable (particulate phosphorus, representing algae), for an input ensemble representing a drastic change in the nutrient loading.

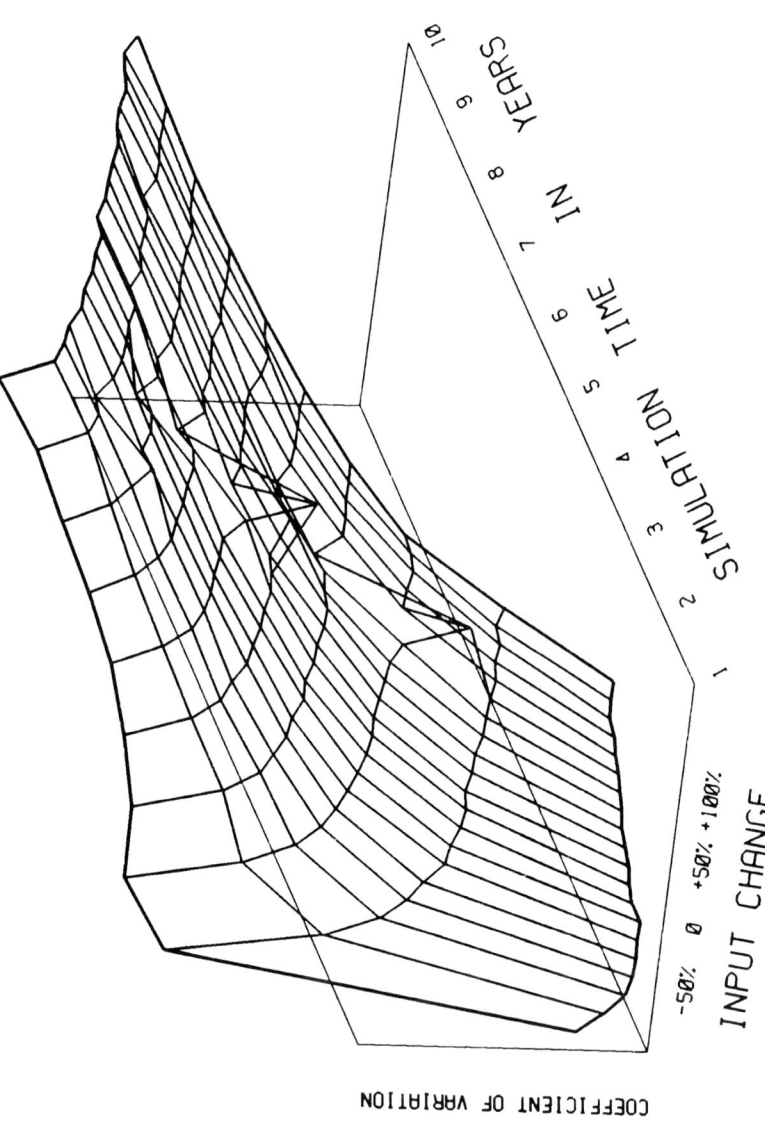

FIGURE 11 Prediction uncertainty versus extrapolation in the input conditions and extrapolation in time (from Fedra, 1981a).

equilibrium because of the continuous application of one and the same set of inputs and parameters. The method thus estimates the uncertainty of the equilibrium state of the model for a given input condition. This equilibrium uncertainty is now found to be related to the degree of change in the input conditions, i.e., the loading-determining coefficients in this example. The larger the change to be simulated, the more uncertainty there will be in the predictions of the final as well as the intermediate states of the system. These results correspond with what one would assume intuitively, and in addition the method allows one to estimate quantitatively prediction uncertainty or the limits of predictability for a given initial uncertainty that stems from data uncertainty as well as system variability.

7 DISCUSSION

To build the complex hypotheses required to describe and explain the structural and behavioral features of ecological systems, a formal approach and rigorous testing procedures are required. As has been demonstrated, parts of the observed behavior of a system may easily be reproduced. This, however, goes in parallel with unrealistic behavior in other parts of the system. A complex hypothesis or model, however, can only be accepted as a valuable working tool with explanatory value and predictive capabilities if it fulfills all the constraints one formulates as defining the observed system behavior. Violation of one single condition necessitates the rejection of such a model, which should be just one step in an iterative process of analysis. This represents an alternative concept of "disagreement" under uncertainty, where a gradual "goodness of fit" concept is replaced by the more appropriate test of individual conditions.

A basic idea of the approach is to use the available information according to its relevance for the model's (in other words, the theory's) level of abstraction. Obviously, the states of a system can be described much more easily on the appropriate level than can the process rates and controls (just think in terms of phytoplankton biomass versus production rate). Consequently, the argument of the hypothesis-testing process is turned around: instead of putting the "known" initial conditions (the rates, among other factors) into the model structure and deriving the response for comparison, the allowable response is used as a constraint to identify possible initial conditions. This is to say, a given region in the response hyperspace of a model is mapped back into the input hyperspace.

The test is then as follows: does this region in the input space exist within the specified possible or plausible bounds? In addition, several other features of the input space can be used as a basis for either rejecting or corroborating a given hypothesis, for example, the uniqueness of the input-space region, whether or not it is closed, and its structure, which is determined by the interdependence of the individual input values. In addition, all these features, including the relationship or correlation of input and output space, allow us to learn something about the way the proposed system structure functions. The method facilitates understanding of system behavior on the appropriate level of abstraction (in terms of the input and output of the model) and it also provides diagnostic information for hypothesis generation.

The same underlying idea of inverse mapping of response space into the input space of the model is used in the parameter-estimation procedure. A formal discussion of this procedure is given in Fedra et al. (1981). Thus the approach illustrates the intimate coupling between estimation (or calibration) and prediction. At the same time, it points out the importance of data uncertainty and/or systems variability for the estimation process, resulting in nonunique estimates of the parameters or inputs to be estimated, which in turn are reflected in nonunique predictions.

The approach emphasizes the discrepancy between the raw data, measurements or experiments, and the entities conceptualized in the model. Taking into account the sample nature of the data describing the system, it is necessary to derive a description of the system that is meaningful in terms of the model entities, taking into account the sample statistics as well as the problems of interpretation or mapping the data into the (model) system. This is an attempt to model the system, not the data. In addition, the approach tries to capture the full behavioral repertoire of the system rather than to use any (arbitrarily) specific set of observations to test and improve the model's performance. All these attempts to reduce arbitrariness by explicitly accounting for various sources and effects of uncertainty, however, lead to seemingly imprecise, ambiguous results.

The dilemma of an easily defendable, but ambiguous description of a system, and a seemingly precise, but arbitrary one, is irritating (Fedra et al., 1981); but it should also lead to a more critical reconsideration of the basic and implicit assumptions of any model-based analysis, and to a clearer statement of the questions to be addressed and the objectives of the answers to be expected.

Predictions by means of the above methods quickly degenerate into trivial statements about the future of the system modeled. This, however, should be taken as a warning to the analyst that there are obvious limits to predictability. In many cases, the initial information available will not support a quantitative analysis of the response of the system to changes in its external conditions. If "precision" has to be based on arbitrary assumptions which cannot be tested against available empirical evidence or purely logical reasoning based on well-established theories, no useful and reliable forecast is possible. However, the method proposed should also allow identification of data needs and critical gaps in the available knowledge. The analysis of the structure of input or parameter space can be very useful in terms of sensitivity analysis, and could even be used to indicate inadequacies in the model structure and process descriptions.

The probability distributions generated for the model output variables must be understood as the results of the evolution or propagation of the initial uncertainty in the available information. This uncertainty can only in part be attributed to problems of data collection and interpretation; much of the uncertainty may stem from noise in the driving conditions of the system, such as, for example, weather phenomena. However, as this part of the uncertainty at least is inevitable and an essential feature of the system itself, we must learn to live with it and increasingly incorporate it into numerical analysis and modeling, rather than ignore it.

ACKNOWLEDGMENTS

The research described in this paper was carried out within the framework of the Austrian Lake Ecosystems Case Study, supported by the Austrian Fonds zur Förderung der wissenschaftlichen Forschung, grant No. 3905. The author gratefully acknowledges as-yet unpublished data and information made available by the Biological Station Helgoland, Federal Republic of Germany, and numerous Austrian colleagues.

REFERENCES

Argentesi, F. and Olivi, L. (1976). Statistical sensitivity analysis of a simulation model for the biomass–nutrient dynamics in aquatic ecosystems. Proceedings of the Summer Computer Simulation Conference, 4th, pp. 389–393.

Attersee Report (1976). Vorläufige Ergebnisse des OECD-Seeneutrophierungs und des MaB-Programms. Weyregg, Austria.

Attersee Report (1977). Vorläufige Ergebnisse des OECD-Seeneutrophierungs und des MaB-Programms. Weyregg, Austria.

Beck, M.B. (1979a). System Identification, Estimation, and Forecasting of Water Quality: Part I: Theory. WP-79-31. International Institute for Applied Systems Analysis, Laxenburg, Austria.

Beck, M.B. (1979b). Applications of System Identification and Parameter Estimation in Water Quality Modeling. WP-79-99. International Institute for Applied Systems Analysis, Laxenburg, Austria.

Beck, M.B., Halfon, E., and van Straten, G. (1979). The Propagation of Errors and Uncertainty in Forecasting Water Quality – Part I: Method. WP-79-100. International Institute for Applied Systems Analysis, Laxenburg, Austria.

Benson, M. (1979). Parameter fitting in dynamic models. Ecological Modelling, 6: 97–115.

Bierman, V.J., Dolan, D.M., Stoermer, E.F., Gannon, J.E., and Smith, V.E. (1980). The Development and Calibration of a Spatially Simplified Multi-Class Phytoplankton Model for Saginaw Bay, Lake Huron. Contribution No. 33. Great Lakes Environmental Planning Study, US Environmental Protection Agency, Grosse Ile, Michigan.

Biologische Anstalt Helgoland (1964–1979). Jahresberichte. Biologische Anstalt Helgoland, Hamburg.

Conrad, M. (1976). Patterns of biological control in ecosystems. In B.C. Patten (Editor), Systems Analysis and Simulation in Ecology, Volume IV. Academic Press, New York, pp. 431–457.

DiToro, D.M. and van Straten, G. (1979). Uncertainty in the Parameters and Predictions of Phytoplankton Models. WP-79-27. International Institute for Applied Systems Analysis, Laxenburg, Austria.

DiToro, D.M., O'Connor, D.J., and Thomann, R.V. (1971). A dynamic model of phytoplankton populations in the Sacramento–San Joaquin Delta. Advances in Chemistry Series, 106: 131–180.

Fedra, K. (1979a). Modeling Biological Processes in the Aquatic Environment; With Special Reference to Adaptation. WP-79-20. International Institute for Applied Systems Analysis, Laxenburg, Austria.

Fedra, K. (1979b). A Stochastic Approach to Model Uncertainty: a Lake Modeling Example. WP-79-63. International Institute for Applied Systems Analysis, Laxenburg, Austria.

Fedra, K. (1980a). Mathematical modeling – a management tool for aquatic ecosystems? Helgoländer Wissenschaftliche Meeresuntersuchungen, 34(2):221–235.

Fedra, K. (1980b). Austrian Lake Ecosystem Case Study: Achievements, Problems and Outlook After the First Year of Research. CP-80-41. International Institute for Applied Systems Analysis, Laxenburg, Austria.

Fedra, K. (1981a). Estimating model prediction accuracy: a stochastic approach to ecosystem modeling. In D.M. Dubois (Editor), Progress in Ecological Engineering and Management by Mathematical Modelling. Cebedoc, Liège.

Fedra, K. (1981b). Pelagic foodweb analysis: hypothesis testing by simulation. Proceedings of the 15th European Marine Biology Symposium. Kieler Meeresforschung, Sonderheft 5:240–258.

Fedra, K. (1982). Environmental Modeling Under Uncertainty: Monte Carlo Simulation. WP-82-42. International Institute for Applied Systems Analysis, Laxenburg, Austria.

Fedra, K., van Straten, G., and Beck, B. (1981). Uncertainty and arbitrariness in ecosystems modeling: a lake modeling example. Ecological Modelling, 13: 87–110.
Flögl, H. (1974). Die Reinhaltung der Salzkammergutsee. Österreichische Wasserwirtschaft, 26:1–11.
Gjessing, D.T. (1979). Environmental remote sensing. Physics in Technology, 10: 266–271.
Greve, W. (1981). Invertebrate predator control in a coastal marine ecosystem: the significance of Beroe gracilis (Ctenophora). Proceedings of the 15th European Marine Biology Symposium. Kieler Meeresforschung, Sonderheft 5:211–217.
Hagmeier, E. (1978). Variations in phytoplankton near Helgoland. Rapports et Procès Verbaux, Conseil International pour l'Exploration de la Mer, 172: 361–363.
Halfon, E. (Editor) (1979). Theoretical Systems Ecology. Academic Press, New York.
Hornberger, G.M. and Spear, R.C. (1980). Eutrophication in Peel Inlet – I. Problem-defining behavior and a mathematical model for the phosphorus scenario. Water Research, 14: 29–42.
Imboden, D.M. and Gächter, R. (1978). A dynamic lake model for trophic state prediction. Ecological Modelling, 4: 77–98.
Kremer, J.N. and Nixon, S.W. (1978). A Coastal Marine Ecosystem. Ecological Studies 24. Springer, New York.
Lewis, S. and Nir, A. (1978). A study of parameter estimation procedures of a model of lake phosphorus dynamics. Ecological Modelling, 4: 99–118.
Lucht, F. and Gillbricht, M. (1978). Long-term observations on nutrient contents near Helgoland in relation to nutrient input of the river Elbe. Rapports et Procès Verbaux, Conseil International pour l'Exploration de la Mer, 172: 358–360.
Moog, O. (1980). Jahresbericht 1979. Arbeiten aus dem Labor, Weyregg, 4/1980.
Müller, G. (1979). Jahresbericht 1978. Arbeiten aus dem Labor, Weyregg, 3/1979.
Nihoul, J.C.J. (1975). Modelling of Marine Systems. Elsevier Oceanography Series 10. Elsevier, Amsterdam.
O'Neill, R.V. and Gardner, R.H. (1979). Sources of uncertainty in ecological models. In B.P. Zeigler, M.S. Elzas, G.J. Klir, and T.I. Oren (Editors), Methodology in Systems Modelling and Simulation. North-Holland, Amsterdam, pp. 447–463.
O'Neill, R.V. and Rust, B. (1979). Aggregation error in ecological models. Ecological Modelling, 7: 91–105.
Park, R.A. et al. (1974). A generalized model for simulating lake ecosystems. Simulation, August: 33–56.
Pielou, E.C. (1975). Ecological Diversity. Wiley, New York.
Popper, K.R. (1959). The Logic of Scientific Discovery. Hutchinson, London.
Reckhow, K.H. (1979). The use of a simple model and uncertainty analysis in lake management. Water Resources Bulletin, 15: 601–611.
Scavia, D. (1980a). An ecological model of Lake Ontario. Ecological Modelling, 8: 49–78.
Scavia, D. (1980b). The need for innovative verification of eutrophication models. In R.V. Thomann and T.O. Barnwell, Jr. (Editors), Workshop on Verification of Water Quality Models. EPA–600/9-80-016. US Environmental Protection Agency, Athens, Georgia.
Spear, R.C. and Hornberger, G.M. (1980). Eutrophication in Peel Inlet – II. Identification of critical uncertainties via generalized sensitivity analysis. Water Research, 14: 43–49.
Steele, J.H. (1962). Environmental control of photosynthesis in the sea. Limnology and Oceanography, 7: 137–150.
Steele, J.H. (1974). The Structure of Marine Ecosystems. Harvard University Press, Cambridge, Massachusetts.
Steele, J.H. (Editor) (1978). Spatial Pattern in Plankton Communities. Plenum Press, New York.
Straskraba, M. (1976). Development of an Analytical Phytoplankton Model with Parameters Empirically Related to Dominant Controlling Variables. Symposium für Umweltbiophysik, Abhandlungen der Akademie der Wissenschaften der DDR, Jg. 1974. Akademieverlag, Berlin, German Democratic Republic.
Straskraba, M. (1979). Natural control mechanisms in models of aquatic ecosystems. Ecological Modelling, 6: 305–321.
Tiwari, J.L., Hobbie, J.E., Reed, J.P., Stanley, D.W., and Miller, M.C. (1978) Some stochastic differential equation models for an aquatic ecosystem. Ecological Modelling, 4: 3–27.

van Straten, G. (1980). Analysis of Model and Parameter Uncertainty in Simple Phytoplankton Models for Lake Balaton. WP-80-139. International Institute for Applied Systems Analysis, Laxenburg, Austria.

van Straten, G. and de Boer, B. (1979). Sensitivity to Uncertainty in a Phytoplankton–Oxygen Model for Lowland Streams. WP-79-38. International Institute for Applied Systems Analysis, Laxenburg, Austria.

Vollenweider, R.A. (1969). Possibilities and limits of elementary models concerning the budget of substances in lakes. Archiv für Hydrobiologie, 66 (1): 1–36.

Vollenweider, R.A. (1975). Input–output models with special reference to the phosphorus loading concept in limnology. Schweizerische Zeitschrift für Hydrologie, 37: 53–84.

THE NEED FOR SIMPLE APPROACHES FOR THE ESTIMATION OF LAKE MODEL PREDICTION UNCERTAINTY

Kenneth H. Reckhow*
Michigan State University, East Lansing, Michigan (USA)

Steven C. Chapra**
National Oceanic and Atmospheric Administration, Ann Arbor, Michigan (USA)

1 INTRODUCTION

A variety of water-quality models have been developed to help assess the impact of land use on water quality. Initially, most models proposed were deterministic. However, as modelers acquired more information on the functioning of lake–watershed systems, and as engineers and planners inquired about the reliability of the models, considerations of uncertainty began to appear. Modelers who examined uncertainty in their models, and planners who demanded an estimate of the uncertainty in the techniques that they used, realized that they must have a measure of the reliability of their methods. Without this, there was no way to assess the value of the information provided by a model. Under those conditions, inefficient or incorrect decisions were more apt to be made because the model results were given too much or too little weight.

Despite the fact that many water-quality models exist and more are being developed, this does not necessarily represent a significant duplication of effort. Models are needed for a range of problems, and thus they are developed to address a variety of issues at different levels of mathematical complexity and for different degrees of spatial and temporal resolution. Thus, for a model user, the choice of model to be applied will depend upon: (1) the issue of concern; (2) the level of spatial and temporal aggregation appropriate to the issue; (3) the familiarity of the users with a particular model, or the mathematical sophistication of the user; (4) the cost and time required for acquisition of data necessary to run the model; and (5) the cost of model acquisition and model runs.

In the field of lake trophic-quality modeling, ecosystem models (Thomann et al., 1975; Scavia and Robertson, 1979) have been developed to address the problem of

---
* Present address: Duke University, Durham, North Carolina, USA.
** Present address: Texas A & M University, College Station, Texas, USA.

eutrophication in a multidimensional manner, often with a fairly high degree of spatial and temporal resolution. In order to make these models more useful in the planning process, modelers have begun to quantify the error terms for ecosystem models. As this occurs, lake ecosystem models will become even more useful for the evaluation of lake-management strategies.

At the other end of the lake-model complexity spectrum, black-box nutrient models have been proposed for the assessment of certain lake-quality issues where considerable spatial and temporal aggregation is permissible. These models are attractive to many planners and engineers because they are often more compatible with the position of the planner/engineer on the model-selection criteria mentioned above (particularly with regard to mathematical background and financial support). Since it has been shown that uncertainty analysis is relatively easily applied to the black-box model, modeling with error analysis is now being undertaken by a group of model users who might otherwise work strictly with deterministic methods.

This is not to say that all lake-model users addressing management concerns should be applying black-box models. On the contrary, model-selection criteria (1) and (2) above clearly state that the model chosen should be appropriate to the issue of concern. Certainly there are many issues of importance in lake quality that are not addressed well with a black-box model. Yet, at the same time, there are issues, and potential model users, who need simple, aggregated models, because of model-selection criteria (3), (4), and (5). Some of these users may demand an estimate of the model uncertainty. It is more likely, however, that many of these users may not have thought a great deal about uncertainty. A procedure that allows these users to calculate a numerical value for an estimate of prediction uncertainty can be a powerful tool for convincing engineers, planners, and decision-makers of the importance of uncertainty.

2 SIMPLE APPROACHES TO LAKE MODELING UNDER UNCERTAINTY

Most simple lake models have dealt with nutrient enrichment or eutrophication and stem from the work of Vollenweider (1968, 1975, 1976). While there have been a variety of expressions of Vollenweider's approach, which is generally referred to as the "phosphorus-loading concept", the models are typically designed to predict a measure of trophic state (such as total phosphorus concentration) as a function of a small number of variables representing the lake's loading, morphometry, and hydrology. Although some of the models have a theoretical basis, they are generally statistical in the sense that the functional relationships are derived from data for large numbers of lakes; in other words, they are the result of regression analysis. However, possibly because some of the seminal contributions were developed in an informal or intuitive fashion, little was done to quantify the errors associated with these models. Recently, attempts have been made to estimate the uncertainty of these models as well as to develop methods for presenting the probabilistic information in an easily utilizable format (Reckhow, 1977, 1979a,b,c; Chapra and Reckhow, 1979; Reckhow and Chapra, 1979; Reckhow and Simpson, 1980). In the present report, three of these papers, which illustrate the progress as well as the deficiencies of efforts to date, are reviewed.

2.1 Chapra and Reckhow (1979)

One of the initial attempts to quantify the uncertainty of the phosphorus-loading concept took the obvious step of determining the standard error of the regression used to derive one of Vollenweider's (1976) most widely applied models. This model is expressed mathematically as

$$[P]_\lambda = [P]_i/(1 + \tau_w^{1/2}) \tag{1}$$

where $[P]_\lambda$ is the lake's total phosphorus concentration in mg m^{-3}, $[P]_i$ is the phosphorus inflow concentration (which represents the loading divided by the water flow into the lake) in mg m^{-3}, and τ_w is the lake's water residence time in years. Data from 117 lakes in the Northern hemisphere, temperate climatic zone, were used to corroborate this model form and to estimate the standard error. Figure 1 presents a comparison of predictions, using eqn. (1) ($[\hat{P}]_{\lambda,\text{VILCM}}$), with the measured values ($[P]_\lambda$) from the 117 lakes. (The subscript VILCM denotes Vollenweider's Improved Loading Criteria Model.) Also presented in Figure 1 is the standard error expressed in terms of prediction intervals. The logarithmic transform was used to maintain homoscedacity (constant variance).

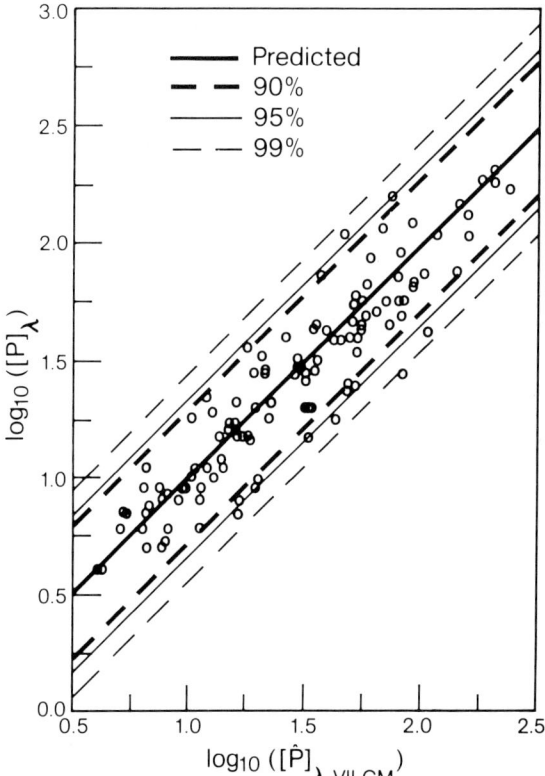

FIGURE 1 Comparison of model and data on a logarithmic scale with prediction intervals.

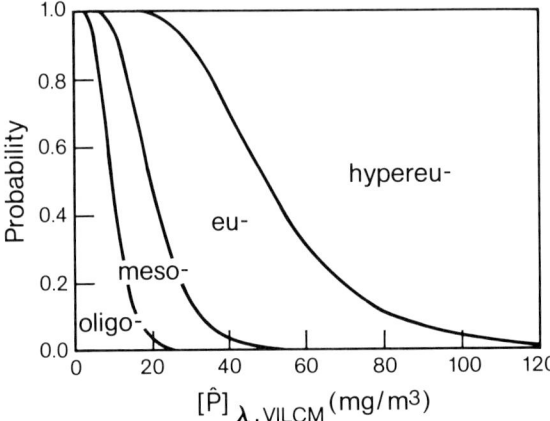

FIGURE 2 Probability of a prediction falling within a particular trophic class.

By assuming that the standard error was a valid estimate of the uncertainty of the model's predictions, two methods were presented to suggest how such information might be structured for use by a lake manager. First, limnological studies were used to express trophic state in terms of the model's dependent variable (total phosphorus concentration). For example, a lake was classified as mesotrophic when $[P]_\lambda$ fell between 10 and 20 mg m^{-3}. Then, by assuming that the standard error was normally distributed, the prediction was expressed in terms of the probability that a lake will fall within the bounds of a particular trophic state (as illustrated in Figure 2). For example, if eqn. (1) was used to predict that the lake in question would have a phosphorus concentration of 20 mg m^{-3}, Figure 2 could be used to estimate that the lake would have approximately a 4%, 46%, 49%, and 1% chance of being oligotrophic, mesotrophic, eutrophic, and hypereutrophic, respectively.

Another way in which models such as eqn. (1) are used is to determine the loading that is required to maintain a lake's trophic state at a prespecified level. Figure 3, which was also developed using the model's standard error, is designed for that application. With knowledge of a lake's residence time, this plot may be used to determine the loading level (as reflected by the inflow concentration $[P]_i$) that is needed to insure that the lake would be at a particular trophic state with a specified degree of certainty. For example, it might be desired that a lake with a residence time of 16 years be 95% certain of being better than eutrophic (< 20 mg m^{-3}). Figure 3 can then be used to estimate that the inflow concentration would have to be set at approximately 50 mg m^{-3} to attain this goal.

Obviously, the foregoing exercise entails assumptions and limitations. Some of these, such as divergence from normality and the effect of parameter error, were tested and judged to have negligible effect on the plots. Others represent important questions regarding the efficacy of the technique. The primary deficiency relates to the fact that there are errors in the estimates of both the dependent and independent variables in the model-development data set which tend to inflate the model's standard error. One way to circumvent this shortcoming is to stipulate that the model's application be

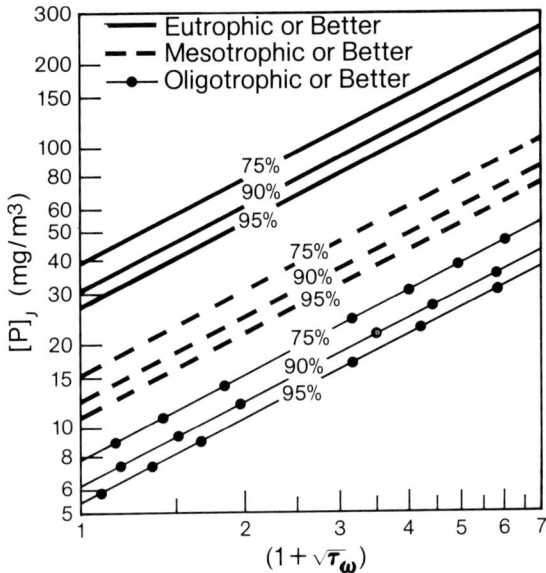

FIGURE 3 Probabilistic loading plot showing the logarithm of the predicted inflow concentration as a function of the water residence time. Percentages represent the certainty of the effectiveness of the inflow concentration achieving the expected trophic state.

limited to lakes where observations were gathered at about the same level of uncertainty as those for the model-development data set. This is obviously a severe limitation on the model's use since each lake could have a different level of uncertainty in the variables representing its condition. In particular, nutrient loadings from projected land uses cannot be measured and must be indirectly estimated, frequently from values in the literature. These extrapolated literature values may be quite uncertain. An attempt to account for this is described in the next section.

2.2 Reckhow (1979c)

In this publication, the standard error of another of Vollenweider's models was estimated. The model, which was derived from 47 north, temperate lakes, can be expressed mathematically as

$$[P]_\lambda = L/(11.6 + 1.2q_s) = yL \tag{2}$$

where L is the lake's areal phosphorus loading in $g\,m^{-2}\,year^{-1}$, and q_s is the areal water load in $m\,year^{-1}$.

In this case, the total prediction uncertainty, s_T, was expressed as

$$s_T = f(s_m, s_L) \tag{3}$$

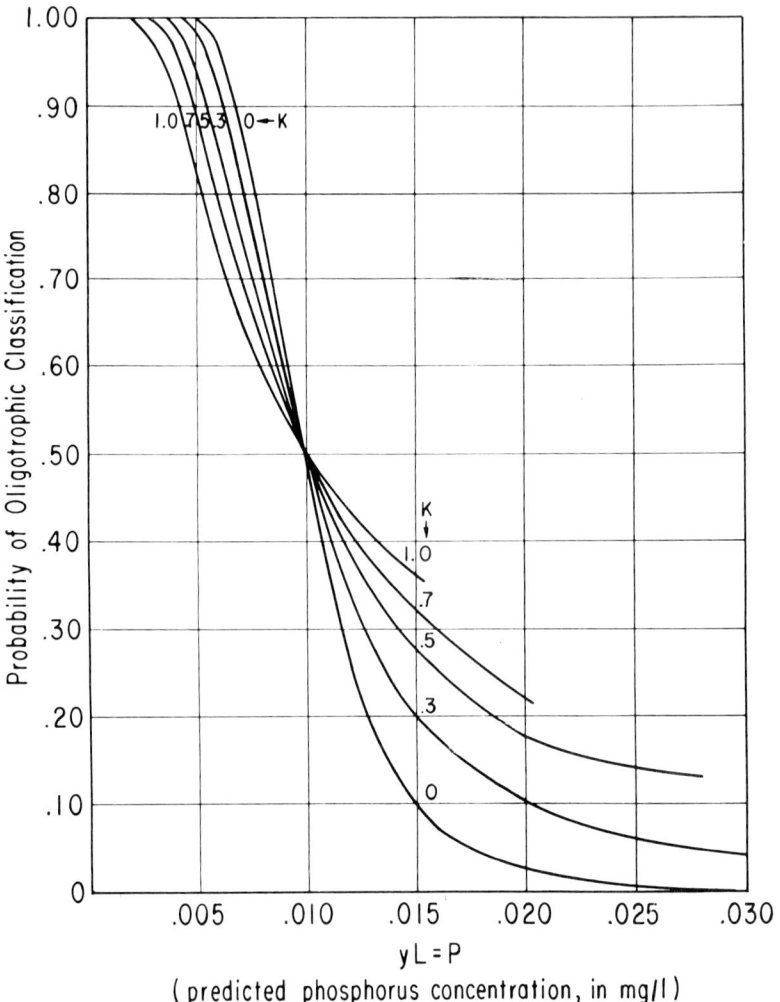

FIGURE 4 Probability of oligotrophic classification as a function of phosphorus loading, model uncertainty, and phosphorus-loading uncertainty.

where s_m is the model standard error, and s_L is the uncertainty of the loading. The error terms were converted to mutually consistent units and then added together in variance form. For mathematical convenience, s_L is expressed as a fraction, k, of the loading

$$s_L = kL \qquad (4)$$

As in Chapra and Reckhow (1979), this information can be expressed graphically (Figures 4–6). For example, for Lake Charlevoix, Michigan: $L = 0.12 \, \text{g m}^{-2} \text{year}^{-1}$, $q_s = 5.25 \, \text{m year}^{-1}$, and s_L is assumed to be $0.5L$. Using Figures 4–6, and the $k = 0.5$ line, it is then estimated that the probability of Lake Charlevoix being oligotrophic is

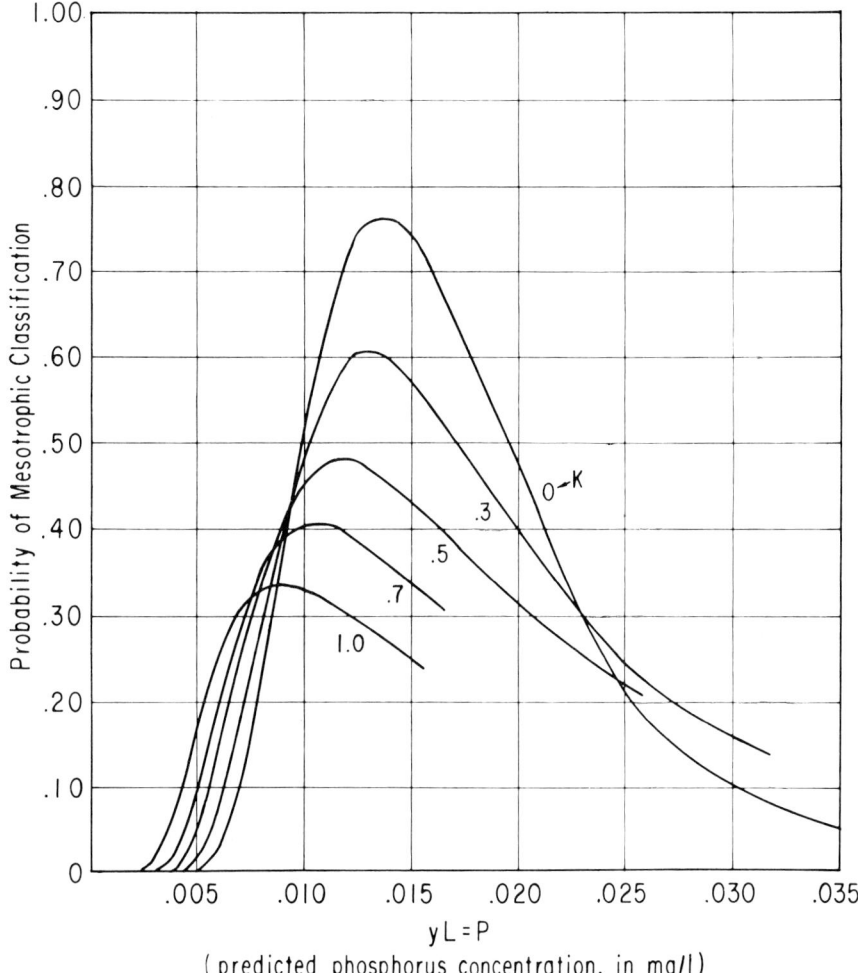

FIGURE 5 Probability of mesotrophic classification as a function of phosphorus loading, model uncertainty, and phosphorus-loading uncertainty.

approximately 0.80, the mesotrophic probability approximately 0.20, and the eutrophic probability approximately zero.

Such a procedure is valid only if certain assumptions are made. First, parameter error, dependent-variable error, and error in q_s are assumed to be negligible in comparison to the standard error of the model and the loadings. Second, when the loading is measured directly it is assumed that the error is approximately the same as for lakes in the model-development data set. In this case, the uncertainty of the loading is incorporated in the model standard error, and the trophic-state probability is estimated using Figures 4–6 with the $k = 0$ lines. In cases where the loading is approximated from the literature, k may be estimated, and then the appropriate k-curve in Figures 4–6 is used to assign trophic-state probabilities. While this involves a double counting of loading

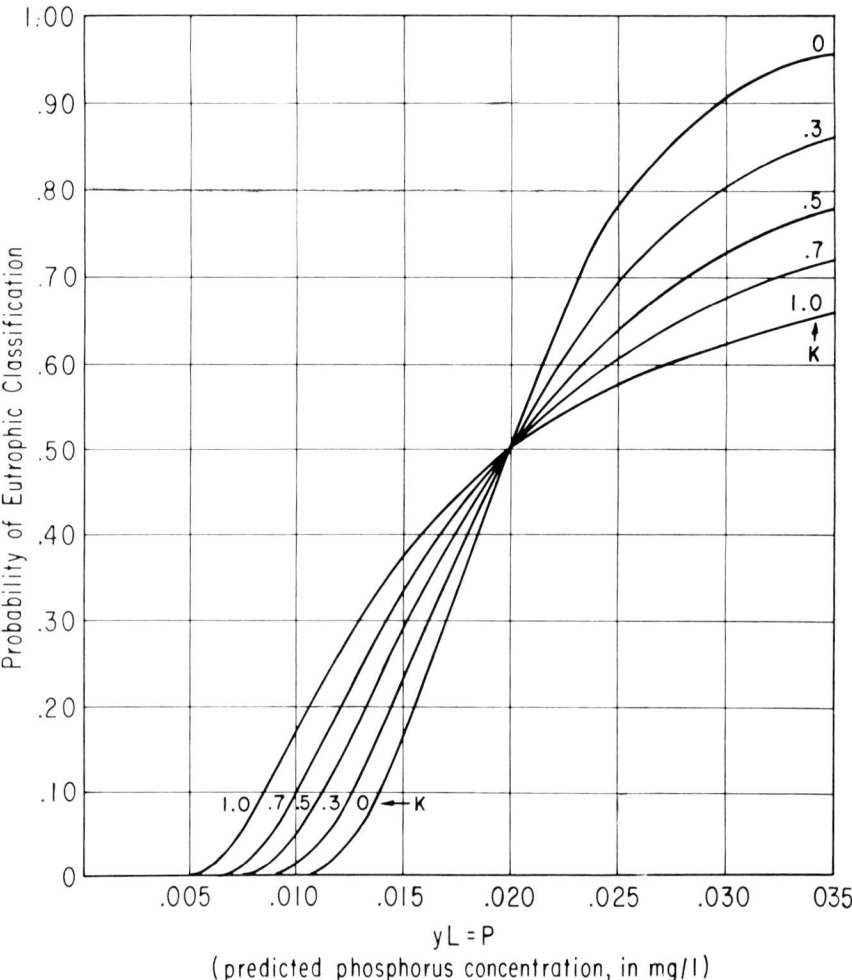

FIGURE 6 Probability of eutrophic classification as a function of phosphorus loading, model uncertainty, and phosphorus-loading uncertainty.

error, it is assumed that the component of the standard error contributed by the loading estimates in the model-development data set is small in comparison to the uncertainty associated with the literature loading estimates. The next section describes an analysis of the error terms, including a procedure for reducing double counting of errors.

2.3 Reckhow and Simpson (1980)

It should be apparent that in order to make these simple phosphorus lake models truly useful for planners and engineers examining lake-management policy, the model developer must clearly describe the interpretation of the error terms. This is important since, as noted above, a portion of the independent-variable (L and q_s) uncertainty is incorporated in the standard error for the model. This incorporated portion of variable error represents the variance and bias inherent in these variables in the data set used to construct the model. Many applications of the models (including all those for which future projection of water quality is the goal of the analysis) will be undertaken using data that were acquired in a manner different from the techniques used to acquire the model-development data set. This means that the model developer must understand the methods used to collect his data, and he must further understand what these methods mean in terms of variability and bias. This information is then conveyed to the model users as model documentation.

The model user in turn, must add, or remove, terms from the prediction error that represent the difference in variable errors between the model-development data set and the application lake. As a first attempt at suggesting how this might be accomplished, Reckhow and Simpson (1980) have taken the work of Reckhow (1979c) a bit further by proposing a step-by-step procedure for modeling the impact of watershed land use on lake trophic quality using Reckhow's model. In their paper, Reckhow and Simpson attempt to specify the issues of concern when error terms are estimated. By determining the approximate sources and levels of phosphorus-loading errors in the model-development data set, Reckhow and Simpson are able to estimate the additional loading error needed when the model is applied to other lakes. For example, they recommend that the error term for septic-system phosphorus loading be a function of: uncertainty in the projection of future population and future occupancy rate; and the estimated retention of phosphorus by the soil. This latter becomes important only when the estimated phosphorus loading from septic systems is substantial or when the average soil retention is thought to be quite different from 90%. These recommendations for the error term are based on the methods of data collection for the model-development data set, and the uncertainty in phosphorus loading expected as a result of those methods. Thus, it was determined that additional loading error, beyond that specified above, was not necessary for septic-tank phosphorus loading, because of the error already inherent in the model standard error.

Reckhow and Simpson describe the composition of the other error terms associated with variable error so that the double counting, or neglecting, of errors is minimized. This aspect of the error-term definition is most effective when the model developer has control over the design of the sampling program used to acquire the model-development data set, or when the modeler at least has intimate knowledge of this data set. Unfortunately, Reckhow and Simpson relied on data from several sources with different sampling schemes, some of which probably led to biased estimates of the independent variables. Therefore, at best, they were able only to suggest the composition of the application lake error terms based on limited knowledge and some speculation on the errors in the model-development data set.

3   WHERE DO WE GO FROM HERE?

Experience to date with the black-box nutrient model and error analysis has firmly convinced the authors that this approach is useful for lake management. Yet some issues clearly exist that need to be addressed before this modeling approach can achieve its potential effectiveness. Some suggestions are presented below for future directions of research and development with the black-box lake model.

(1) A better understanding of the data in the model-development data sets is critical. It is probably not worthwhile examining the data used to develop most existing models because of the poor sampling design used to acquire those data. However, for data gathered under a good sampling scheme, data analyses should be undertaken prior to model development. In particular, the modeler should study the cross-sectional (multilake) data distributions and the single-lake data distributions. This will help in the identification of appropriate summary statistics, data transformations, and error terms.

(2) Once the data have been thoroughly studied, a generally-applicable model may be developed if the independent-variable error in the model-development data set is removed from the model standard error. This, of course, can be undertaken only if the modeler can identify these error terms and estimate their magnitudes.

(3) An alternative is to develop a group of data set-specific models. Under this approach, different models would be proposed for data sets acquired under different conditions, so that error levels would be homogeneous within data sets. Then, essentially all of the prediction error would be contained within the model standard error, and additional error-accounting processes would not be necessary. For example, a model should be developed using indirect estimates of nutrient loading (e.g., literature nutrient-export coefficients) for use in the prediction of future trophic quality from projected land use. In that case, the only additional error term needed would be the uncertainty in the future land-use projections.

(4) Since the distributional aspects of the data and the prediction errors have been inadequately characterized, a nonparametric approach should be adopted. Existing work (Reckhow, 1979b; Reckhow and Simpson, 1980) is based on an assumption of normality, but that cannot be justified at this time. Therefore, error bounds should be constructed using nonparametric methods until the distributions are studied and identified.

(5) To examine the effectiveness of the first-order error-propagation equation (the method used here to combine errors from different sources) for this lake-modeling application, a Monte Carlo simulation should be run, after the distributions are characterized. This may be useful for a number of purposes in model development (including the analysis of a time-dependent model), and in the evaluation of lake-management strategies when the necessary resources exist.

In summary, there is a need for a variety of lake water-quality models and for an analysis of the uncertainty associated with the application of these models. Particular concerns and circumstances will favor different modeling approaches. For example, the black-box nutrient model has been found to be useful when the issue of interest may be analyzed on an aggregated basis, or when the resources of the model user preclude any other type of mathematical model. Regardless of the modeling/error analysis procedures employed, however, it is imperative that the communication between the model developer and the model user be explicit. Thus, for example, when the model documentation includes a step for "estimation of the phosphorus-loading error term", the model user must be told exactly what is meant and the exact intended composition of this term. Without this clear communication, model-prediction error may ironically be increased by a well-intentioned but ill-constructed uncertainty analysis.

REFERENCES

Chapra, S.C. and Reckhow, K.H. (1979). Expressing the phosphorus loading concept in probabilistic terms. Journal of the Fisheries Research Board of Canada, 36(2): 225–229.

Reckhow, K.H. (1977). Phosphorus Models for Lake Management. Ph.D. dissertation, Harvard University, Cambridge, Massachusetts.

Reckhow, K.H. (1979a). Quantitative Techniques for the Assessment of Lake Quality. EPA-440/5-79-015. U.S. Environmental Protection Agency, Washington, D.C.

Reckhow, K.H. (1979b). The use of a simple model and uncertainty analysis in lake management. Water Resources Bulletin, 15(3): 601–611.

Reckhow, K.H. (1979c). Uncertainty analysis applied to Vollenweider's phosphorus loading criterion. Journal of the Water Pollution Control Federation, 51(8): 2123–2128.

Reckhow, K.H. and Chapra, S.C. (1979). Error analysis for a phosphorus retention model. Water Resources Research, 15(6): 1643–1646.

Reckhow, K.H. and Simpson, J.T. (1980). A procedure using modeling and error analysis for the prediction of lake phosphorus concentration from land use information. Canadian Journal of Fisheries and Aquatic Sciences, 37(9): 1439–1448.

Scavia, D. and Robertson, A. (Editors) (1979). Perspectives on Lake Ecosystem Modeling. Ann Arbor Science Publishers, Ann Arbor, Michigan.

Thomann, R.V., DiToro, D.M., Winfield, R.P., and O'Connor, D.J. (1975). Mathematical Modeling of Phytoplankton in Lake Ontario, Part 1. Model Development and Verification. EPA-660/3-75-005. U.S. Environmental Protection Agency, Corvallis, Oregon.

Vollenweider, R.A. (1968). The Scientific Basis of Lake and Stream Eutrophication, with Particular Reference to Phosphorus and Nitrogen as Eutrophication Factors. OECD Technical Report DAS/DSI/68. OECD, Paris.

Vollenweider, R.A. (1975). Input–output models with special reference to the phosphorus loading concept in limnology. Schweizerische Zeitschrift für Hydrologie, 37: 53–84.

Vollenweider, R.A. (1976). Advances in defining critical loading levels for phosphorus in lake eutrophication. Memorie dell'Istituto Italiano di Idrobiologia, 33: 53–83.

STATISTICAL ANALYSIS OF UNCERTAINTY PROPAGATION AND MODEL ACCURACY

Dennis B. McLaughlin
Resource Management Associates, 3738 Mt. Diablo Blvd., Suite 200, Lafayette, California 94549 (USA)

1 INTRODUCTION

Until recently the subjects of model uncertainty and prediction accuracy were largely ignored by water-quality modelers. There were many reasons for this, including a widespread conviction that model predictions could be made as accurate as desired simply by increasing the detail and complexity of the governing equations. Enthusiasm for complex model structures led to a proliferation of sophisticated ecosystem models, which grew larger and larger and included more and more biological compartments, chemical interactions, etc. Unfortunately, increases in model size and complexity did not necessarily provide the expected improvements in prediction accuracy. If anything, they made the models more difficult to use and the results harder to interpret. It became apparent that the primary factor limiting model performance in many applications was not lack of detail but rather insufficiently accurate model inputs.

Classical statistics provides a convenient means for analyzing the effects of input errors on prediction accuracy when a model's basic structure is adequate, i.e., when the model has the inherent complexity and flexibility needed to reproduce observed behavior patterns. In such cases it is possible to quantify propagation of uncertainty from the inputs of the model (e.g., its coefficients, initial conditions, source terms, etc.) forward to its predictions of particular water-quality constituent concentrations. Each model input can be assigned an *a priori* covariance which measures its uncertainty or probable level of error. The error-propagation equations then predict the accuracy that would be achieved if the *a priori* levels of uncertainty were actually encountered in practice. One of the major advantages of this approach is its emphasis on "what if" questions, which focus on the relative sensitivity of each error source rather than on a single aggregated measure of accuracy.

This paper presents an analysis of uncertainty propagation which is particularly appropriate for distributed-parameter water-quality models satisfying the "structural-adequacy" requirement mentioned above. A typical example is the familar conservative constituent mass-transport equation

$$\partial C/\partial t = -\nabla \cdot VC + \nabla \cdot K\nabla C + S \tag{1}$$

where

$C(x, t)$ = concentration of the conservative constituent,
$V(x)$ = advective vector velocity (assumed, for simplicity, to be time-invariant),
$K(x)$ = time-invariant dispersivity (diffusivity) tensor,
$S(x, t)$ = source term,
x = location vector, and
t = time.

This equation is based on fundamental principles of conservation-of-mass and continuum fluid flow and should be "structurally adequate" for most water-quality transport problems. The difficulties in actually applying it result from the need for accurate specification of the following inputs: the velocity field $V(x)$ and the dispersivity tensor $K(x)$ (model parameters); the initial concentration field $C(x, t_0)$ (initial conditions); values for $C(x, t)$ or $\nabla C(x, t)$ on the boundaries of the region of interest (boundary conditions); and the source term $S(x, t)$. Error-propagation analysis provides a straightforward way to determine how errors in such inputs influence a model's predictive accuracy.

The paper begins with a review of numerical methods for solving distributed-parameter model equations such as eqn. (1). This review is followed by a derivation of the error-propagation equations and a discussion of the *a priori* covariances required in the error analysis. The final section provides a brief summary of the major results described in the paper and an assessment of the practical potential of error-propagation analysis.

2 STATE-SPACE FORMULATION OF DISTRIBUTED-PARAMETER MODELS

Most distributed-parameter water-quality models are originally formulated as partial-differential equations similar to eqn. (1). Although such equations may be solved analytically in certain special cases, numerical solution methods are required for problems having irregular geometries and/or inhomogeneities. Numerical solutions generally proceed in two phases. First, the partial-differential equation is discretized in space. Next, the resulting set of ordinary-differential equations is discretized in time to provide a set of recursive "state-space" equations. These state-space equations are the starting point for our statistical analysis of uncertainty propagation.

This section briefly reviews numerical procedures for converting the linear conservation-of-mass transport model of eqn. (1) into vector state-space form. The procedures described here apply to any linear distributed-parameter model and may, in fact, be readily extended to nonlinear models as well (see McLaughlin, 1979). Further details on numerical methods for solving partial-differential equations may be found in texts such as Zienkiewicz (1977).

The dependent variable C and the independent variables V, K, and S defined in eqn. (1) are generally continuous functions of location. In order for this equation to be solved numerically, the spatial derivatives of these functions — the terms $\nabla \cdot VC$ and

$\nabla \cdot K \nabla C$ — must be approximated or discretized. Two of the more general methods for accomplishing this discretization are the finite-difference and finite-element solution techniques. Both of these techniques conceptually superimpose a grid of N discrete "node points" on the geographical region of interest. An unknown time-dependent concentration value $\bar{C}_j(t)$ is assigned to each node ($j = 1, 2, \ldots, N$) and the continuous function $C(x, t)$ is approximated by a weighted sum of the $\bar{C}_j(t)$ values:

$$C(x, t) \simeq \eta^T(x)\bar{C} \tag{2}$$

where

$\bar{C}$ = an unknown state vector of nodal concentration values ($\bar{C}_1(t) \ldots \bar{C}_N(t)$), and

$\eta(x)$ = an unknown interpolation vector of simple location-dependent polynomials which weight the various components of $\bar{C}$.

Similar methods may be used to discretize the independent variables u, K, and S. Generally speaking, these variables need not be discretized as finely as $C(x, t)$.

When the nodal weighting technique of eqn. (2) is used to approximate spatial variations in the model's variables, eqn. (1) may be transformed into the following ordinary-differential equation:

$$\partial \bar{C}/\partial t = A(\bar{\alpha})\bar{C} + D(\bar{\alpha})\bar{u} + E(\bar{\alpha})\bar{\beta} \tag{3}$$

where

$A(\bar{\alpha})$ = an $N \times N$ matrix of coefficients which depend on the geometry of the spatial discretization grid,

$D(\bar{\alpha})$ = an $N \times M$ matrix of grid-dependent coefficients which weight the influence of the M discretized source terms,

$E(\bar{\alpha})$ = an $N \times P$ matrix of grid-dependent coefficients which weight the influence of the P discretized boundary conditions,

$\bar{u}$ = a vector of M discretized source terms derived from $S(x, t)$,

$\bar{\beta}$ = a time-invariant vector of P discretized boundary-condition terms derived from specified values of C or ∇C along the regional boundaries, and

$\bar{\alpha}$ = a vector of discretized parameter (i.e., V and K) values derived by applying an approximation similar to eqn. (2) to the functions $V(x)$ and $K(x)$.

Note that the matrices $A(\bar{\alpha})$, $D(\bar{\alpha})$, and $E(\bar{\alpha})$ are nonlinear functions of the model's discretized parameters. These parameters are lumped, for convenience, into a single vector $\bar{\alpha}$. Both $\bar{\alpha}$ and $\bar{\beta}$ are assumed to be time-invariant in order to simplify subsequent statistical expressions. These assumptions are not essential and may, in fact, need to be relaxed in certain practical applications.

A number of different temporal discretization techniques are typically used to integrate eqn. (3). Most of these approximate $\partial \bar{C}/\partial t$ and $\bar{C}$ at time t_{k+1} with functions of $\bar{C}_k$ and $\bar{C}_{k+1}$, the nodal concentration vectors at t_k and t_{k+1}, respectively. The resulting ordinary-difference equation for $\bar{C}_{k+1}$ may be written as

$$\bar{C}_{k+1} = \Phi(\bar{\alpha})\bar{C}_k + \Gamma(\bar{\alpha})\bar{u}_k + \Lambda(\bar{\alpha})\bar{\beta} \qquad (4)$$

where

$\Phi(\bar{\alpha}), \Gamma(\bar{\alpha}), \Lambda(\bar{\alpha}) = N \times N, N \times M,$ and $N \times P$ matrices that depend on $\mathbf{A}(\bar{\alpha}), \mathbf{D}(\bar{\alpha}), \mathbf{E}(\bar{\alpha}),$ and Δt (the time step); and

$\bar{u}_k = $ source term vector $\bar{u}$ evaluated at time t_k.

Equation (4) can be conveniently generalized if the coefficient matrices are allowed to vary with time. This can be indicated symbolically with appropriate subscripts:

$$\bar{C}_{k+1} = \Phi_k(\bar{\alpha})\bar{C}_k + \Gamma_k(\bar{\alpha})\bar{u}_k + \Lambda_k(\bar{\alpha})\bar{\beta} \qquad (5)$$

Equation (5) is solved recursively, starting with a vector of initial concentrations $\bar{C}_0$ defined at each of the N node points. On each time step, all time-dependent terms are updated and the next concentration vector $\bar{C}_{k+1}$ is computed.

In some applications, the variables of most interest are not the nodal concentrations but concentrations at other locations such as monitoring sites. The nodal weighting technique of eqn. (2) provides a way to predict concentrations at any desired set of locations. The method is conveniently summarized by the following equation:

$$\bar{Z}_{k+1} = \mathbf{H}\bar{C}_{k+1} \qquad (6)$$

where

$\bar{Z}_{k+1} = $ a vector of predicted concentrations (at time t_{k+1}) at L specified points $(x_1, x_2, \ldots, x_L)$ in the region of interest, and

$\mathbf{H} = $ an $L \times N$ matrix of coefficients constructed from the weighting functions $\eta(x)$ evaluated at $x_1, x_2, \ldots, x_L$.

The rows of the matrix $\mathbf{H}$ weight appropriate nodal concentrations to give interpolated concentration values at the locations $x_1, x_2, \ldots, x_L$. The resulting model output vector $\bar{Z}_{k+1}$ depends, directly or indirectly, on the initial condition, parameter, source rate, and boundary-condition values specified by the modeler. Errors in these values clearly have an effect on the accuracy of the model's predictions.

3 FIRST-ORDER ANALYSIS OF MODEL ERROR PROPAGATION

The preceding section demonstrates how a typical distributed-parameter water-resource model may be formulated in "state-space" terms, i.e., as a set of linear-difference equations. This set of equations summarizes the modeler's deterministic view of reality.

The equations for a deterministic model for simulation and prediction are

$$\bar{C}_{k+1} = \Phi_k(\bar{\alpha})\bar{C}_k + \Gamma_k(\bar{\alpha})\bar{u} + \Lambda_k(\bar{\alpha})\bar{\beta} \tag{7}$$

$$\bar{Z}_{k+1} = H\bar{C}_{k+1} \tag{8}$$

In order to perform an *a priori* analysis of the potential accuracy of this model, it is convenient to postulate a stochastic (nondeterministic) model of the real world. This stochastic model has an overall structure similar to the deterministic model of eqns. (7) and (8) but includes a number of random terms and coefficients.

The equations for a stochastic model for accuracy analysis are as follows

$$C_{k+1} = \Phi_k(\alpha)C_k + \Gamma_k(\alpha)u_k + \Lambda_k(\alpha)\beta \tag{9}$$

$$Z_{k+1} = HC_{k+1} + \omega_{k+1} \tag{10}$$

where

- α = random parameter vector,
- u_k = random source–sink vector,
- β = random boundary-condition vector,
- ω_{k+1} = random sampling-error vector (measurement noise),
- C_k = random vector of "true" nodal concentrations,
- Z_{k+1} = random vector of concentration measurement,
- C_0 = random initial condition vector, and
- $\Phi_k, \Gamma_k, \Lambda_k, H$ = system matrices defined previously.

This stochastic model can be viewed as an attempt to explain the quasirandom (never completely predictable) behavior of the real-world system. In a sense, field observations (the Z_k values of eqn. (10)) behave as if they were produced by complicated noise generators which add random variations to the "known" deterministic inputs $\bar{\alpha}, \bar{\beta}, \bar{u}_k$, and $\bar{C}$. The detrimental effects of these hypothetical noise generators can be derived from a statistical analysis of the model state and output errors.

Since $\bar{C}_{k+1}$ is the deterministic model's simulated state and C_{k+1} is the supposed "true" state, the model's state error may be measured by the difference between C_{k+1} and $\bar{C}_{k+1}$. Inspection of eqns. (7)–(10) shows that this difference is:

$$\delta_{k+1} = C_{k+1} - \bar{C}_{k+1} = \Phi_k(\alpha)C_k - \Phi_k(\bar{\alpha})\bar{C}_k + \Gamma_k(\alpha)u_k - \Gamma_k(\bar{\alpha})\bar{u}_k$$
$$+ \Lambda_k(\alpha)\beta - \Lambda_k(\bar{\alpha})\bar{\beta} \tag{11}$$

The statistical properties of the state error δ_{k+1} can be better explored if the nonlinear functions of α are expanded. This can be done with the Vetter calculus (see, for example, Dettinger and Wilson, 1979) or a simple perturbation approach may be used. The nonlinear functions of interest may be written as:

$$\Phi_k(\alpha) = \Phi_k(\bar{\alpha}) + \delta\Phi(\alpha, \bar{\alpha})$$

$$\Gamma_k(\alpha) = \Gamma_k(\bar{\alpha}) + \delta\Gamma(\alpha, \bar{\alpha})$$

$$\Lambda_k(\alpha) = \Lambda_k(\bar{\alpha}) + \delta\Lambda(\alpha, \bar{\alpha})$$

where the $\delta\Phi$, $\delta\Gamma$, $\delta\Lambda$ are random (matrix) error terms caused by differences between the parameter vectors of the deterministic and stochastic models. Similarly, the true state, source rate, and boundary-condition vectors may be written as:

$$C_k = \bar{C}_k + \delta_k$$

$$u_k = \bar{u}_k + v_k$$

$$\beta = \bar{\beta} + \delta\beta$$

where the δ_k, v_k, $\delta\beta$ are random error terms.

With these definitions, eqn. (11) may be written as a function of the known system matrices $\Phi_k(\bar{\alpha})$, $\Gamma_k(\bar{\alpha})$, and $\Lambda_k(\bar{\alpha})$ and random error terms:

$$\begin{aligned}\delta_{k+1} = {}& \Phi_k(\bar{\alpha})\delta_k + \delta\Phi(\alpha, \bar{\alpha})\bar{C}_k + \delta\Phi(\alpha, \bar{\alpha})\delta_k \\ & + \Gamma_k(\bar{\alpha})v_k + \delta\Gamma(\alpha, \bar{\alpha})\bar{u}_k + \delta\Gamma(\alpha, \bar{\alpha})v_k \\ & + \Lambda_k(\bar{\alpha})\delta\beta + \delta\Lambda(\alpha, \bar{\alpha})\bar{\beta} + \delta\Lambda(\alpha, \bar{\alpha})\delta\beta \end{aligned} \quad (12)$$

The last term on each line of this expression is a second-order error term (the product of two errors) while the other terms are first order. Note that eqn. (12) does not depend on any approximations — it is exact.

Some approximations must now be introduced if the effects of parameter errors are to be included explicitly in the error analysis. One alternative is to define two non-linear vector functions of α

$$\text{first order:} \quad f(\alpha) = \delta\Phi(\alpha, \bar{\alpha})\bar{C}_k + \delta\Gamma(\alpha, \bar{\alpha})\bar{u}_k + \delta\Lambda(\alpha, \bar{\alpha})\bar{\beta} \quad (13)$$

$$\text{second order:} \quad g(\alpha) = \delta\Phi(\alpha, \bar{\alpha})\delta_k(\alpha) + \delta\Gamma(\alpha, \bar{\alpha})v_k + \delta\Lambda(\alpha, \bar{\alpha})\delta\beta \quad (14)$$

Each of these functions may be expanded in a Taylor series about $\bar{\alpha}$, as follows:

$$f(\alpha) = f(\bar{\alpha}) + \mathbf{D}_{f\alpha}(\bar{\alpha})\delta_\alpha + \text{higher-order terms} \quad (15)$$

where

$\mathbf{D}_{f\alpha}(\bar{\alpha}) = $ matrix of $f(\alpha)$ gradients, taken with respect to α and evaluated at $\bar{\alpha}$, and
$\delta_\alpha = \alpha - \bar{\alpha}$.

The derivatives in the second-order ($g(\alpha)$) expansion are fairly difficult to evaluate, either numerically or analytically. Since we wish to illustrate the concepts of error propagation with a minimum amount of mathematical complexity, we will retain only the first-order ($f(\alpha)$) terms of eqn. (12). A first-order analysis can be informative and revealing, even though it is admittedly approximate. Further details on second-order analysis of the state error (specifically, second-order mean analysis) are presented in Dettinger and Wilson (1979).

The first-order $f(\alpha)$ expansion of eqn. (15) may be written as:

$$f(\alpha) = f(\bar{\alpha}) + (\partial/\partial\xi)[\Phi(\xi)\bar{C}_k + \Gamma(\xi)\bar{u}_k + \Lambda(\xi)\bar{\beta}]|_{\xi=\bar{\alpha}}\,\delta_\alpha$$

$$- (\partial/\partial\xi)[\Phi(\bar{\alpha})\bar{C}_k + \Gamma(\bar{\alpha})\bar{u}_k + \Lambda(\bar{\alpha})\bar{\beta}]\delta_\alpha$$

$$+ \text{second-order and other higher-order terms ignored in first-order analysis}$$

(16)

Since $f(\bar{\alpha})$ is zero and the second bracketed term in eqn. (16) is independent of ξ, $f(\alpha)$ is

$$f(\alpha) = \mathbf{D}_{\alpha k}\delta_\alpha \qquad (17)$$

where $\mathbf{D}_{\alpha k}$ is a sensitivity matrix with element ij given by

$$(\partial/\partial\xi_j)\{\Phi(\xi)\bar{C}_k + \Gamma(\xi)\bar{u}_k + \Lambda(\xi)\bar{\beta}\}_i|_{\xi=\bar{\alpha}}$$

The sensitivity matrix may be evaluated numerically at each step of the simulation, either with simple perturbations or with the Vetter calculus (see Dettinger and Wilson, 1979).

Substitution of eqn. (17) into eqn. (12) gives, when second-order terms are omitted, the following expression for the state error:

$$\delta_{k+1} = \Phi_k(\bar{\alpha})\delta_k + \Gamma_k(\bar{\alpha})\mathbf{v}_k + \Lambda_k(\bar{\alpha})\beta + \mathbf{D}_{\alpha k}\delta_\alpha \qquad (18)$$

This equation contains only deterministic (known) coefficient matrices ($\Phi_k(\bar{\alpha})$, $\Gamma_k(\bar{\alpha})$, $\Lambda_k(\bar{\alpha})$, and $\mathbf{D}_{\alpha k}$) and random error vectors ($\delta_k, \mathbf{v}_k, \beta, \delta_\alpha$). The statistical properties of δ_{k+1} may be recursively derived from eqn. (18) if the properties of all the noise-generating random vectors are defined. In particular, the first and second moments of each random error source must be specified, together with appropriate information on cross-correlations between errors and on the temporal correlation properties of $\mathbf{v}_k$. Although many alternatives are possible, the following seem reasonable:

First moments (means)

$$E[\delta_0] = E[\beta] = E[\delta_\alpha] = 0$$

$$E[\mathbf{v}_k] = 0 \quad \text{(for all } k\text{)}$$

Second moments (covariances)

$$\text{cov}[\boldsymbol{\delta}_0] = \mathbf{P}_0$$

$$\text{cov}[\boldsymbol{\beta}] = \mathbf{P}_\beta$$

$$\text{cov}[\boldsymbol{\delta}_\alpha] = \mathbf{P}_\alpha$$

$$\text{cov}[\mathbf{v}_k] = \mathbf{Q}_k$$

Cross-correlations

$$E[\boldsymbol{\delta}_0 \boldsymbol{\beta}] = E[\boldsymbol{\delta}_0 \boldsymbol{\delta}_\alpha] = E[\boldsymbol{\delta}_0 \mathbf{v}_k] = E[\boldsymbol{\beta} \boldsymbol{\delta}_\alpha] = E[\boldsymbol{\beta} \mathbf{v}_k] = E[\boldsymbol{\delta}_\alpha \mathbf{v}_k] = 0$$

Temporal correlation of $\mathbf{v}_k$

$$E[\mathbf{v}_k \mathbf{v}_j] = 0 \quad (\text{for all } j \neq k)$$

The rationales for the assumptions reflected in these definitions are as follows:

First moments. It is reasonable to assume that all error means are zero since if any error had a known nonzero mean, this mean would be incorporated into the deterministic model as a compensation factor. Of course, the errors may have unknown nonzero means but there is no way to deal with this without complicating the error analysis further.

Cross-correlation. It is reasonable to suppose that the fundamental error sources — initial condition errors, parameter errors, source–sink errors, and boundary-condition errors — arise from different statistically independent effects (this is one of the reasons the errors are classified into four separate categories). Statistical independence implies zero correlation.

Temporal correlation of $\mathbf{v}_k$. In many situations the errors influencing source–sink terms are statistically independent from time to time, location to location, and sample to sample. When available field information indicates this is not the case, the error equations may be modified to account for temporal correlation of $\mathbf{v}_k$, provided that good estimates of the correlation function can be obtained. Otherwise, it is best to assume that the source–sink errors are completely uncorrelated.

With the above assumptions, the first and second moments of $\boldsymbol{\delta}_{k+1}$ may be obtained directly from eqn. (18):

$$E[\boldsymbol{\delta}_{k+1}] = 0 \tag{19}$$

$$\text{cov}[\boldsymbol{\delta}_{k+1}] = \mathbf{P}_{k+1} = \boldsymbol{\Phi}_k \mathbf{P}_k \boldsymbol{\Phi}_k^T + \boldsymbol{\Gamma}_k \mathbf{Q}_k \boldsymbol{\Gamma}_k^T + \boldsymbol{\Lambda}_k \mathbf{P}_\beta \boldsymbol{\Lambda}_k^T + \mathbf{D}_{\alpha k} \mathbf{P}_\alpha \mathbf{D}_{\alpha k}^T$$

$$+ \boldsymbol{\Phi}_k \mathbf{P}_{k\beta} \boldsymbol{\Lambda}_k^T + \boldsymbol{\Lambda}_k \mathbf{P}_{\beta k} \boldsymbol{\Phi}_k^T + \boldsymbol{\Phi}_k \mathbf{P}_{k\alpha} \mathbf{D}_{\alpha k}^T + \mathbf{D}_{\alpha k} \mathbf{P}_{\alpha k} \boldsymbol{\Phi}_k^T \tag{20}$$

where

$$P_k = \text{cov}[\delta_k], \text{ and}$$
$P_{k\beta}, P_{\beta k}, P_{k\alpha}, P_{\alpha k}$ = cross-covariances defined below.

The dependence of the coefficient matrices on $\bar{\alpha}$ is now assumed rather than indicated explicitly. Also, note that δ_k and v_k are uncorrelated because δ_k depends only on v_0 through v_{k-1}, which are uncorrelated with v_k.

The first-order state error mean of eqn. (19) is zero because all of the component error source means are zero (the second-order state error mean is not zero but, instead, depends on the error source second moments). The first-order state error covariance of eqn. (20) depends on three types of terms: the propagated error covariances, the propagated cross-covariances of δ_k and β ($P_{k\beta}$ and $P_{\beta k}$), and the propagated cross-covariances of δ_k and α ($P_{k\alpha}$ and $P_{\alpha k}$). The error covariances P_0, Q_k, P_β, and P_α are specified (i.e., assumed *a priori*). The cross-covariances $P_{k\beta}$, $P_{\beta k}$, $P_{\alpha k}$, and $P_{k\alpha}$ are, on the other hand, obtained from separate recursive equations. This is illustrated by the derivation of $P_{k+1,\beta}$ from eqn. (18):

$$P_{k+1,\beta} = E[\delta_{k+1}\beta^T] = \Phi_k E[\delta_k \beta^T] + \Lambda_k E[\beta\beta^T]$$

or

$$P_{k+1,\beta} = \Phi_k P_{k\beta} + \Lambda_k P_\beta; \quad P_{0\beta} = 0 \tag{21}$$

Note that the cross-terms containing $E[v_k \beta^T]$ and $E[\delta_\alpha \beta^T]$ are zero because β is assumed uncorrelated with v_k and δ_α. Also, $P_{0\beta}$ is zero because the initial condition error δ_0 and boundary condition error β are assumed to be uncorrelated. Similar reasoning may be used to show that:

$$P_{k+1,\alpha} = \Phi_k P_{k\alpha} + D_{\alpha k} P_\alpha; \quad P_{0\alpha} = 0 \tag{22}$$

The other cross-covariances $P_{\beta,k+1}$ and $P_{\alpha,k+1}$ are the transposes of $P_{k+1,\beta}$ and $P_{k+1,\alpha}$.

Equations (20)–(22) constitute a coupled set of recursive algorithms for computing the error covariance of the model state vector at any time. The first term ($\Phi_k P_k \Phi_k^T$) of eqn. (20) describes the evolution of uncertainty in C_k forward to C_{k+1}. If the linear system is stable, scalar metrics of $\Phi_k P_k \Phi_k^T$ will be smaller than P_k, indicating that the system dynamics tend to damp initial-condition uncertainty (in a sense, C_k is the initial condition for C_{k+1}). The next three terms of eqn. (20) describe the incremental addition of uncertainty between times t_k and t_{k+1} due to random source, boundary condition, and parameter errors. Since the boundary condition and parameter errors are assumed to be time-invariant, their incremental uncertainty contributions (from time step to time step) will be correlated. The last four terms of eqn. (20) are "correction factors" which account for this correlation effect. The correction factors depend on cross-covariances such as $P_{k\beta}$ which are, in turn, derived recursively (in eqns. 21 and 22) from the parameter and boundary-condition covariances, P_α and P_β.

The state error covariance is a measure of the "closeness" of the simulated state $\bar{C}_k$ to the true state C_k. A similar error covariance can be derived to measure the closeness of the simulated model output $\bar{Z}_k$ to the field measurement vector Z_k. Equations (8) and (10) imply that:

$$\epsilon_{k+1} = Z_{k+1} - \bar{Z}_{k+1} = H\delta_{k+1} + \omega_{k+1} \tag{23}$$

The statistical properties of the measurement error ω_{k+1} may be defined by analogy with the model error sources discussed earlier:

$$E[\omega_{k+1}] = 0$$

$$\text{cov}[\omega_{k+1}] = R_{k+1}$$

$$E[\omega_{k+1}\delta_0] = E[\omega_{k+1}\delta_\alpha] = E[\omega_{k+1}v_k] = E[\omega_{k+1}\beta] = 0$$

$$E[\omega_k \omega_j] = 0 \quad \text{for all } j \neq k$$

It is reasonable to assume that the measurement error mean is zero since any known nonzero mean (bias) would be subtracted from the measurement during or subsequent to field sampling. It is also reasonable to assume that the measurement error during the simulation period (i.e., after model calibration/parameter estimation has been completed) is uncorrelated with the other random errors included in our analysis. Output measurement errors during this period generally arise independently of any errors influencing the model's initial conditions, source terms, boundary conditions, or parameters. It is worth noting, however, that the parameter error may be correlated with output measurements taken prior to the simulation period if these earlier measurements were used to estimate the parameter vector $\bar{\alpha}$. This prior parameter/measurement correlation does not influence $E[\omega_{k+1}\delta_\alpha]$ because the measurement error sequence is assumed to be temporally uncorrelated, i.e., ω_{k+1} is not correlated with errors in the measurements used to estimate $\bar{\alpha}$.

With the statistical properties of the measurement error defined, the first and second moments of ϵ_{k+1} may be obtained directly from eqn. (23):

$$E[\epsilon_{k+1}] = 0 \tag{24}$$

$$\text{cov}[\epsilon_{k+1}] = HP_{k+1}H^T + R_{k+1} \tag{25}$$

Also of interest is the mean of the scalar sum-squared output error, which is often used as an empirical measure of model accuracy:

$$E[S] = \sum_k E[\epsilon_{k+1}^T \epsilon_{k+1}] = \sum_k \{E[\delta_{k+1}^T H^T H \delta_{k+1}] + E[\omega_{k+1}^T \omega_{k+1}]\}$$

This expression reduces (see Kendall and Stuart, 1973) to:

$$E[S] = \sum_k \{\text{Tr}[H^T P_{k+1} H] + \text{Tr}[R_{k+1}]\} \tag{26}$$

where $\text{Tr}[\cdot]$ is the scalar trace of the matrix in brackets. Equation (26) suggests that the sum-squared error computed from a comparison of model outputs with field observations will depend on both model errors ($\mathbf{H}_k^T \mathbf{P}_{k+1} \mathbf{H}_k$) and measurement errors ($\mathbf{R}_{k+1}$). The sum-squared error computed from any particular set of field observations is, of course, a single sample from a random population having a mean $\mathbf{E}[S]$. The variance of this population depends on higher-order moments of $\boldsymbol{\delta}_{k+1}$ and $\boldsymbol{\omega}_{k+1}$ and generally decreases as the number of measurements increases.

Equations (24)–(26) complete our analysis of uncertainty propagation. Although the error covariance equations derived in this analysis appear complex at first glance, they may be readily incorporated into a model's normal computational cycle. The only complication is evaluation of the parameter sensitivity matrix $\mathbf{D}_{\alpha k}$, which adds somewhat to the overall computation cost since an additional (perturbed) simulation must be performed at each time step. In return, the modeler obtains a step-by-step, node-by-node breakdown of his model's performance which provides estimates of the relative influence of each potential error source. Needless to say, this information can be valuable not only to the model-user concerned with prediction accuracy but also to the model-developer seeking to reduce the detrimental effects of input errors.

4 SPECIFICATION OF *A PRIORI* COVARIANCES

The error-propagation analysis presented in the preceding section is implicitly based on a Bayesian viewpoint, which assumes that an *a priori* probability density function can be assigned to each error source. The *a priori* covariances $\mathbf{P}_0, \mathbf{P}_\beta, \mathbf{P}_\alpha, \mathbf{Q}_k$, and $\mathbf{R}_k$ are the covariances of these density functions. If no measurements of input or sampling error are available, the *a priori* density functions reflect the modeler's own subjective uncertainty about the "true" values of the model inputs and field measurements. In some applications, the *a priori* densities must be based entirely on intuition and common sense. But often enough data are available to suggest at least a range of reasonable values for *a priori* statistics. The types of data likely to be used to help define *a priori* covariances in water-resource applications are discussed briefly below.

4.1 Errors in Initial Conditions and Boundary Conditions

The initial conditions and boundary conditions needed to solve eqn. (1) or other similar distributed-parameter model equations are often estimated from field samples taken at scattered monitoring sites. Even if field-measurement procedures are perfect (which they rarely are in water-quality applications), errors arise when the measurements are extrapolated over an entire nodal grid. The magnitudes and distribution of these errors depend, of course, on the extrapolation/interpolation method used. Automated interpolation methods such as spline-fitting routines or Kriging algorithms are particularly convenient from an error analysis point of view because they provide estimates of their own accuracy (see Delhomme, 1976; Moore and McLaughlin, 1980). Most Kriging programs supply contour maps of estimation error variance which define the initial-condition error covariance $\mathbf{P}_0$ and boundary-condition error covariance $\mathbf{P}_\beta$ used in eqns. (20) and (21). These maps can also help the modeler decide where to draw regional boundaries and where to propose additional sampling sites.

4.2 Source Rate Errors

In most water-quality simulation studies, source rates such as pollutant-loadings are either hypothesized or inferred from extrapolations of historical measurements. In the first case, the source rate errors depend on the accuracy of the modeler's hypotheses and must generally be estimated subjectively. In the second case, the errors depend on the accuracy of the historical measurements and the extrapolation procedure used to extend these measurements forward in time. If the extrapolation procedure can be described statistically it may be possible to assign approximate variances to the source rate errors associated with a particular simulation. For example, the mean, variance, and autocorrelation function of pollutant-loading rates extrapolated from a least-squares trend line may be readily derived if certain assumptions are made about the statistical properties of the residual errors. Such statistical measures can help define the elements of the source rate error covariance $\mathbf{Q}_k$ used in eqn. (20).

4.3 Measurement Errors

Measurement errors depend on both field-sampling and laboratory-analysis procedures. The statistical properties of such errors can be derived from a careful analysis of each stage of the sampling procedure, using manufacturer's specifications or published guidelines to establish the accuracy of measuring instruments or laboratory tests. Alternatively, replicate samples can be collected in the field and the pooled statistics of these replicates used to define the error covariance matrix $\mathbf{R}_{k+1}$ required by eqns. (25) and (26).

4.4 Parameter Errors

Model parameters are frequently estimated indirectly from field measurements of related variables, often with the aid of the simulation model itself. The diffusion coefficients of a transport model such as eqn. (1) may, for example, be estimated from historical measurements of pollutant concentration during periods when loading rates and boundary conditions were well known. A wide variety of parameter-estimation procedures have been proposed in the literature, including such statistically-oriented techniques as least-squares, Kalman filtering, instrumental variables, and maximum likelihood (see Beck and Young, 1976; Lettenmaier and Burges, 1976; Moore, 1978). Most of these procedures include algorithms for computing the covariances of their parameter estimates, given certain assumptions about the model and measurement errors acting when the estimates were derived. These covariances may be used to define the matrix $\mathbf{P}_\alpha$ required in eqns. (20) and (22).

It should be emphasized that the covariance values supplied by statistical analyses such as those outlined above are approximate and can be modified by the modeler if desired. The Bayesian approach does not require that the *a priori* covariances in eqns. (20)–(26) be estimated from field data. Derived estimates of such covariances serve merely to refine the modeler's subjective assessment of input and measurement uncertainties.

5 SUMMARY AND CONCLUSIONS

The error-propagation analysis presented in this paper provides a set of recursive equations for computing the mean and covariance of the states of a linear distributed-parameter simulation model. The equations, which are summarized in the Appendix, depend on two types of matrix variables:

- *Coefficient matrices* – the matrices $\mathbf{\Phi}_k$, $\mathbf{\Gamma}_k$, $\mathbf{\Lambda}_k$, $\mathbf{D}_{\alpha k}$, and $\mathbf{H}$, which may be numerically or analytically derived from the model's governing equations, discretization procedures, and interpolation algorithms.
- *A priori covariance matrices* – the matrices $\mathbf{P}_0$, $\mathbf{Q}_k$, $\mathbf{P}_\alpha$, $\mathbf{P}_\beta$, and $\mathbf{R}_k$, which are measures of the modeler's *a priori* uncertainty about the magnitudes of errors in initial conditions, source rates, parameters, boundary conditions, and field measurements.

In practice, the modeler assigns *a priori* covariance matrices which reflect the levels of input uncertainty likely to be encountered during a model simulation run. The error-propagation equations then predict the accuracy (error covariance) that would be achieved if these *a priori* levels of uncertainty were actually attained. This approach to error analysis allows the modeler to evaluate the effects of a range of input uncertainties, from very optimistic levels to very pessimistic levels. The "most likely" input covariances falling in the middle of this range can be based on quantitative measures of input accuracy derived from statistical analyses of field data, if sufficient data are available.

The error-propagation algorithms derived in this paper are easy to program and can, in fact, be incorporated into most simulation models with a minimal amount of effort. Although there are few examples of formal error-propagation studies in the water-quality field, it is likely that the number of practical applications will increase dramatically in the near future. Related studies of groundwater flow models have already been reported by Dettinger and Wilson (1979). As such applications become more common, water-quality modelers will undoubtedly begin to appreciate the advantages of quantitative error analysis and will rely on it increasingly, both as a research tool and as a practical method for evaluating model performance.

REFERENCES

Beck, M.B. and Young, P.C. (1976). Systematic identification of a DO–BOD model structure. Journal of the Environmental Engineering Division, American Society of Civil Engineers, 102 (EE5): 909–927.

Delhomme, J.P. (1976). Kriging in Hydrosciences. Condensed version of Ph.D. Thesis, University of Paris, Ecole Nationale Supérieure, Des Mines de Paris.

Dettinger, M.D. and Wilson, J.L. (1979). Numerical Modeling of Aquifer Systems under Uncertainty: A Second Moment Analysis. MIT/Cairo University Technological Planning Program Working Paper No. 1. Department of Civil Engineering, Massachusetts Institute of Technology, Cambridge, Massachusetts.

Kendall, M.G. and Stuart, A. (1973). The Advanced Theory of Statistics. Vol. 2, Inference and Relationship. Hafner Publishing Co., New York.

Lettenmaier, D.P. and Burges, S.J. (1976). Use of state estimation techniques in water resources system modeling. Water Resources Bulletin, 12(1): 88–99.

McLaughlin, D.B. (1979). Hanford Groundwater Modeling – A Numerical Comparison of Bayesian and Fisher Parameter Estimation Techniques. Rockwell Hanford Operations Consultants Report RHO-C-24 (available from Resource Management Associates, 3738 Mt. Diablo Blvd., Suite 200, Lafayette, California 94549, USA).

Moore, S.F. (1978). Applications of Kalman filter to water quality studies. In C.L. Chiu (Editor), Applications of Kalman Filter to Hydrology, Hydraulics and Water Resources. Stochastic Hydraulics Program, Department of Civil Engineering, University of Pittsburgh, Pittsburgh, Pennsylvania, pp. 485–499.

Moore, S.F. and McLaughlin, D.B. (1980). Computer Mapping of Contaminant Plumes in an Arid Site Vadose Zone. Final report for Rockwell Hanford Operations, Richland, Washington (available from Resource Management Associates, 3738 Mt. Diablo Blvd., Suite 200, Lafayette, California 94549, USA).

Zienkiewicz, O.C. (1977). The Finite Element Method. McGraw-Hill, London.

APPENDIX: SUMMARY OF FIRST-ORDER ERROR-PROPAGATION EQUATIONS

Error Means

State error mean (eqn. 19): $\quad E[\delta_{k+1}] = 0 \quad$ (for all k)

Measurement error mean (eqn. 24): $\quad E[\varepsilon_{k+1}] = 0 \quad$ (for all k)

Recursive Error Covariance Equations

State error covariance (eqn. 20):

$$P_{k+1} = \underbrace{\Phi_k P_k \Phi_k^T}_{\substack{\text{state} \\ \text{error} \\ \text{propagation}}} + \underbrace{\Gamma_k Q_k \Gamma_k^T}_{\substack{\text{source} \\ \text{error} \\ \text{propagation}}} + \underbrace{\Lambda_k P_\beta \Lambda_k^T}_{\substack{\text{boundary} \\ \text{condition} \\ \text{error} \\ \text{propagation}}} + \underbrace{D_{\alpha k} P_\alpha D_{\alpha k}^T}_{\substack{\text{parameter} \\ \text{error} \\ \text{propagation}}}$$

$$+ \underbrace{\Phi_k P_{k\beta} \Lambda_k^T + \Lambda_k P_{\beta k} \Phi_k^T}_{\text{state-boundary condition correlations}}$$

$$+ \underbrace{\Phi_k P_{k\alpha} D_{\alpha k}^T + D_{\alpha k} P_{\alpha k} \Phi_k^T}_{\text{state-parameter correlations}}$$

State-boundary condition cross-covariance (eqn. 21): $\mathbf{P}_{k+1,\beta} = \mathbf{\Phi}_k \mathbf{P}_{k\beta} + \mathbf{\Lambda}_k \mathbf{P}_\beta$
State-parameter cross-covariance (eqn. 22): $\mathbf{P}_{k+1,\alpha} = \mathbf{\Phi}_k \mathbf{P}_{k\alpha} + \mathbf{D}_{\alpha k} \mathbf{P}_\alpha$
Measurement-error covariance (eqn. 25): $\text{cov}[\mathbf{\epsilon}_{k+1}] = \mathbf{H} \mathbf{P}_{k+1} \mathbf{H}^T + \mathbf{R}_{k+1}$

Error Covariance Initial Conditions

Initial state error covariance: $\mathbf{P}_0 \triangleq \mathbf{P}_0$
Initial state boundary condition cross-covariance: $\mathbf{P}_{0\beta} = 0$
Initial state parameter cross-covariance: $\mathbf{P}_{0\alpha} = 0$

Variable Definitions

$\mathbf{\Phi}_k, \mathbf{\Gamma}_k, \mathbf{\Lambda}_k$ = Time-dependent coefficient matrices derived from the model's governing equations and discretization procedures,

$\mathbf{D}_{\alpha k}$ = Parameter-sensitivity matrix evaluated at each time step with element ij given by

$$(\partial/\partial \xi_j)\{\mathbf{\Phi}(\xi)\bar{C}_k + \mathbf{\Gamma}(\xi)\bar{u}_k + \mathbf{\Lambda}(\xi)\bar{\beta}\}_i|_{\xi=\alpha}$$

$\mathbf{H}$ = Matrix of coefficients constructed from spatial interpolation functions relating nodal locations to measurement locations,

$\mathbf{P}_0, \mathbf{Q}_k, \mathbf{P}_\beta, \mathbf{P}_\alpha, \mathbf{R}_k$ = *A priori* covariance matrices for errors in initial conditions, source rates, boundary conditions, parameters, and measurements, respectively,

$\mathbf{P}_{k+1}$ = Covariance of the error between true and simulated nodal concentrations at time $k+1$,

$\text{cov}[\mathbf{\epsilon}_{k+1}]$ = Covariance of the error between model outputs and field measurements at time $k+1$,

$\mathbf{P}_{k\beta}, \mathbf{P}_{k\alpha}, \mathbf{P}_{\alpha k}, \mathbf{P}_{\beta k}$ = Cross-covariances computed as intermediate variables in the error-propagation equations.

Detailed definitions of these variables are provided in Sections 2 and 3 of this paper.

MODELING AND FORECASTING WATER QUALITY IN NONTIDAL RIVERS: THE BEDFORD OUSE STUDY

P.G. Whitehead
Institute of Hydrology, Wallingford, Oxfordshire (UK)

1 INTRODUCTION

In addition to being the major sources of water, river systems are used as the principal disposal pathways for waste material from man's activities. Such waste material alters the concentration of many chemical substances in water and impairs the quality and thus the usefulness of that water. Moreover, the variety of pollutants generated by a highly industrial society appears to grow continuously and as discussed by Stott (1979), "the problems of water quality are now more difficult and demanding than water quantity".

While, in general, average water quality in the UK has tended to improve, in certain respects there have been grounds for concern. For example, some water authorities have been observing progressively increasing levels of nitrates in their system. The mechanisms governing these increases are not wholly understood and, as a result, strategies for the management of nitrate levels have not been fully identified. In particular, nitrate levels in the River Thames and the River Lea have increased dramatically over the past ten years with the average concentration increasing from $4\,\text{mg}\,\text{N}\,\text{l}^{-1}$ in 1968 to an average of $11.1\,\text{mg}\,\text{N}\,\text{l}^{-1}$ in 1977 in the River Lea (Thames Water Statistics, 1978). This level is close to the World Health Organization (WHO) limit of $11.3\,\text{mg}\,\text{N}\,\text{l}^{-1}$ and at certain times of the year nitrate levels in the River Lea have in fact exceeded the WHO limit, thereby preventing the abstraction of water for potable supply. Moreover, the observation that certain acceptable limits of quality are exceeded from time to time indicates that desirable stream quality is not only quantified in terms of, say, yearly average indices; transient, intermittent deterioration of quality is also important, and may be of growing concern for the future.

In this paper water-quality models developed during the recent Bedford Ouse Study (Bedford Ouse Study, 1979; Whitehead et al., 1979, 1981) are briefly described and applied to assess the impact of effluent on the river system. Concern over the future water quality in the Bedford Ouse has led to the development of an extensive automatic water-quality monitoring and computer-controlled telemetry system. Water-quality models are included in the mini/microcomputer system and provide forecasts for operational management. In this paper, models of ammonia and dissolved oxygen are developed using the extended Kalman filter (EKF) technique applied to data obtained from the automatic monitors and the utility of such forecasting schemes is discussed.

2 MODELING FOR WATER-QUALITY MANAGEMENT

There has been a tendency in recent years to categorize water-quality models as either planning or operational management aids. However, such a breakdown is not strictly correct since planning models provide the "steady state" or annual average water-quality conditions and identify measures which alter the natural distribution of water quality in time and space in accordance with an overall development objective. Steady-state planning models do not account for the uncertainties in the system such as errors associated with sampling measurement and the imprecise knowledge of system mechanisms; they provide only a rough guide to likely future water-quality levels.

By contrast operational management is concerned with the short-term (hourly or daily) behavior of water quality and models are required for selecting optimal operating rules and control procedures and providing real-time forecasts of water quality in river systems.

A third intermediate stage between planning and operational models is required during the detailed design of a water-resource system. Here, there must be some consideration of risk and information on the day-to-day changes in river quality is required, since it is the transient violation of water-quality standards that creates particular problems. The approach of digital simulation provides a convenient method of analyzing systems during this design phase and historic and synthetic inputs can be simulated and information on the distributions of water quality used to assess risk.

If the model is to be useful for the purpose of design it should possess the following properties:

1. It should be a truly dynamic model, capable of accepting time-varying input (upstream) functions of water quality which are used to compute time-varying output (downstream) responses.
2. The model should be as simple as possible yet consistent with the ability to characterize adequately the important dynamic and steady-state aspects of the system behavior.
3. It should provide a reasonable mathematical approximation of the physicochemical changes occurring in the river system and should be calibrated against real data collected from the river at a sufficiently high frequency and for a sufficiently long period of time.
4. It should account for both the inevitable errors associated with laboratory analysis and sampling, and the uncertainty associated with imprecise knowledge of the pertinent physical, chemical, and biological mechanisms.

3 AN INTEGRATED MODEL OF FLOW AND WATER QUALITY

Mathematical models which satisfy these four properties have been developed during the recent Bedford Ouse Study (Whitehead et al., 1979, 1981) and the principal

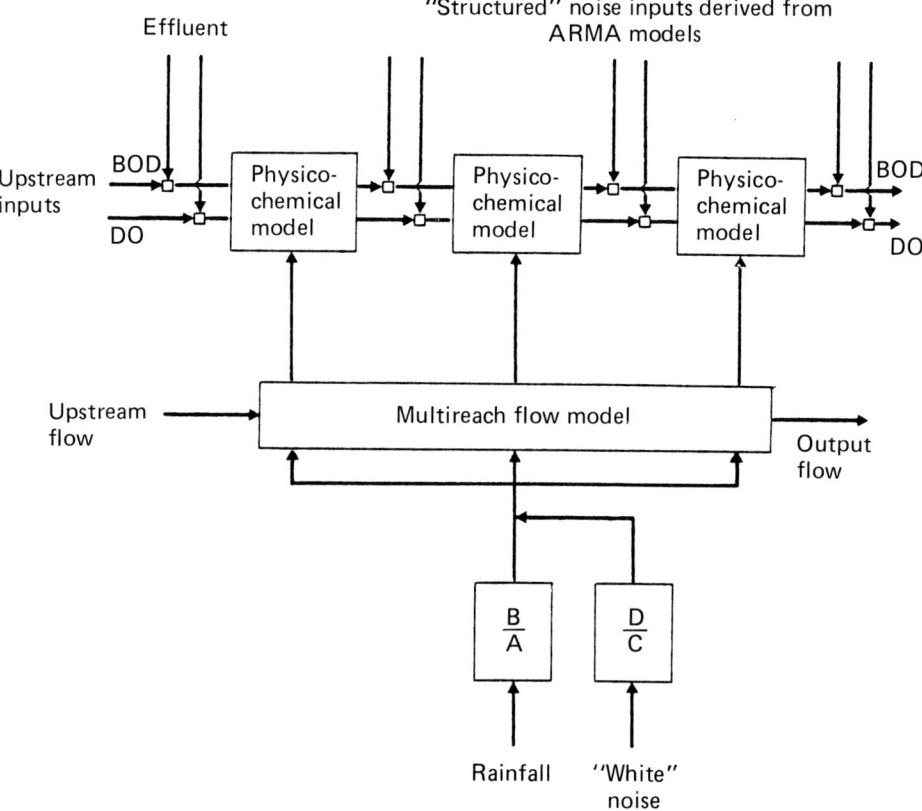

FIGURE 1 Interaction between hydrological and water-quality models.

interactions between flow and water-quality components of the model are illustrated in Figure 1. The underlying hydrology of a river system is modeled using a deterministic nonlinear storage model to relate flow variations at downstream points on the system to input flows at the upstream system boundaries. Having accounted for most of the flow variations with the deterministic streamflow model, the residual between the deterministic model output and the observed downstream flow is modeled using stochastic methods of time-series analysis (Whitehead, 1979). The stochastic time-series models represent the residual flow variations due to rainfall and runoff effects. As shown in Figure 1, information on flow is transferred to physicochemical models of water quality which contain the principal mechanisms governing water-quality behavior, based on a mass balance over the reach.

The structure of these models is based on a transportation delay/continuously stirred reactor (CSTR) idealization of a river (Beck and Young, 1976). The mathematical formulation of this model is in terms of lumped-parameter, ordinary-differential equations and draws upon standard elements of chemical engineering reactor analysis (see, for example, Himmelblau and Bischoff, 1968). As indicated by Whitehead et al. (1979), this idealization can be shown to approximate the analytical properties of the distributed-parameter, partial-differential equation representations of advection dispersion mass transport in addition to the experimentally observed transport and dispersion mechanisms (Whitehead, 1980).

The principal advantages of this model over the equivalent partial-differential equation descriptions are:

(a) the simplified computation required to solve the lumped-parameter differential equations;
(b) the availability of statistically efficient algorithms for model identification and parameter estimation, which can only be readily applied to the lumped-parameter form;
(c) the availability of extensive control system methods which may be used for management purposes and are most suited to the ordinary-differential equation model.

The mathematical form of the model is derived from a component mass balance. For the CSTR

$$(dx/dt)(t) = (Q/V)(t)\tilde{u}(t) - (Q/V)(t)x(t) + S(t) + \zeta(t) \tag{1}$$

and for the transportation delay

$$\tilde{u}(t) = u(t - \tau(t))$$

where

$u(t)$ = the vector of input, upstream component concentration (mg l^{-1}),
$\tilde{u}(t)$ = the vector of time-delayed input, upstream component concentration (mg l^{-1}),
$x(t)$ = the vector of output, downstream component concentration (mg l^{-1}),
$S(t)$ = the vector of component source and sink terms (mg l^{-1}),
$\zeta(t)$ = the vector of chance, random disturbances affecting the system (mg l^{-1}),
$Q(t)$ = the stream discharge (m^3 day^{-1}),
V = the reach volume (m^3),
$\tau(t)$ = the magnitude of the transportation-delay element (day), and
t = the independent variable of time.

The errors associated with the laboratory analysis and sampling are included in the observation equation

$$y(t) = x(t) + \eta(t) \tag{2}$$

where

$y(t)$ = the vector of observed (measured) downstream component concentration (mg l^{-1}), and
$\eta(t)$ = the vector of the chance measurement error.

Equations (1) and (2) provide the basic description of the conceptual water-quality model. The identification and estimation of these models against water-quality data is given in detail elsewhere (Beck and Young, 1976; Whitehead et al., 1979, 1981).

4 THE BEDFORD OUSE STUDY

The Bedford Ouse Study was initiated in 1972 by the Great Ouse River Division of the Anglian Water Authority and the Department of the Environment. The objective of the study was to develop and utilize water-quality models in the planning, design, and operational management of the Bedford Ouse River system in central eastern England. In particular the development of the new city of Milton Keynes (see Figure 2) is likely to have a considerable impact, and effluent from the city is discharged about 55 km upstream of an abstraction plant supplying water to Bedford.

The research has therefore been directed towards obtaining models of water quality which could be used to investigate the impact of effluent on the aquatic environment. Details of the Bedford Ouse Study are given elsewhere (Bedford Ouse Study, 1979; Whitehead et al., 1979, 1981) and the integrated models of flow and water quality discussed in the previous section have been extensively applied to the Bedford Ouse River system. For example, a typical simulation of flow based on data from the upstream flow-gauging stations and the daily rainfall in the area is given in Figure 3. This shows the simulated river flow superimposed on the observed flows together with a plot of the residual error. The mean percentage error of 8.6% is within the accuracy of the flow-gauging stations estimated at 10% by the Great Ouse River Division. In addition, the model explains 99% of the variance of the original flow series and the errors are

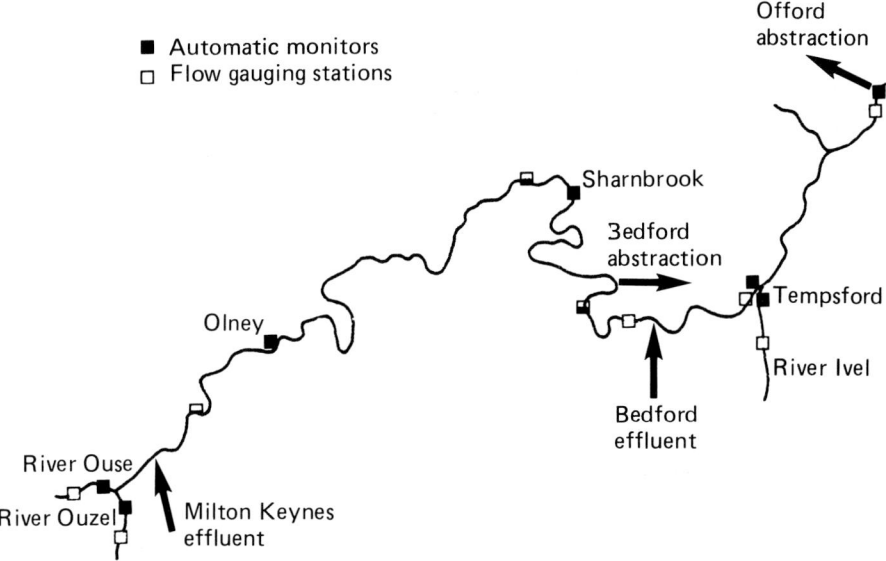

FIGURE 2 The Bedford Ouse River system.

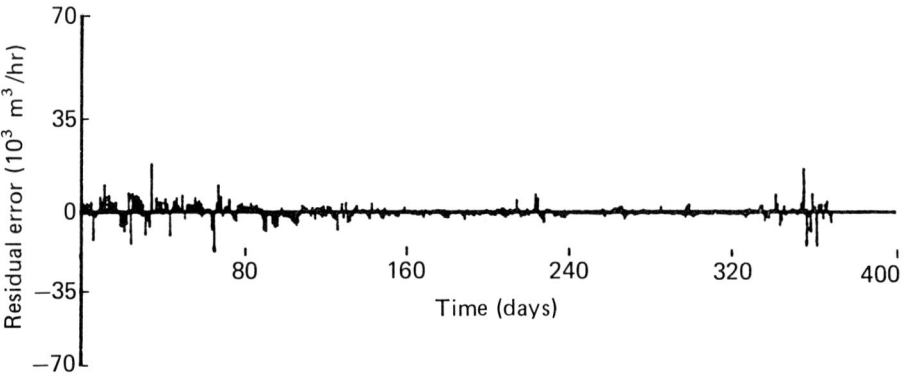

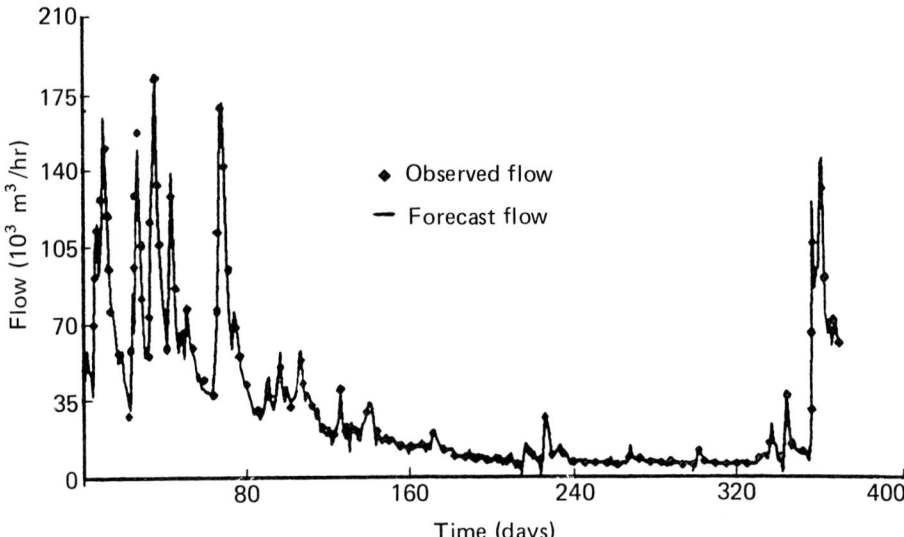

FIGURE 3 Simulated, observed, and residual flows on the Bedford Ouse over 1972.

within 10% of the observed flow for 70% of the time. The model has been validated using several years data and it appears that the combination of a deterministic flow-routing model and the stochastic rainfall–runoff model provides a satisfactory representation of the system.

4.1 Assessing the Impact of Effluent on River Water Quality

Water-quality models for the Bedford Ouse have been developed for chloride, dissolved oxygen (DO), biochemical oxygen demand (BOD), total oxidized nitrogen (TON), and ammonia.

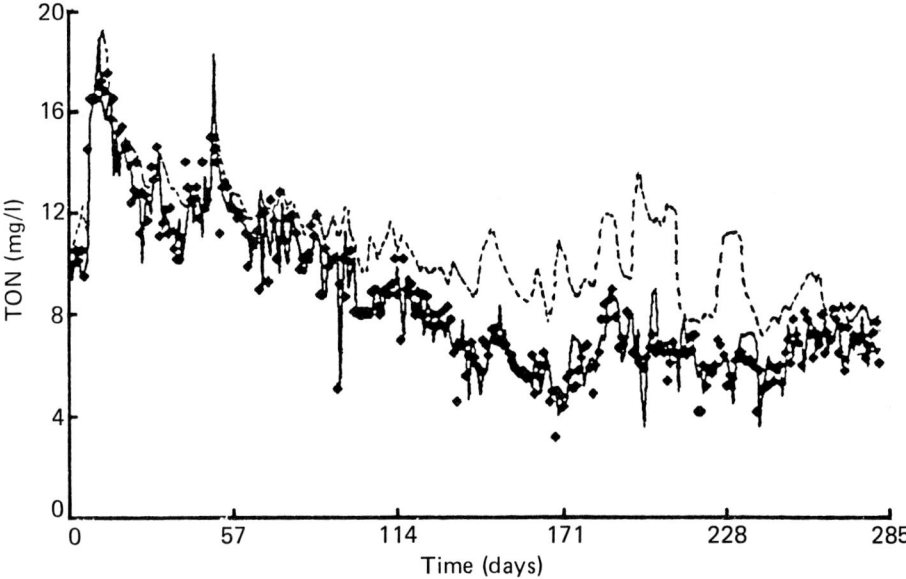

FIGURE 4 Simulated and observed daily TON concentrations on the Bedford Ouse during 1974.

A typical simulation for nitrate over 1974 is given in Figure 4 and again upstream water-quality information can be used to reliably simulate downstream behavior. In addition, since the models are based on mass-balance principles, it is possible to assess the impact of effluent in the river system. Figure 4 shows the effect on downstream nitrate levels assuming an effluent flow from Milton Keynes of 114,000 $m^3 day^{-1}$ with nitrate levels of $10 mg l^{-1}$. During high flow conditions the impact of the effluent is minimal because of dilution effects, and upstream sources of nitrogen and runoff effects predominate. In this situation nitrate treatment at Milton Keynes would have relatively little effect and alternative methods of overcoming the high nitrate levels are required, such as blending with groundwater or reservoir water at the abstraction plant at Bedford. During low flow conditions and increased temperature levels during summer, the background levels of nitrogen fall, and the effluent effect is more significant.

In addition to providing time-varying concentrations at the downstream point, the models may be used in a Monte Carlo simulation study to provide predictions directly in terms of probability distributions rather than exact values (Whitehead and Young, 1979). The stochastic simulation approach is extremely useful where analytical solutions are difficult or even impossible to obtain, as is often the case with reasonably complicated dynamic systems. The system calculations (usually simulations) are performed a large number of times, each time with the values for the stochastic inputs or uncertain parameters selected at random from their assumed (i.e., estimated) parent probability distributions. Each such random experiment or simulation yields a different result for any variable of interest and when all these results are taken together the required probability distribution can be ascertained to any required degree of accuracy from the sample statistics. The degree of accuracy of the probability distribution function estimated in this manner is, of course, a function of the number of random simulations used to calculate the sample statistics, but it is possible to quantify the degree of uncertainty on the distribution using nonparametric statistical tests such as the Kolmogorov–Renyi statistics.

Monte Carlo simulation is a flexible, albeit computationally expensive tool with which to investigate certain design problems. For example, the water-quality standards that are proposed in the Bedford Ouse Study (1979) are presented in terms of the percentage of time that a water-quality level is exceeded, and therefore, provide a reference against which the water quality can be tested. It would be possible to perform Monte Carlo simulation analysis using the water-quality models developed for the study section of the Bedford Ouse and making various assumptions about future levels of effluent input. The outcome of such an analysis would be probability density functions for the water-quality states, which could be compared directly with the water-quality standards. Such information would be extremely useful in assessing the impact of effluent on the system and determining the degree of treatment necessary at Milton Keynes in order to ensure satisfactory water quality at the abstraction point.

TABLE 1 Effluent conditions used to assess the impact on aquatic environments.

	Flow rate $(m^3 s^{-1})$	BOD concentration in effluent $(mg\,l^{-1})$	Variance of BOD levels
Case 1	0.1	5	1
Case 2	0.4	10	4
Case 3	1.0	10	4

An initial assessment of the impact of Milton Keynes effluent on the aquatic environment may now be obtained using Monte Carlo simulation; details are given in Whitehead and Young (1979). Altogether three effluent conditions were considered at different flow rates and BOD levels, as shown in Table 1. It was assumed that the effluent has no dissolved oxygen present; this condition represents the worst situation but is not unrealistic, since the effluent is to be pumped directly from the treatment works via a 4-km pipe into the river. Effluent BOD levels fluctuate in practice and a stochastic component defined by a noise signal of variance of 1, 4, and 4 $mg\,l^{-1}$, respectively, was added to the three BOD levels shown in Table 1. The distributions of BOD and DO at Bedford given these three effluent conditions are compared with the present situations in Figure 5. At low discharge conditions there is relatively little effect on the aquatic environment. At the $1\,m^3 s^{-1}$ condition, however, the mean BOD level has risen to 4.5 $mg\,l^{-1}$, the mean DO level has fallen to 6.5 $mg\,l^{-1}$, and the DO distribution ranges from 4.5 to 9 $mg\,l^{-1}$. These distributions represent only an initial assessment of the impact of Milton Keynes effluent, and an updated prediction based on a reestimated model in two years time may indicate an improved situation. On the other hand the DO levels may be adversely affected by the changing biological nature of the river and some form of control action may be necessary to improve the DO distribution.

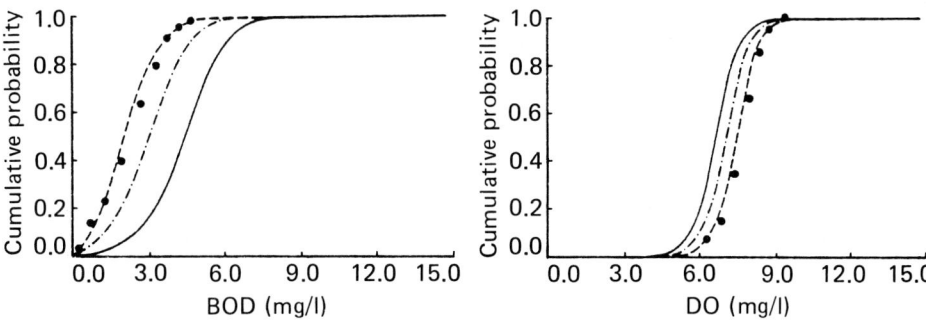

FIGURE 5 Distributions of DO and BOD at Clapham obtained from the Monte Carlo simulation study.

5   THE REAL-TIME MONITORING SCHEME FOR THE BEDFORD OUSE

In the short-term operational management of water-resource systems, a major requirement is for information on the present condition of the river system and on future changes in water quality. Operational managers must be able to respond quickly to emergency situations in order to protect and conserve the river and maintain adequate water supplies for public use. Moreover, the costs of water treatment and bankside storage are particularly high and there are therefore considerable benefits to be gained from the efficient operational management of river systems from the viewpoint of water quality (Young and Beck, 1974; Whitehead, 1978; Beck, 1979; Rinaldi et al., 1979).

In recent years there has been some progress towards providing more efficient operational management by the installation of automatic, continuous water-quality monitors on river systems. These measure such water-quality variables as dissolved oxygen, ammonia, and temperature, and, if combined with a telemetry scheme relaying information to a central location, provide immediate information on the state of the river for pollution control officers. Whilst the reliability of such schemes is still rather poor, there is now an opportunity to use this information together with mathematical models for making real-time forecasts of water quality.

The practical problems associated with the continuous field measurement and telemetry of water quality have largely limited the application of on-line forecasting and control schemes. Continuous flow of water past sensors for measuring water quality gives rise to severe fouling of optical and membrane surfaces, thereby drastically reducing the accuracy of the data produced. In recent years, however, there have been several studies and applications of continuous water-quality monitors (Briggs, 1975; Kohonen et al., 1978). Most UK water authorities have established monitoring and telemetry schemes (Cooke, 1975; Hinge and Stott, 1975; Caddy and Akielan, 1978) and report reasonable reliability provided the monitors are regularly maintained. More recently, Wallwork (1979) has described an application on the River Wear in northeast England where a continuous monitor is used to protect an abstraction point.

The application of particular interest in this paper is an extensive monitoring and telemetry scheme which has been developed along the Bedford Ouse River system.

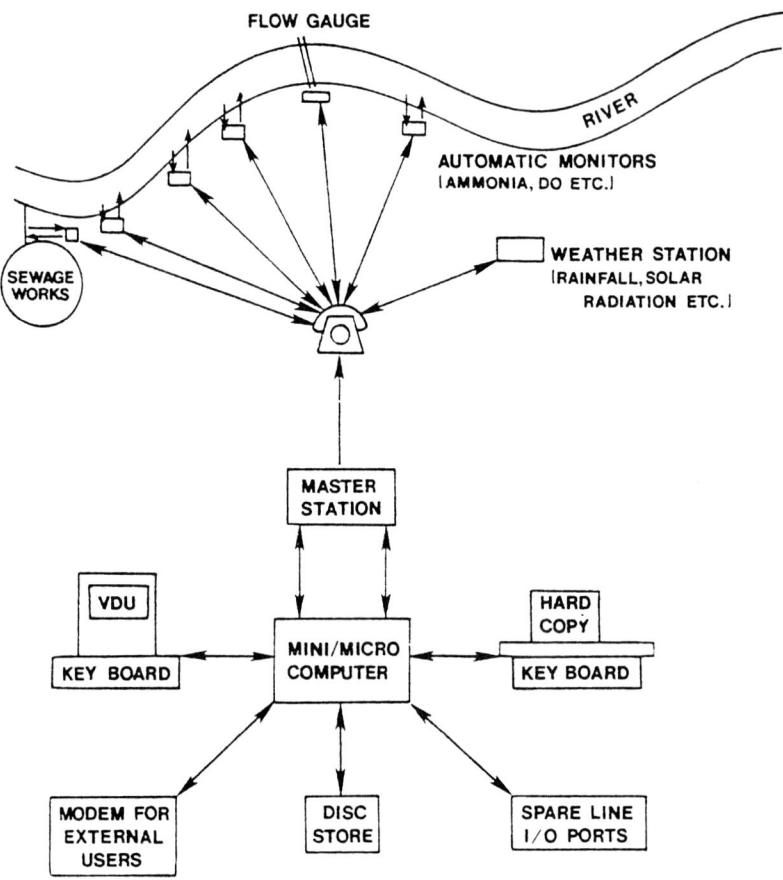

FIGURE 6 Monitoring, telemetry, and mini/microcomputing system operational management.

As indicated in Figure 6, automatic water-quality monitors have been installed at several sites along the river and data on dissolved oxygen, pH, ammonia, and temperature are telemetered at four-hourly intervals to the central control station located in Cambridge. It is proposed to extend this telemetry scheme to include information on flow and such variables as rainfall and solar radiation, and to use a mini/microcomputer located in Cambridge to analyze the data on-line. The system will provide rapid information on the present state of the river and will incorporate a dynamic water-quality model for making real-time forecasts of flow and quality at key locations along the river system.

The data from the automatic monitors are telemetered at four-hourly intervals to the central master station in Cambridge and, in order to assess and model the short-term behavior, data have been obtained for the monitoring stations located at Sharnbrook and Tempsford (see Figure 2) for the period from July to November 1978. The stretch of river between these two sites is of particular interest to the Anglian Water Authority because of the location of the Bedford Water Division abstraction plant at Clapham,

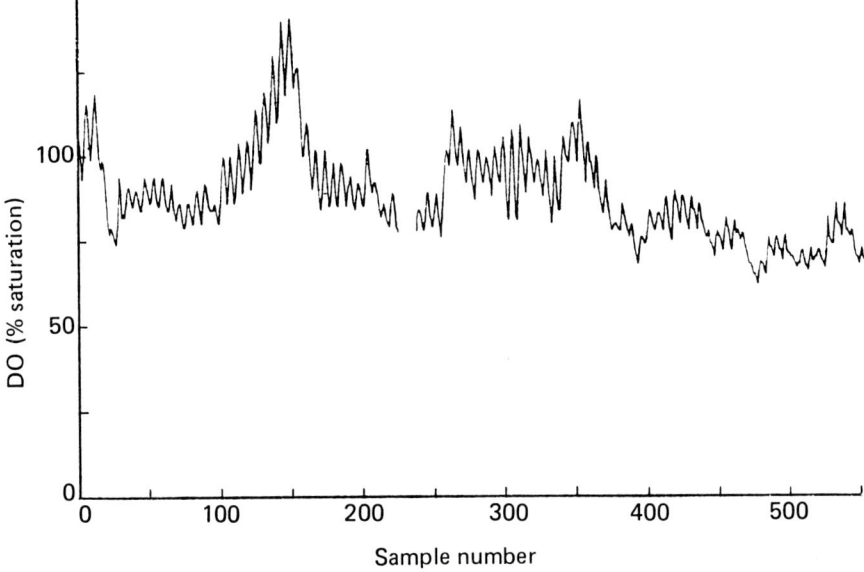

FIGURE 7   Continuous DO data at Tempsford on the Bedford Ouse.

the discharge of effluent from Bedford Sewage Works, and the abstraction of water at Offord just downstream of Tempsford.

Data have been obtained for dissolved oxygen, ammonia, flow, temperature, and solar radiation together with data on the quality and quantity of effluent from Bedford Sewage Works. A plot of dissolved oxygen at the upstream site is given in Figure 7 and shows clearly the daily oscillations caused by oxygen production and consumption processes, and the longer-term fluctuations which are due to other variables such as temperature and streamflow. Initially, mathematical analysis of these data has been restricted to the first 108 samples (18 days) since this period corresponds with a major storm event and high levels of ammonia in the river downstream of the sewage works.

6   AMMONIA AND DISSOLVED OXYGEN MODELS

The model of ammonia and dissolved oxygen is based on the mass-balance description of eqn. (1) but contains additional terms to describe source and sink processes such as the nitrification of ammonia and the production of oxygen by photosynthesis. The river between Sharnbrook and Tempsford has been divided into four reaches with reach boundaries corresponding to the abstraction plant at Bedford, the Bedford Sewage Works, and an intermediate point between the sewage works discharge and the Tempsford monitor. The upstream ammonia concentrations are particularly low ($< 0.05 \, \mathrm{mg\,l^{-1}}$) and therefore the ammonia model has been formulated for just the two reaches below the sewage works. The models identified using the EKF are as follows:

Dissolved oxygen

$$dx_1/dt = (Q/V_1)u_1 - (Q/V_1)x_1 + k_1 S - k_2 \tag{3}$$

$$dx_2/dt = (Q/V_2)x_1 - (Q/V_2)x_2 + k_1 S - k_2 \tag{4}$$

$$dx_3/dt = (Q/V_3)x_2 - (Q/V_3)x_3 + k_1 S - k_3 - 4.33(k_4/Q)x_5 \tag{5}$$

$$dx_4/dt = (Q/V_4)x_3 - (Q/V_4)x_4 + k_1 S - k_3 - 4.33(k_4/Q)x_6 \tag{6}$$

Ammonia

$$dx_5/dt = (Q/V_3)U_e - (Q/V_3)x_5 - (k_4/Q)x_5 \tag{7}$$

$$dx_6/dt = (Q/V_4)x_5 - (Q/V_4)x_6 - (k_4/Q)x_5 \tag{8}$$

where

x_1, x_2, x_3, x_4 = DO at the downstream boundary of the four reaches (mg l^{-1}),
x_5, x_6 = the ammonia concentrations at the downstream boundary of the third and fourth reaches,
u_1 = the upstream DO concentration entering the first reach at Sharnbrook (mg l^{-1}),
U_e = the ammonia in the effluent discharge calculated as the effective instream ammonia level (mg l^{-1}),
Q = the flow rate measured at Bedford (m^3 day^{-1}),
S = a sunlight term to account for addition of oxygen by photosynthesis,
V_1, V_2, V_3, V_4 = volumes of the reaches (m^3),
k_1 = the rate constant associated with oxygen production by photosynthesis (day^{-1}),
k_2 = the loss of DO caused by BOD upstream of Bedford (mg l^{-1} day^{-1}),
k_3 = the loss of DO caused by BOD downstream of Bedford (mg l^{-1} day^{-1}), and
k_4 = the nitrification rate (day^{-1}).

The sunlight term, S, is a function of solar radiation, S_r (see Water Research Centre Annual Report, 1968) and is determined as

$$S = S_r^{0.28}$$

The constant 4.33 in eqns. (5) and (6) represents the mass of oxygen removed from the water for each unit mass of ammonia nitrified.

One feature of particular interest in this model is the inclusion of the flow term, Q, in the ammonia nitrification expression in eqns. (5)–(8). The flow is included to account for the lower nitrification rate occurring under high flow conditions (Garland,

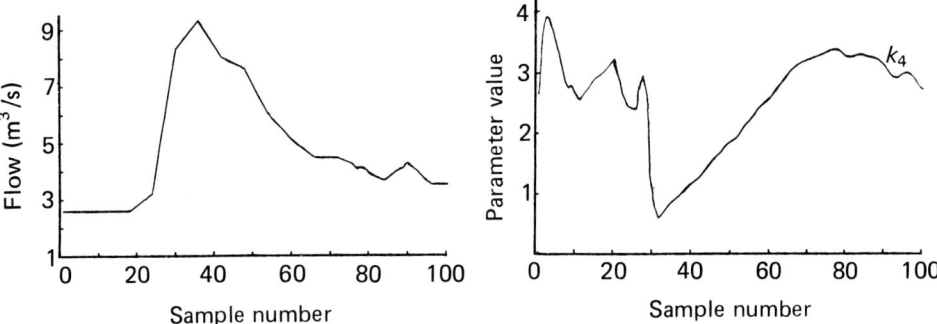

FIGURE 8 Recursive estimate of ammonia-decay coefficient and measured flow at Tempsford.

1978). During the initial EKF runs the flow term was not included and the parameter k_4, as shown in Figure 8, is estimated recursively and appears to be inversely proportional to the flow, Q. Inclusion of the flow term and reestimation of k_4 produced an essentially constant or slowly varying parameter, as shown in Figure 9. The higher flows tend to flush the reach of the nitrifying bacteria which are responsible for the conversion of ammonia to nitrite and nitrate and hence reduce the nitrification processes. The EKF is particularly useful in identifying this behavior and reducing an essentially time-varying parameter model to a model which is time-invariant (Whitehead, 1979).

The other parameters in the dissolved oxygen model do not vary significantly over the sampling period, as shown in Figure 9, although the parameter k_1 increases slightly during the estimation. This is most probably due to the presence of large algal populations in the river, which have not been explicitly included in the model. During the course of the Bedford Ouse Study (1979) the sunlight term was modified to account for the algal populations using chlorophyll-a concentrations as a measure of the oxygen-producing matter in the river. In the present study, chlorophyll-a data are not available and the sunlight term is therefore dependent on solar radiation only. As shown in Figure 10, there are large diurnal variations in dissolved oxygen which are indicative of algal activity, and further work incorporating the algal components is therefore required.

The simulation of dissolved oxygen and ammonia, as shown in Figure 10, is reasonable although the peak of the ammonia is considerably underestimated. This may be due to the inaccurate measurement of effluent flow from the sewage plant during the peak of the storm and the additional inputs along the reach from agricultural and urban runoff.

7 CONCLUSIONS

The design of a water-resource system from the viewpoint of water quality has conventionally been based on "steady-state" models which provide information on annual average conditions. However, for many design problems, detailed information on the transient behavior of water quality is required together with a description of the

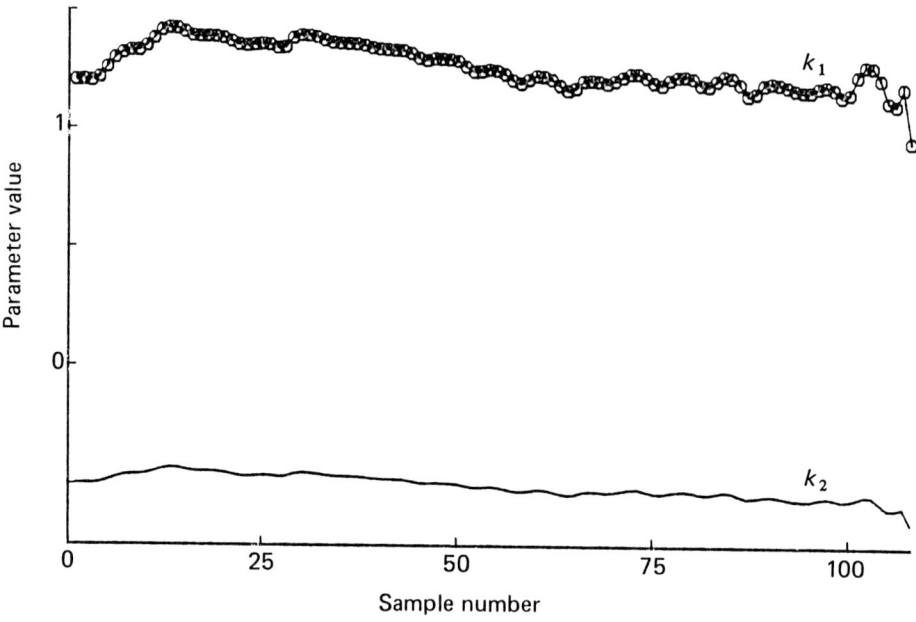

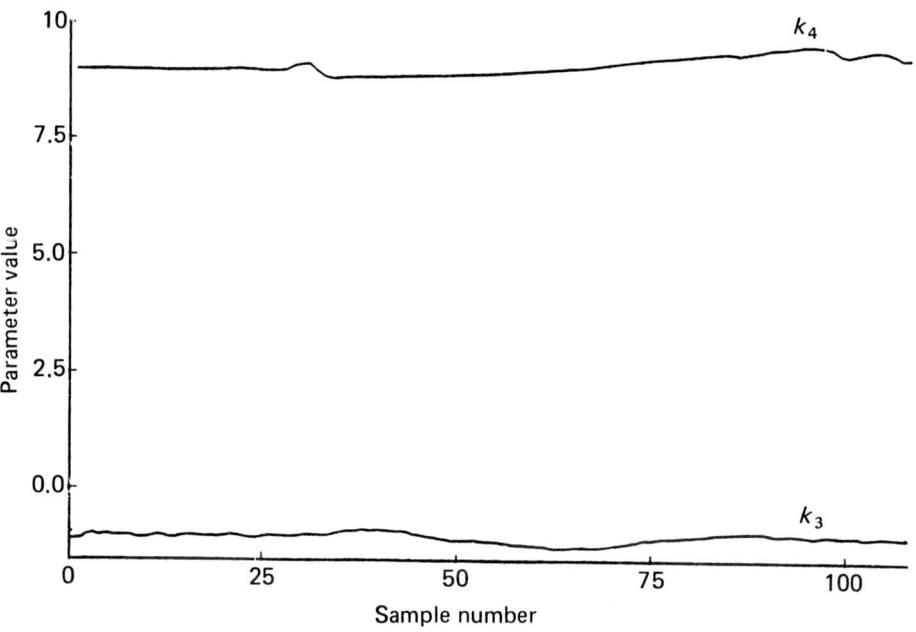

FIGURE 9 Recursive estimates of parameters with inclusion of flow term.

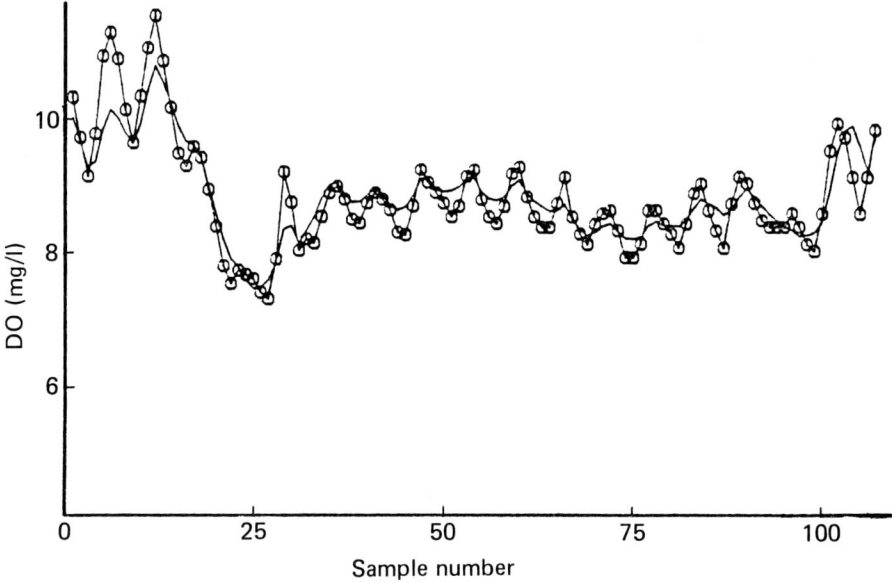

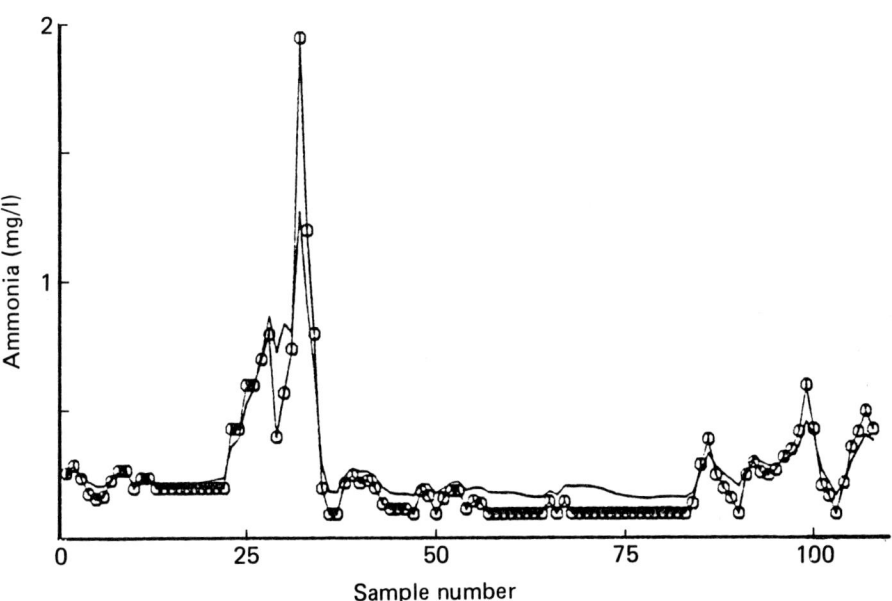

FIGURE 10 Simulated and observed DO and ammonia concentrations at Tempsford.

stochastic aspects of water quality. Such information can be obtained using the integrated models of flow and water quality developed during the Bedford Ouse Study. In this paper the models have been used to assess the impact of effluent on the Bedford Ouse River.

In recent years continuous water-quality monitoring schemes have been developed in conjunction with telemetry systems to provide real-time information for operational management. The rapid development in microcomputers has enhanced such schemes by providing considerable analytical power for on-line data-processing at relatively low cost. The application of real-time forecasting and control of water quality along critical stretches of river systems is therefore an option available to operational management. Such an application has been considered for the Bedford Ouse river system and this scheme is currently being implemented by the Anglian Water Authority and the Institute of Hydrology.

ACKNOWLEDGMENTS

The results presented in this paper reflect the opinions of the author and are not necessarily the views of the Anglian Water Authority. The author is particularly grateful to Dr. D. Caddy of the Great Ouse Division of the Anglian Water Authority for the provision of water-quality data and collaboration in establishing the water-quality study.

REFERENCES

Beck, M.B. (1979). The Role of Real Time Forecasting and Control in Water Quality Management. WP-79-1. International Institute for Applied Systems Analysis, Laxenburg, Austria.

Beck, M.B. and Young, P.C. (1976). Systematic identification of DO–BOD model structure. Journal of the Environmental Engineering Division, American Society of Civil Engineers, 102 (EE5): 909–927.

Bedford Ouse Study (1979). Conference Proceedings and Final Report, Anglian Water Authority, Huntingdon, UK.

Briggs, R. (1975). Instrumentation for monitoring water quality. Journal of the Society for Water Treatment and Examination, 24: 23–45.

Caddy, D.E. and Akielan, A.W. (1978). Management of river water quality. International Environment and Safety, 1: 18–26.

Cooke, G.H. (1975). Water Quality Monitoring in the River Trent System. Instruments and Control Systems Conference, Water Research Centre, Medmenham, Bucks., UK.

Garland, J.H.N. (1978). Nitrification in the River Trent. In A. James (Editor), Mathematical Models in Water Pollution Control. Wiley, Chichester.

Himmelblau, D.M. and Bischoff, K.B. (1968). Process Analysis and Simulation. Wiley, New York.

Hinge, D.C. and Stott, D.A. (1975). Experience in the Continuous Monitoring of River Water Quality. Instruments and Control Systems Conference, Water Research Centre, Medmenham, Bucks., UK.

Kohonen, T., Hell, P., Muhonen, J., and Vuolas, E. (1978). Automatic Water Quality Monitoring Systems in Finland. Report 153. National Board of Waters, Helsinki, Finland.

Rinaldi, S., Soncini-Sessa, R., Stehfest, H., and Tamura, H. (1979). Modelling and Control of River Quality. McGraw-Hill, New York.

Stott, P. (1979). Opening Address to the Conference on Water Resources – A Changing Strategy. Institution of Civil Engineers, London.

Thames Water Statistics (1978). Thames Water Authority, Nugent House, Reading, UK.

Wallwork, J.F. (1979). Protecting a Water Supply Intake – River Water Data Collection and Pollution Monitoring. River Pollution Control Conference, Water Research Centre, Medmenham, Bucks., UK.

Water Research Centre Annual Report (1968). Water Research Centre, Medmenham, Bucks., UK.

Whitehead, P.G. (1978). Modelling and operational control of water quality in river systems. Water Research, 12: 377–384.

Whitehead, P.G. (1979). Applications of recursive estimation techniques to time variable hydrological systems. Journal of Hydrology, 40: 1–16.

Whitehead, P.G. (1980). An instrumental variable method of estimating differential-equation models of dispersion and water quality in non-tidal rivers. Ecological Modelling, 9: 1–14.

Whitehead, P.G. and Young, P.C. (1979). Water quality in river systems: Monte Carlo analysis. Water Resources Research, 15(2): 451–459.

Whitehead, P.G., Young, P.C., and Hornberger, G. (1979). A systems model of flow and water quality in the Bedford Ouse River system: part I. Streamflow modelling. Water Research, 13: 1155–1169.

Whitehead, P.G., Beck, M.B., and O'Connell, P.E. (1981). A systems model of flow and water quality in the Bedford Ouse River system: part II. Water Research, 15: 1157–1171.

Young, P.C. and Beck, M.B. (1974). The modelling and control of water quality in a river system. Automatica, 10: 455–468.

ADAPTIVE PREDICTION OF WATER QUALITY IN THE RIVER CAM

H.N. Koivo and J.T. Tanttu
Tampere University of Technology, Tampere (Finland)

1 INTRODUCTION

Modeling can be used for two main purposes — prediction or control. This must be kept in mind when methods for either, or both, of the purposes are chosen. In this paper the prediction of water quality in a river is investigated.

Prediction or forecasting has received much attention, especially in the time-series literature (Brown, 1963; Coutie, 1964; Box and Jenkins, 1970). Applications of forecasting methods to water-pollution research have been described by, for example, Kashyap and Rao (1973) and Beck (1977). In addition, a very thorough account of related modeling aspects has been given by Beck (1978), in which river water quality is specifically discussed. In this volume, the papers by Ikeda and Itakura (1983) and by Tamura and Kondo (1983) describe the application of the group method of data handling to forecasting.

In this paper a recently developed adaptive self-tuning predictor for multivariable stochastic processes (Tanttu, 1980) is used for real-time prediction of water quality in the River Cam.

The field data are given in Beck (1978). The basic idea of a self-tuning predictor is easy to explain: instead of identifying plant parameters and then constructing a multivariable predictor, predictor parameters are identified and updated at each step. This technique was suggested by Wittenmark (1974) and Holst (1977) for the scalar case. It was motivated by the success of self-tuning controllers (Åström and Wittenmark, 1974; Koivo, 1980) and, especially in the multivariate case, by the ease with which computations can be performed compared with the explicit scheme.

2 REVIEW OF THE SELF-TUNING PREDICTOR

Many time-series can be described by an autoregressive moving average (ARMA) model (Box and Jenkins, 1970), which in the multivariable case takes the form

$$y(t) + A_1 y(t-1) + \ldots + A_{n_A} y(t-n_A) = e(t) + C_1 e(t-1) + \ldots$$
$$+ C_{n_C} e(t - n_C) \quad (t = 0, 1, \ldots) \tag{1}$$

where $y(t) \in \mathcal{R}^p$ is the (measured) output, $e(t) \in \mathcal{R}^p$ and $\{e(t), t = 0, 1, \ldots\}$ is a sequence of independent, equally distributed random vectors with zero mean and covariance R. Introducing the backward shift operator q^{-1}, $q^{-1} y(t) := y(t-1)$, eqn. (1) can be written in the more compact form

$$A(q^{-1}) y(t) = C(q^{-1}) e(t) \tag{2}$$

where the $p \times p$ polynomial matrices are given by

$$A(z) = I + A_1 z + \ldots + A_n z^n$$
$$C(z) = I + C_1 z + \ldots + C_n z^n$$

The matrices A_i, C_i ($i = 1, 2, \ldots, n$) are assumed to be time-invariant and they may also be zero. It is further required that det $A(z)$ and det $C(z)$ have all their zeroes outside the unit disk, i.e., the process is stationary and invertible (Box and Jenkins, 1970).

The following notations will be used below. $\hat{y}(t+k|t)$ is the k-step-ahead prediction of output $y(t+k)$, $k \in \mathcal{N}$. The prediction error at time $t+k$ is

$$\varepsilon(t+k) = y(t+k) - \hat{y}(t+k|t) \tag{3}$$

It can be shown (Tanttu, 1980) that model (2) can be written in the form

$$\varepsilon(t+k) = A(q^{-1}) \varepsilon(t) + B(q^{-1}) y(t+k|t) + E(q^{-1}) e(t+k) \tag{4}$$

where $\hat{y}(t+k|t)$ is the prediction that minimizes the loss function (the minimum variance prediction):

$$V = E\{\varepsilon^T(t+k) \varepsilon(t+k)\}$$

the prediction error is then given by

$$\varepsilon(t+k) = E(q^{-1}) e(t+k)$$

The key aim is now to estimate recursively the parameters of the model

$$\varepsilon(t) = \hat{A}(q^{-1}) \varepsilon(t-k) + \hat{B}(q^{-1}) \hat{y}(t|t-k) + w(t) \tag{4'}$$

so that the estimation error $w(t)$ is minimized. The matrices $A(q^{-1})$ and $B(q^{-1})$ can be calculated if matrices $A(q^{-1})$ and $C(q^{-1})$ of model (2) are known. Then according to eqn. (4) the minimum variance prediction is obtained from

$$B(q^{-1})\hat{y}(t+k|t) + A(q^{-1})\epsilon(t) = 0 \tag{5}$$

If model (4') is used, however, and the matrices $\hat{A}(q^{-1})$ and $\hat{B}(q^{-1})$ are estimated recursively, the new prediction is calculated from

$$\hat{B}(q^{-1})\hat{y}(t+k|t) + \hat{A}(q^{-1})\epsilon(t) = 0 \tag{6}$$

Now the prediction algorithm can be presented.

2.1 The Prediction Algorithm

2.1.1 Step 1: Choosing the Model and Initial Values
Choose the integers m and l in eqns. (7) and (8):

$$\hat{A}(q^{-1}) = \hat{A}_0 + \hat{A}_1 q^{-1} + \ldots + \hat{A}_m q^{-m} \tag{7}$$

$$\hat{B}(q^{-1}) = \hat{B}_0 + \hat{B}_1 q^{-1} + \ldots + \hat{B}_l q^{-l} \tag{8}$$

and the initial values of the matrices $\hat{A}_0, \hat{A}_1, \ldots, \hat{A}_m, \hat{B}_0, \hat{B}_1, \ldots, \hat{B}_l$.

2.1.2 Step 2: Estimation
Choose the matrices $\hat{A}_i$ ($i = 0, 1, \ldots, l$), $\hat{B}_j$ ($j = 0, 1, \ldots, m$) in

$$\epsilon(t) = \hat{A}(q^{-1})\epsilon(t-k) + \hat{B}(q^{-1})\hat{y}(t|t-k) + w(t)$$

so that the estimation error $\{w(t)^T w(t)\}$ is minimized.

2.1.3 Step 3: Prediction
Compute the predicted value of $y(t+k)$ from

$$\hat{B}(q^{-1})\hat{y}(t+k|t) = -\hat{A}(q^{-1})\epsilon(t)$$

The well-known recursive parameter-estimation algorithms, in particular the recursive least-squares (RLS) method (see Appendix) and its variant the square-root algorithm (Peterka, 1975; Koivo, 1980) can be used in step 2 if the data vector, parameter matrix, and "measurement vector" are defined as follows:

$$x(t) = [\epsilon^T(t-k), \epsilon^T(t-k-1), \ldots, \epsilon^T(t-k-m);$$
$$y^T(t|t-k), y^T(t-1|t-k-1), \ldots, y^T(t-l|t-k-l)] \tag{9}$$

$$\Theta = [\theta_1, \theta_2, \ldots, \theta_p] = [\hat{A}_0, \hat{A}_1, \ldots, \hat{A}_m; \hat{B}_0, \hat{B}_1, \ldots, \hat{B}_l]^T \tag{10}$$

$$z(t) = \epsilon(t) \tag{11}$$

Using the above notation the prediction algorithm takes the following form.

2.1.4 Algorithm 1
(1) Read the new output $y(t)$.
(2) Form the vectors $x(t)$ and $z(t)$ according to eqns. (9) and (11).
(3) Update the parameter matrix

$$\hat{\Theta} = [\hat{A}_0, \hat{A}_1, \ldots, \hat{A}_m; \hat{B}_0, \hat{B}_1, \ldots, \hat{B}_l]^T$$

using the RLS method.
(4) Compute the new prediction from

$$\hat{B}(q^{-1})\hat{y}(t+k|t) = -\hat{A}(q^{-1})\varepsilon(t)$$

(5) Set $t := t+1$ and return to (1).

2.2 A Modified Algorithm

Usually the predicted signal contains deterministic or almost deterministic components. These parts of the signal may be handled by adding an extra measurement vector $v(t)$. Now we estimate the parameters of the model

$$\varepsilon(t) = \hat{A}(q^{-1})\varepsilon(t-k) + \hat{B}(q^{-1})\hat{y}(t|t-k) + \hat{G}(q^{-1})v(t-k) + w(t) \tag{12}$$

where $v(t-k)$ is the additional p-vector and

$$\hat{G}(z) = \hat{G}_0 + \hat{G}_1 z^1 + \ldots + \hat{G}_r z^{+r} \tag{13}$$

is a $p \times p$ polynomial matrix. The new prediction is computed from

$$\hat{B}(q^{-1})\hat{y}(t+k|t) = -\hat{A}(q^{-1})\varepsilon(t) - \hat{G}(q^{-1})v(t)$$

This can be stated in the form of the following algorithm.

2.2.1 Algorithm 2
(1) Read the values of $y(t)$ and $v(t)$.
(2) Form the data vector

$$x(t) = [\varepsilon^T(t-k), \varepsilon^T(t-k-1), \ldots, \varepsilon^T(t-k-m);$$

$$y^T(t|t-k), y^T(t-1|t-k-1), \ldots, y^T(t-l|t-k-l);$$

$$v^T(t-k), v^T(t-k-1), \ldots, v^T(t-k-r)]$$

(3) Update the parameter estimate matrix

$$\hat{\Theta} = [\hat{A}_0, \hat{A}_1, \ldots, \hat{A}_m; \hat{B}_0, \hat{B}_1, \ldots, \hat{B}_l; \hat{G}_0, \hat{G}_1, \ldots, \hat{G}_r]^T$$

(4) Compute the new prediction from

$$\hat{B}(q^{-1})\hat{y}(t+k|t) = -\hat{A}(q^{-1})\varepsilon(t) - \hat{G}(q^{-1})v(t)$$

(5) Set $t := t+1$ and return to (1).

3 A CASE STUDY – THE RIVER CAM

The algorithms developed were applied to the field data presented in Beck (1978) on the DO and BOD concentrations of the River Cam. A restriction was that only 81 samples were available (80 days). This could have caused problems because in the computer simulations it usually took 10–20 steps before the prediction and parameter estimates became satisfactory.

The origin of the data and several models of DO–BOD–algae interaction are discussed in Beck (1978). In the present case the following notation is used. The data obtained at measurement station D (see Figure 1) are denoted by

$$y(t) = \begin{bmatrix} y_1(t) \\ y_2(t) \end{bmatrix} \quad (t = 0, 1, \ldots, 80)$$

where $y_1(t)$ is the measured DO concentration $(g\,m^{-3})$ and $y_2(t)$ the measured BOD concentration $(g\,m^{-3})$, both at time t.

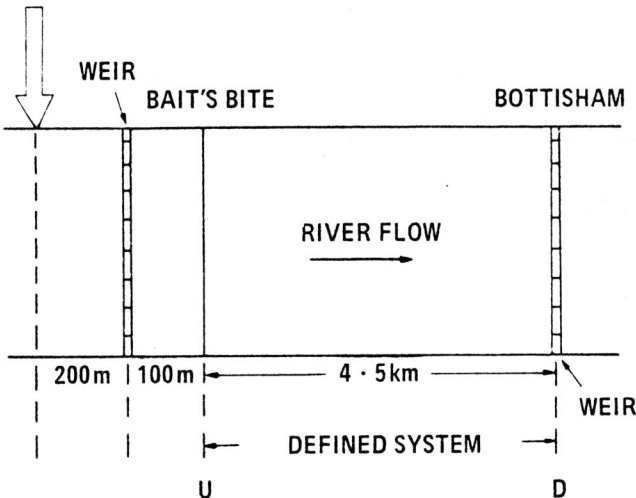

FIGURE 1 Schematic definition of the study reach showing the location of the effluent discharge from Cambridge Sewage Works.

Similarly the data at measurement station U are denoted by

$$v(t) = \begin{bmatrix} v_1(t) \\ v_2(t) \end{bmatrix} \quad (t = 0, 1, \ldots, 80)$$

$v(t)$ is used as an auxiliary variable in Algorithm 2; another auxiliary variable, r, was also used, defined as

$$r(t) = \begin{bmatrix} r_1(t) \\ r_2(t) \end{bmatrix} \quad (t = 0, 1, \ldots, 80)$$

where $r_1(t)$ is the stream temperature (°C) and $r_2(t)$ is the duration of sunlight (h day^{-1}). The aim was to predict $y(t + k)$ based on known data at time t, that is $y(n)$ and $v(n)$ $\{r(n)\}, 0 \leqslant n \leqslant t$. In the following, one-step-ahead predictors are discussed.

When Algorithm 1 is used, only measured values $y(t) \ldots y(0)$ are employed to predict $y(t + 1)$. Though the algorithm is self-tuning, some parameters must be pre-selected: (1) the order of the model, that is, integers m and l in eqns. (7) and (8); (2) the initial values for the parameter estimation algorithm (see Appendix), that is, $\mathbf{P}(0)$ the initial covariance matrix and $\theta(0)$ the initial parameter estimates; (3) $(0 \leqslant \lambda \leqslant 1)$, the "forgetting factor"; and (4) the initial values for the predictor. The effect of $\mathbf{P}(0)$ can be seen from eqns. (A2) and (A3) in the Appendix. It is usually chosen as $r \times \mathbf{I}$, where $\mathbf{I}$ is an identity matrix and r a large positive constant. The forgetting factor λ is used to put more weight on recent data; its value should be chosen as less than 1, especially when the system is time-variant.

Extensive simulations of both self-tuning predictors and controllers have shown that the effects of parameters described in (2) and (4) above are not significant when hundreds or thousands of simulation runs are used.

In all examples, the so-called square-root algorithm (see Appendix) was used, since it is numerically more reliable than the original RLS method. In all the examples that follow, $\mathbf{S}(0) = 1000\mathbf{I}$ (eqn. A6 in the Appendix), and initial parameter estimates are set equal to zero, except for $\mathbf{B_0}$, which is equal to $\mathbf{I}$ ($\mathbf{B_0}$ is inverted in the prediction algorithm).

The initial prediction vector is equal to $[8.0, 2.0]^T$ which is near the true value $[8.0, 2.3]^T = y(0)$. Different values of λ and different predictor structures are compared in Table 1. The loss function V is defined as

$$V = (1/61) \sum_{20}^{80} \varepsilon^T(t) \varepsilon(t) \tag{14}$$

TABLE 1 Values of the loss function using Algorithm 1.

$\lambda = 1.0$			$\lambda = 0.98$		
m	l	V	m	l	V
0	1	0.997	0	1	1.015
1	1	1.034	1	1	1.034

Adaptive prediction of water quality in the River Cam

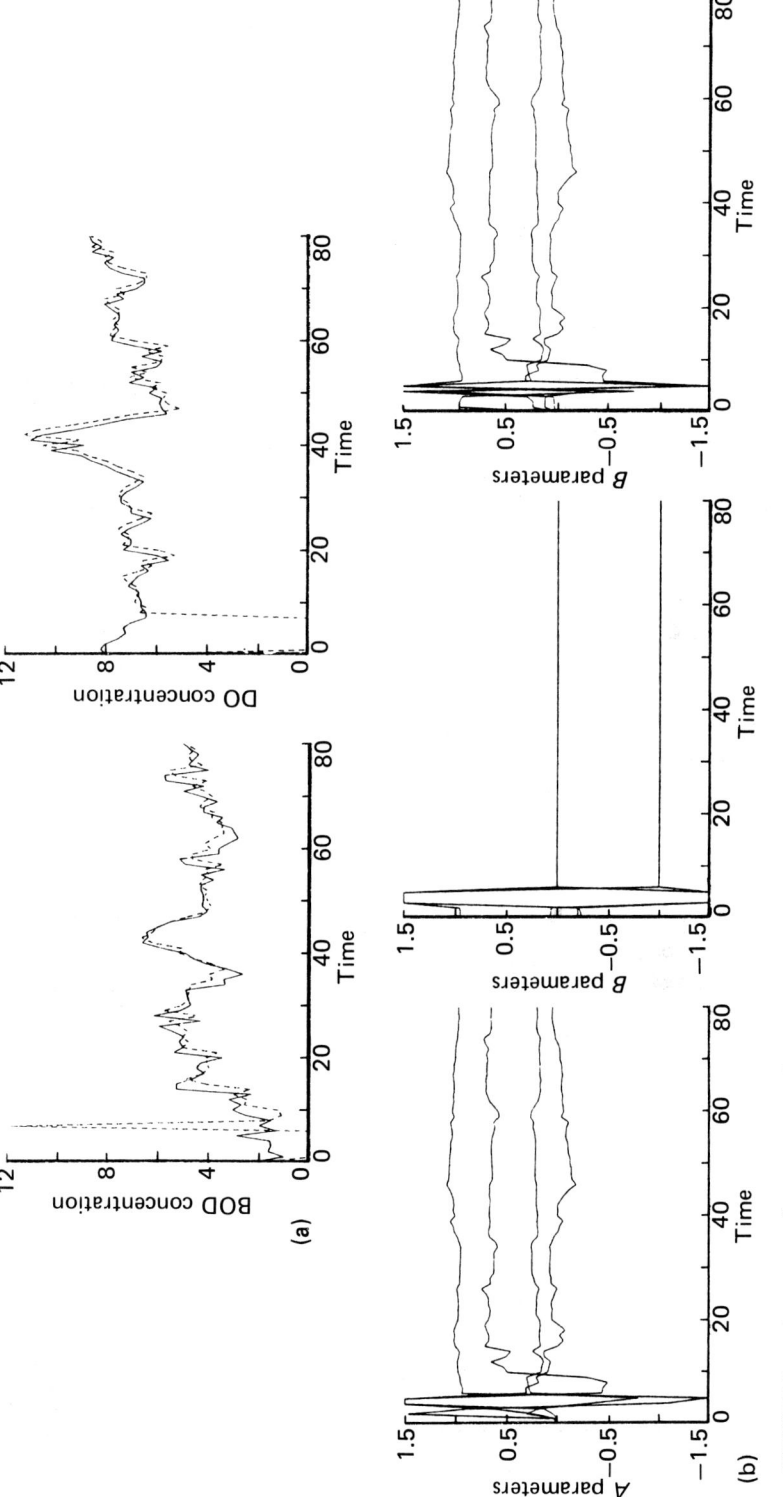

FIGURE 2 (a) Predicted $\hat{y}_1(t+1|t)$ and $\hat{y}_2(t+1|t)$ when $m = 0$, $l = 1$, and $\lambda = 1.0$. (b) Estimated parameters.

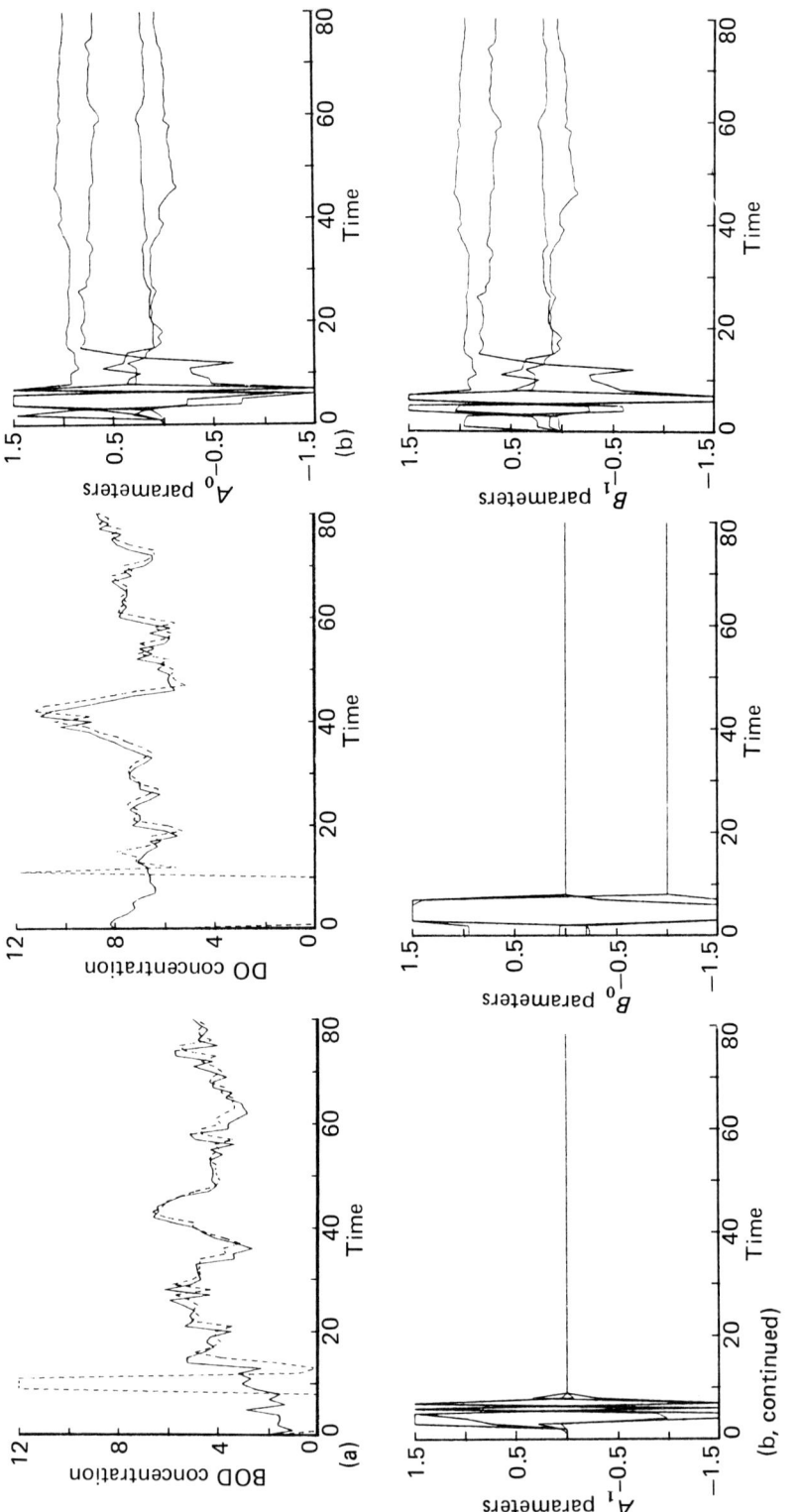

FIGURE 3 (a) Predicted $\hat{y}_1(t+1|t)$ and $\hat{y}_2(t+1|t)$ when $m = 1$, $l = 1$, and $\lambda = 1.0$. (b) Estimated parameters.

Adaptive prediction of water quality in the River Cam

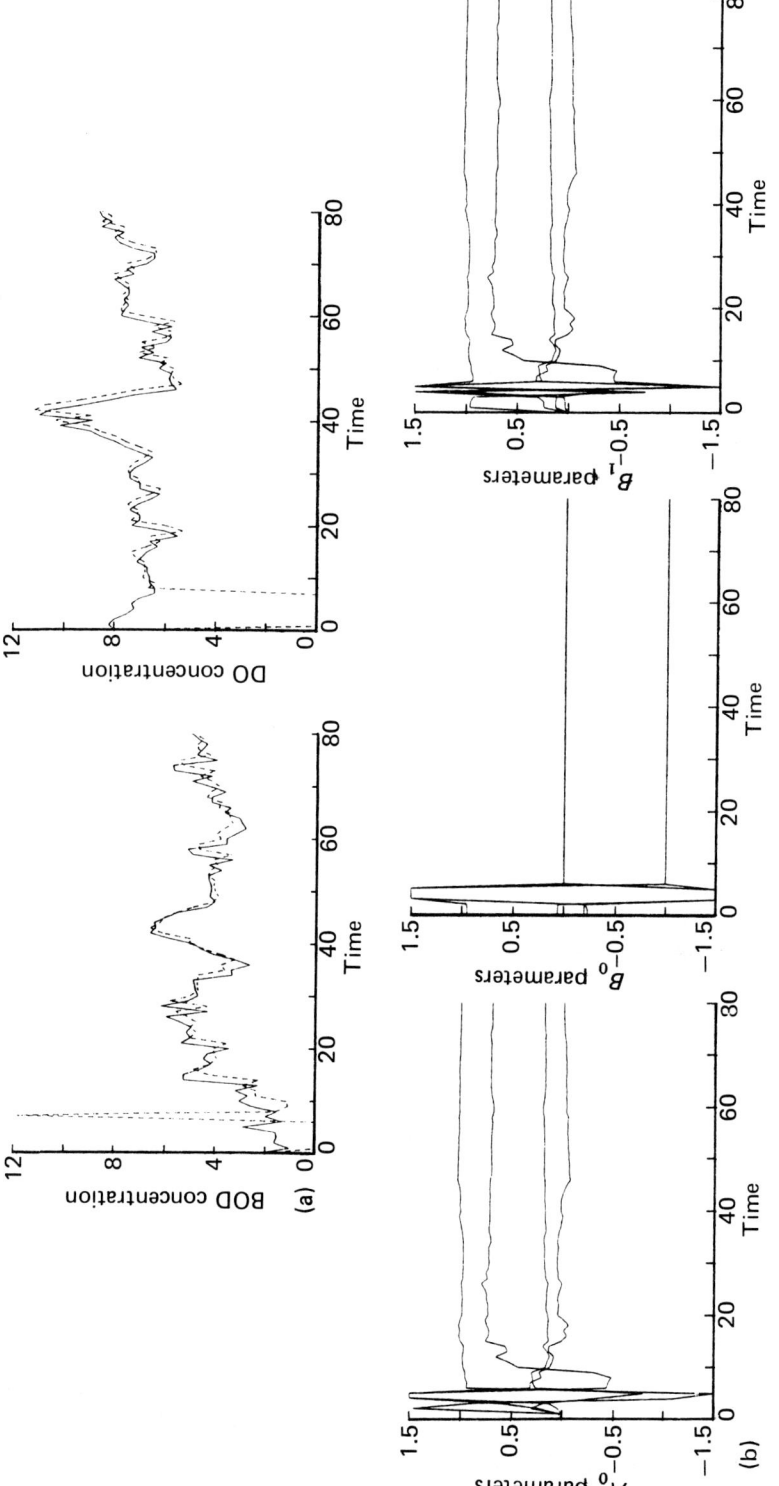

FIGURE 4 (a) Predicted $\hat{y}_1(t+1|t)$ and $\hat{y}_2(t+1|t)$ when $m = 0$, $l = 1$, and $\lambda = 0.98$. (b) Estimated parameters.

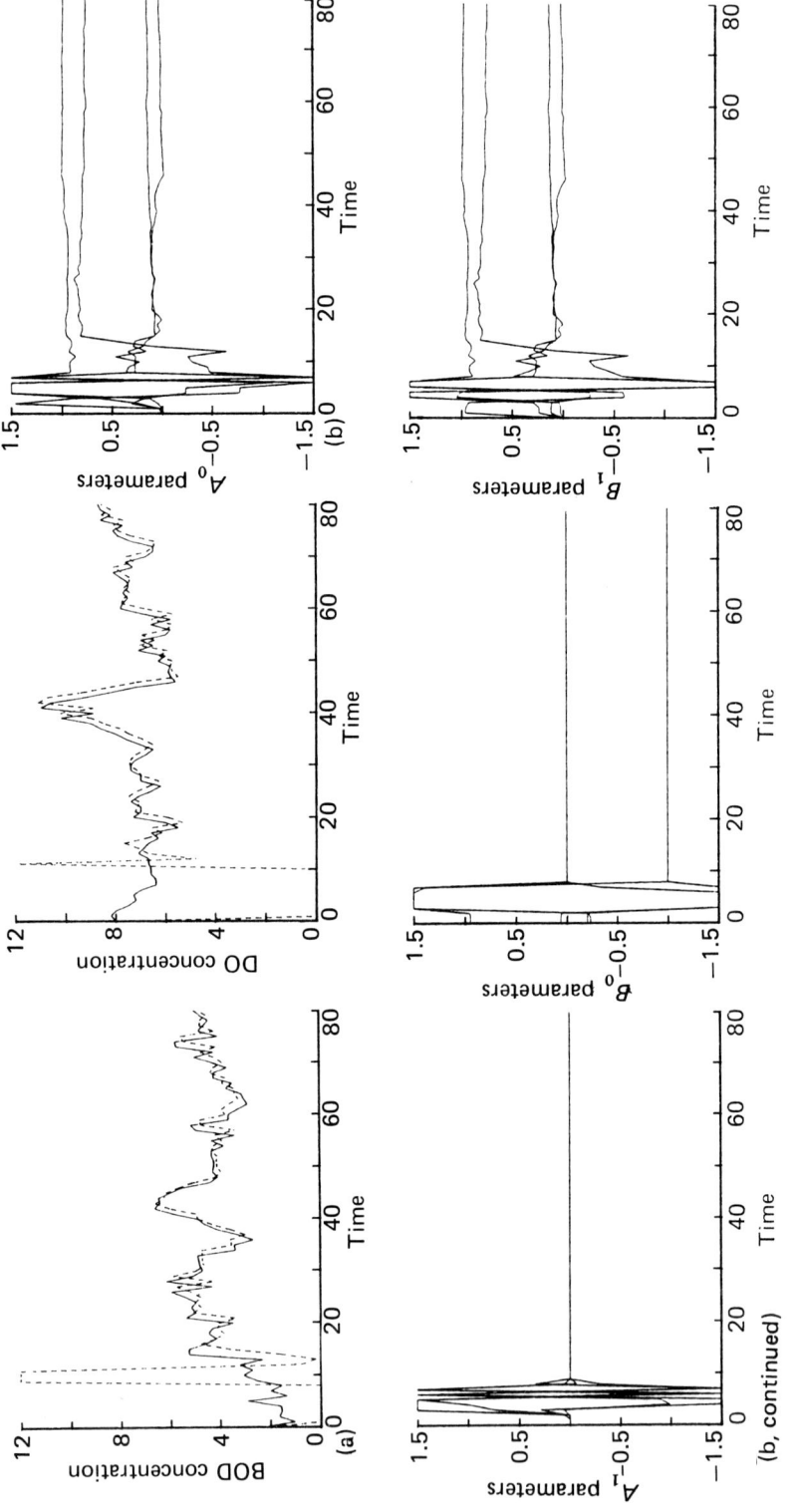

FIGURE 5 (a) Predicted $\hat{y}_1(t+1|t)$ and $\hat{y}_2(t+1|t)$ when $m = 1$, $l = 1$, and $\lambda = 0.98$. (b) Estimated parameters.

The initial value in the calculation of V is 20 because the predictor only "settles" after 10–20 prediction steps (this is due to the parameter estimation algorithm). For comparison, $V = 1.082$ when the "trivial" predictor $\hat{y}(t+1|t) = y(t)$ is used.

The prediction results are shown in Figures 2–5 together with the parameter estimates. The best results are obtained with $m = 0, l = 1$. The final values of the elements of matrix $\mathbf{A}_1$ are about $a \times 10^{-5}$, $0 < |a| < 1$, and thus it is sufficient to estimate matrix $\mathbf{A}_0$ only.

There is also a slight drift in some elements of matrices $\mathbf{A}_0$ and $\mathbf{B}_1$. Using a forgetting factor of less than unity yields, however, slightly worse results. It is also interesting to note the similarity of estimated matrices $\mathbf{A}_0$ and $\mathbf{B}_1$ after about 10 steps. This also holds true for the other cases. The final values of these matrices are as in Figure 2:

$$\mathbf{A}_0 = \begin{bmatrix} 1.00 & -0.00565 \\ 0.184 & 0.698 \end{bmatrix} \quad \mathbf{B}_0 = \begin{bmatrix} -0.999 & -0.00296 \\ 0.00197 & -1.01 \end{bmatrix} \quad \mathbf{B}_1 = \begin{bmatrix} 1.00 & -0.00555 \\ 0.185 & 0.698 \end{bmatrix}$$

So if we assume that $\mathbf{A}_0 = \mathbf{B}_1$ and $\mathbf{B}_0 = -\mathbf{I}$, the model is

$$\varepsilon(t) = \mathbf{A}_0 \varepsilon(t-1) + (-\mathbf{I} + \mathbf{A}_0 q^{-1}) \hat{y}(t|t-1) + w(t)$$

or

$$(\mathbf{I} - \mathbf{A}_0 q^{-1})(\varepsilon(t) + \hat{y}(t|t-1)) - w(t) = 0$$

Noting that $\varepsilon(t) + y(t|t-1)$ equals $y(t)$ we obtain the first-order autoregressive process

$$y(t) = \mathbf{A}_0 y(t-1) + w(t) \tag{15}$$

where

$$\mathbf{A}_0 \approx \begin{bmatrix} 1.00 & -0.0057 \\ 0.19 & 0.70 \end{bmatrix}$$

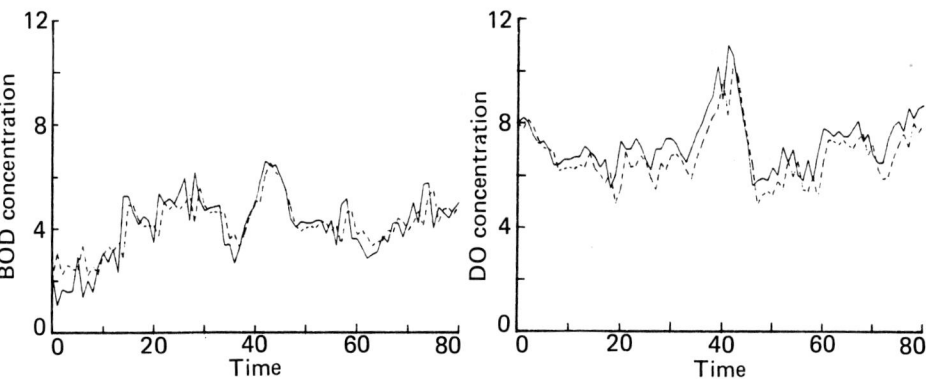

FIGURE 6 Fixed predictor $\hat{y}(t+1|t) = \mathbf{A}_0 y(t)$.

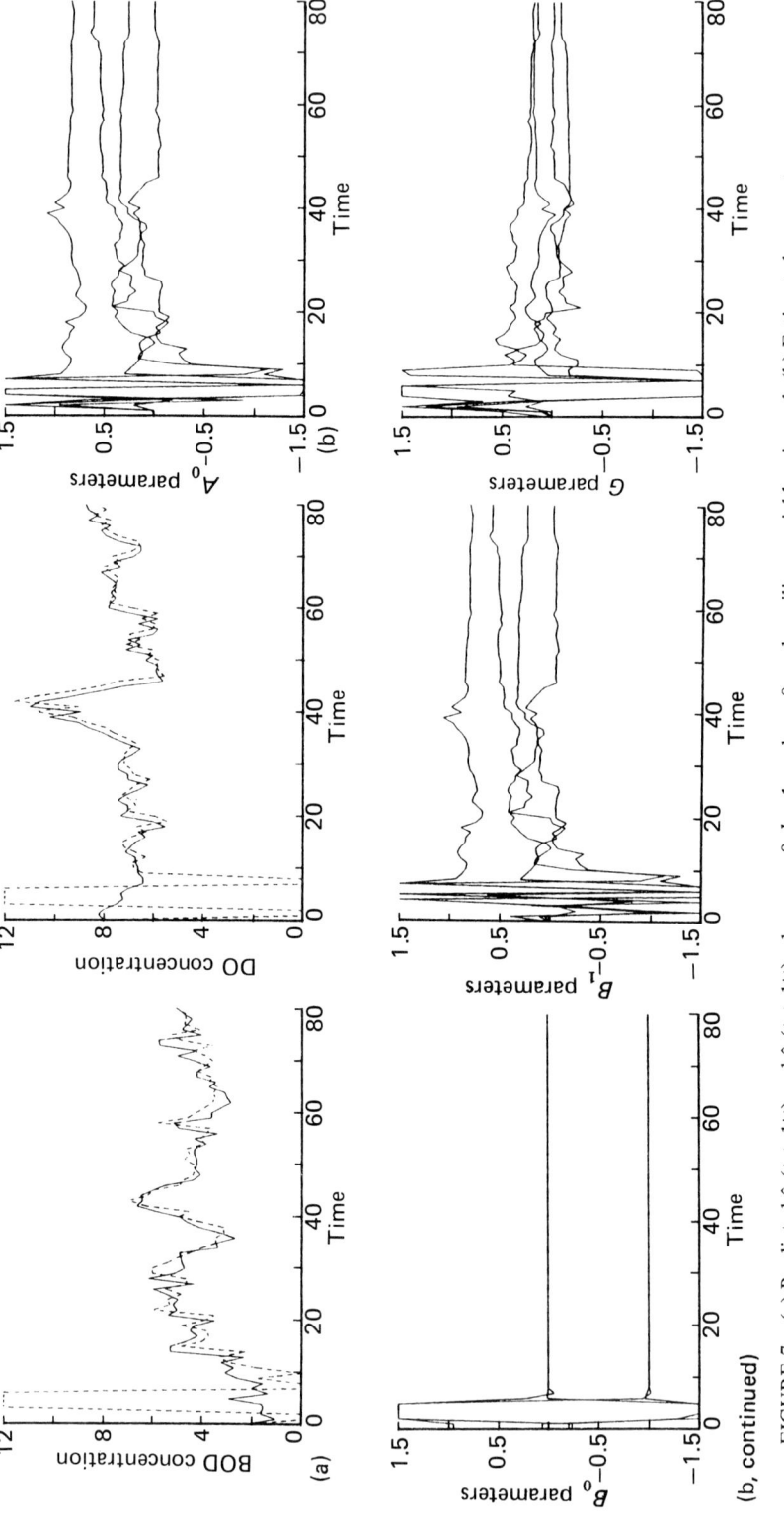

FIGURE 7 (a) Predicted $\hat{y}_1(t+1|t)$ and $\hat{y}_2(t+1|t)$ when $m=0$, $l=1$, and $r=0$, and auxiliary variable v is used. (b) Estimated parameters.

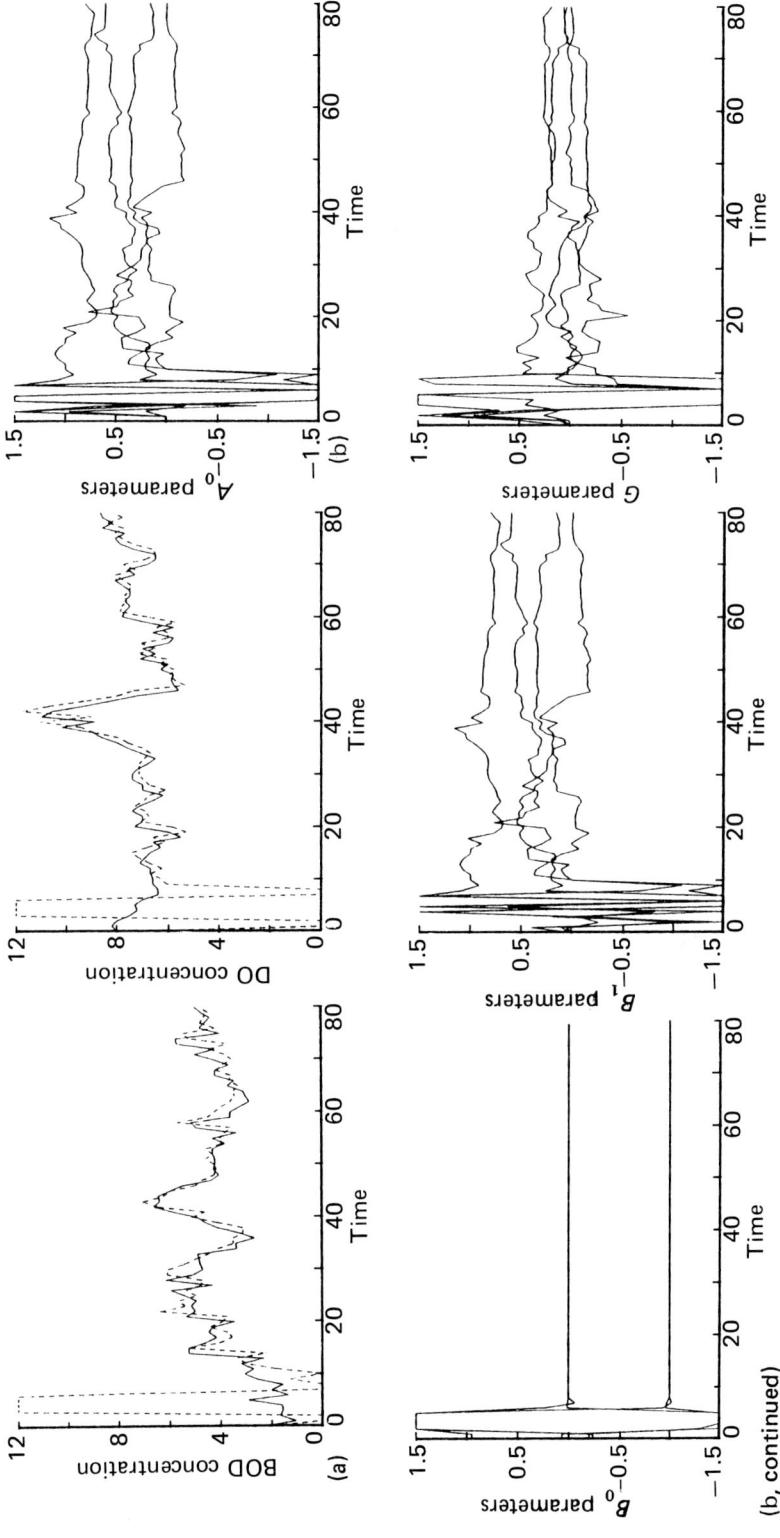

FIGURE 8 (a) Predicted $\hat{y}_1(t+1|t)$ and $\hat{y}_2(t+1|t)$ when $m = 1$, $l = 1$, and $r = 0$, and auxiliary variable v is used. (b) Estimated parameters.

Also the fixed predictor

$$\hat{y}(t+1|t) = \mathbf{A}_0 y(t)$$

was used, and the results are given in Figure 6. The loss function V is equal to 0.209, which is much worse than the values obtained with self-tuning predictors.

When Algorithm 2 is used, the remarks about preselected parameters hold. There is, however, one parameter more to select, that is, integer r in eqn. (13).

The results when auxiliary variable v (upstream DO and BOD concentrations) is used are presented in Table 2, and are also shown in Figures 7–10. It can be seen that Algorithm 1 gives better results. It is again obvious that $\mathbf{A}_1$ parameters are not needed because this matrix is practically zero.

The auxiliary variable r was only used with the structure $m = 0$, $l = 1$, $r = 0$ (Figure 11). The results were almost equal to those shown in Table 1. So the effect of r is negligible, which can also be seen from the values of the $\mathbf{G}_0$ parameters.

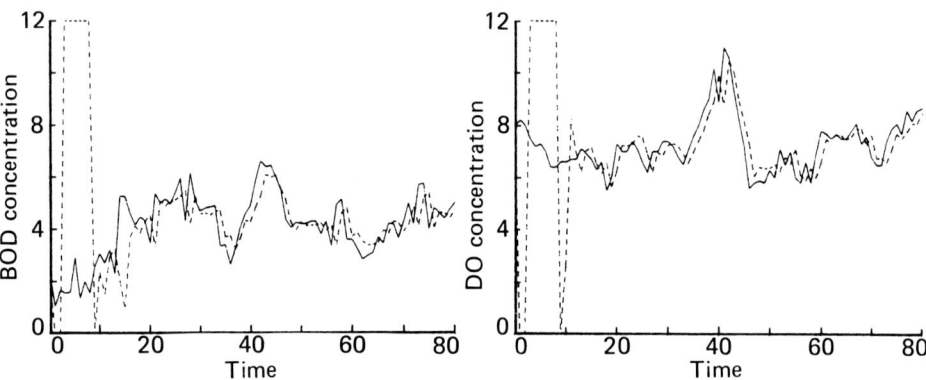

FIGURE 9   Same predictor structure as in Figure 6(a) but λ is 0.98.

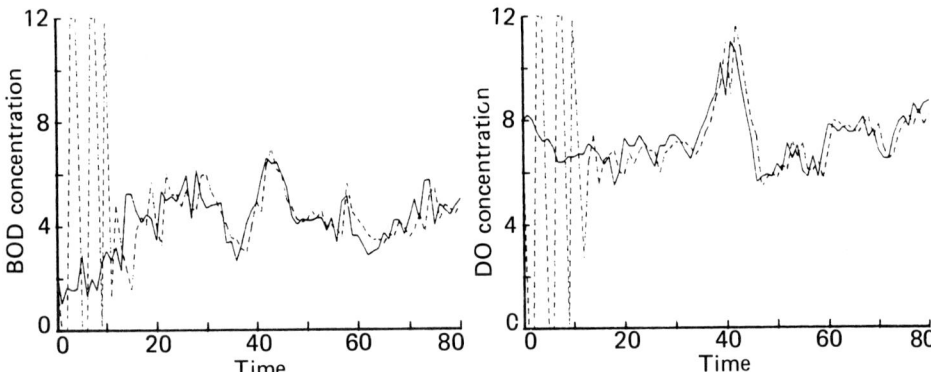

FIGURE 10   Same predictor structure as in Figure 7(a) but λ is 0.98.

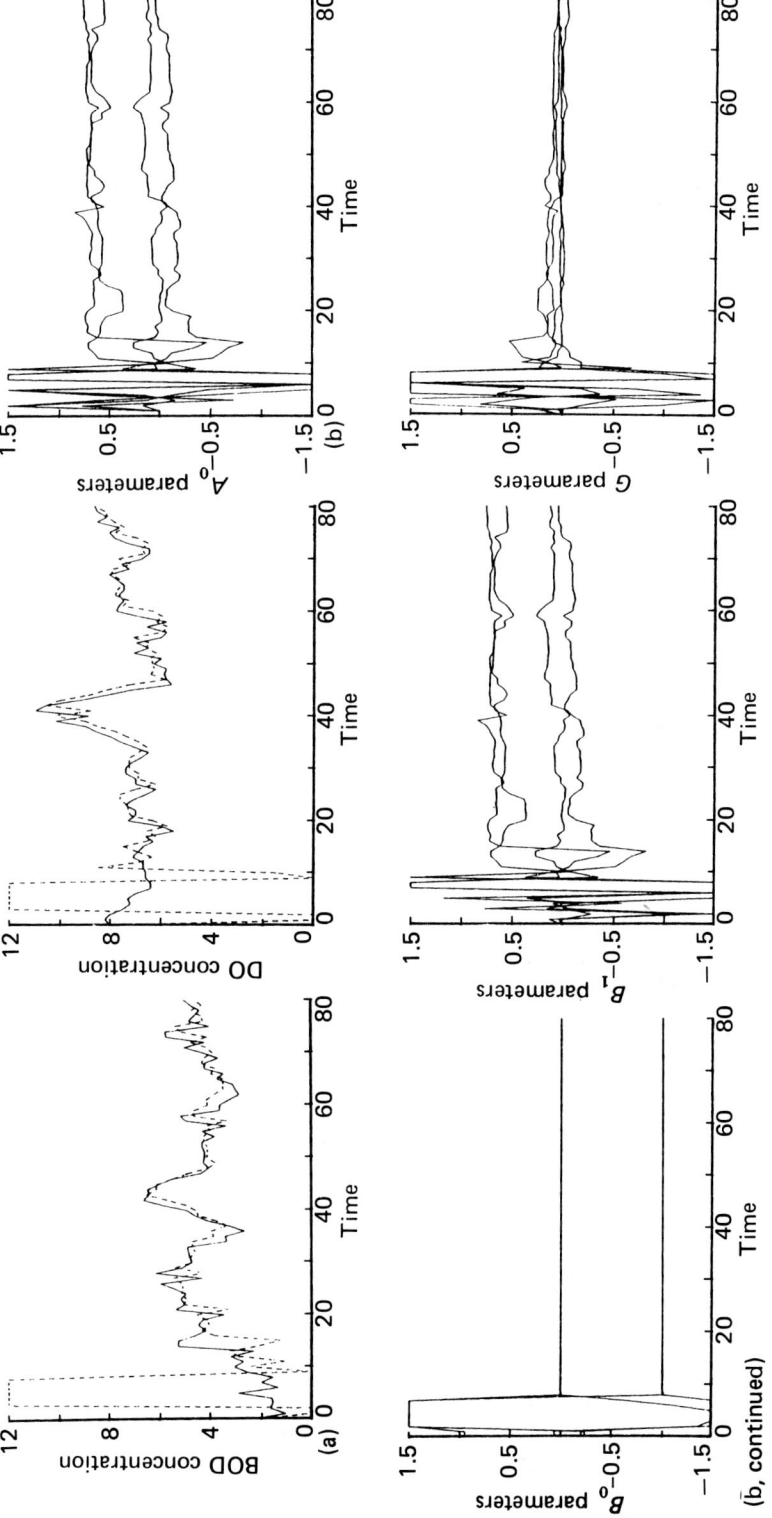

FIGURE 11 (a) Predictor $y_1(t+1|t)$ and $y_2(t+1|t)$ when $m=0$, $l=1$, and $r=0$, and auxiliary variable r is used. (b) Estimated parameters.

TABLE 2 Values of the loss function using Algorithm 2.

$\lambda = 1.0$				$\lambda = 0.98$			
m	l	r	V	m	l	r	V
0	1	0	1.052	0	1	0	1.082
1	1	0	1.182	1	1	0	1.123

It is clear that the prediction method presented cannot be utilized in real time, since BOD measurement requires five days to complete. Thus, at time t, only values $y_2(0), y_2(1), \ldots, y_2(t-5)$ and $v_2(0), v_2(1), \ldots, v_2(t-5)$ are available.

4 CONCLUSIONS

A multivariable self-tuning predictor has been applied to real field data, with encouraging results. The ease of implementation of the self-tuning predictor is obvious. The proposed prediction algorithm is particularly useful when the system model is not well-known or when the system parameters vary slowly with time. The algorithm is suitable for short-term prediction, but should not be used in long-term forecasting.

A more straightforward prediction algorithm would first fit the system model and then recursively identify the parameters, which would then be used in a precalculated optimal predictor. However, computationally this method is more laborious than that proposed, and therefore it was not used here.

ACKNOWLEDGMENT

This research was supported in part by the Finnish Academy.

REFERENCES

Åström, K.J. and Wittenmark, B. (1974). On self-tuning regulators. Automatica, 9: 185–189.
Beck, M.B. (1977). The identification and adaptive prediction of urban sewer flows. International Journal of Control, 25: 425–440.
Beck, M.B. (1978). A Comparative Case Study of Dynamic Models for DO–BOD–Algae Interaction in a Freshwater River. RR-78-19. International Institute for Applied Systems Analysis, Laxenburg, Austria.
Box, G.E.P. and Jenkins, G.M. (1970). Time Series Analysis: Forecasting and Control. Holden-Day, San Francisco, California.
Brown, R.G. (1963). Smoothing, Forecasting and Prediction. Prentice-Hall, Englewood Cliffs, New Jersey.
Coutie, G.A. (1964). Short Term Forecasting. ICI Monograph 2. Oliver and Boyd, Edinburgh.
Holst, J. (1977). Adaptive Prediction and Recursive Estimation. TFRT-1013, Department of Automatic Control, Lund Institute of Technology, Lund, Sweden.

Ikeda. S. and Itakura, H. (1983). Multidimensional scaling approach to clustering multivariate data for water-quality modeling. In M.B. Beck and G. van Straten (Editors), Uncertainty and Forecasting of Water Quality. This volume, pp. 205-223.

Kashyap, R.L. and Rao, A.R. (1973) Real time recursive prediction of river flows. Automatica, 9: 175-183.

Koivo, H.N. (1980). A multivariable self-tuning controller. Automatica, 16: 351-366.

Peterka, V. (1975). A square-root filter for real-time multivariable regression. Kybernetika, 11: 53-67.

Tamura, H. and Kondo, T. (1983). Nonlinear steady-state modeling of river quality by a revised group method of data handling for generating optimal intermediate polynomials. In M.B. Beck and G. van Straten (Editors), Uncertainty and Forecasting of Water Quality. This volume, pp. 225-241.

Tanttu, J.T. (1980). A self-tuning predictor for a class of multivariable stochastic processes. International Journal of Control, 32: 359-370.

Wittenmark, B. (1974). A self-tuning predictor. IEEE Transactions on Automatic Control, AC-19: 848-851.

APPENDIX: RECURSIVE LEAST-SQUARES METHOD

Here the recursive least-squares method is reviewed briefly. We introduce the following notation: a row vector $x(t) \in \mathcal{R}^p$ which contains data known at time t, a parameter matrix $\Theta = [\theta_1, \theta_2, \ldots, \theta_p]$, and a "measurement vector" $z(t) \in \mathcal{R}^p$.

If we try to fit a model of the form

$$z(t) = \Theta^T x(t)^T$$

between z and x, the columns of the parameter matrix can be computed recursively as follows:

$$\hat{\theta}_i(t) = \hat{\theta}_i(t-1) + K(t)[z_i(t) - x(t)\hat{\theta}_i(t-1)] \quad (i = 1, 2, \ldots, p) \quad (A1)$$

$$K(t) = P(t) x(t)^T [1 + x(t) P(t) x^T(t)]^{-1} \quad (A2)$$

$$P(t+1) = \{P(t) - K(t)[1 + x(t) P(t) x^T(t)] K^T(t)\}/\lambda \quad (A3)$$

In eqn. (A1), $z_i(t)$ denotes the ith component of $z(t)$. In eqn. (A3) the scalar λ is the exponential "forgetting factor", which is usually chosen so that $0.9 < \lambda \leq 1.0$.

In the square-root algorithm the main difference is that, instead of the $P(t)$ matrix, its square-root $S(t)$ is updated. Now the equations become

$$\hat{\theta}_i(t) = \hat{\theta}_i(t-1) + K(t)[z_i(t) - x(t) \hat{\theta}_i(t-1)] \quad (i = 1, 2, \ldots, p) \quad (A4)$$

$$K(t) = s^\rho/\sigma_\rho^2 \quad (A5)$$

$$P(t+1) = S(t+1) S^T(t+1) \quad (A6)$$

where

$$S(t+1)_{ij} = \sigma_{j-1}(S(t)_{ij} - f_i s_i^{(j-1)}/\sigma_{j-1}^2)/\sigma_j \lambda^{1/2} \quad (i,j = 1, 2, \ldots, \rho) \quad (\rho = \dim x)$$

$$\sigma_0 = \lambda^{1/2}$$

where λ is the forgetting factor, and

$$\sigma_j = (\sigma_{j-1}^2 + f_j)^{1/2} \quad (j = 1, 2, \ldots, p)$$

$$f_j = \sum_{i=1}^{j} S(t)_{ij} x(t)_i \quad (j = 1, 2, \ldots, p)$$

$$s_i^{(l)} = \begin{cases} \sum_{k_1=1}^{l} S(t)_{ik_1} f_{k_1} & (i = 1, 2, \ldots, l) \\ 0 & (i > l) \end{cases}$$

UNCERTAINTY AND DYNAMIC POLICIES FOR THE CONTROL OF NUTRIENT INPUTS TO LAKES

I.H. Fisher
Resource Engineering Department, University of New England, Armidale, New South Wales (Australia)

1 THE PROBLEM OF NUTRIENT INPUTS TO LAKES

Both natural and man-made lakes are receiving increasing nutrient loads, particularly in the forms of sewage and runoff from agricultural land. These inputs may be treated to a greater or lesser extent prior to discharge into lakes. A lumped representation of this situation is given in Figure 1. The output is neglected in the following discussion on the assumption that its impact on the total nutrient load is small in comparison with the inputs and biological sources–sinks.

The major reason for controlling nutrient inputs is taken to be the prevention of accelerated primary production, usually in the form of algae and especially blue-green algae, and the consequent deterioration in the quality of water for drinking or recreational uses. It is usual to postulate that the maintenance of algal concentration at acceptable levels can be accomplished by maintaining certain nutrient concentrations below specified levels. This is generally an impossible task for two reasons. Firstly, there are usually uncontrolled inputs to the lake of a significance comparable with the amounts removed from the controlled inputs. Secondly, the sedimentary and biological sources–sinks of nutrients within lakes are subject to large, uncontrolled disturbances, for example, by wind-induced currents.

Under these circumstances, the best that can be hoped for is the maintenance of nutrient concentrations below specified levels for a specified proportion of the time. This is in direct contrast to the aim of achieving specified trajectories (or nutrient concentrations), which is so common in the literature of state-variable control theory. It reflects a fundamental philosophical difference between the physical and environmental sciences. Given the lack of understanding of the ecological behavior of lakes, it would be presumptuous, if not dangerous, to aim to achieve particular trajectories of nutrient concentrations over time. Indeed, from the point of view of maintaining resilience (Holling, 1978), the system state should be as unconstrained as management objectives will permit.

The above aim is to be achieved by using an optimal operational policy for the (partial) control of nutrient inputs, within the constraints of the present plant available

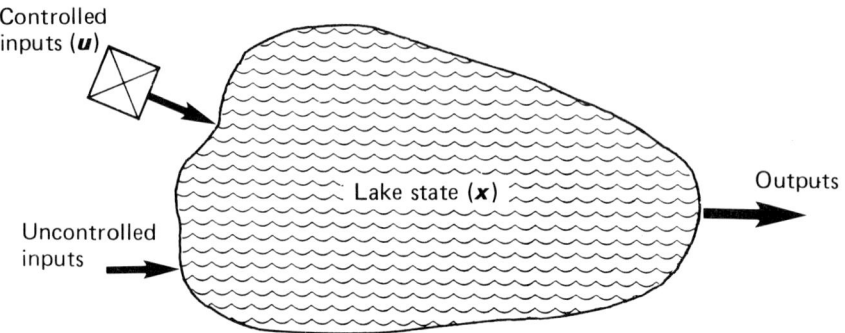

FIGURE 1   Lumped input–output representation of a lake.

for nutrient removal. The term "optimal" is defined here in a welfare sense; that is, an optimal policy is one which results in smaller social net costs, however defined, than do alternative policies. It is quite distinct from the notion of minimizing some squared deviation from a nominal trajectory of state — a notion often criticized in the control theory literature (see, for example, Rosenbrock and McMorran, 1971).

2   DEVELOPMENTAL DECISIONS AND OPERATIONAL POLICIES

Management of lake water quality may be considered to comprise two broad types of decision, namely developmental (long run) and operational (short run). The former type usually involves capital investment, and results in a decrease in social costs, however defined, over a future period of many years. Such decisions are taken infrequently. The installation of equipment to remove phosphates from river water before it enters a lake is an example of a developmental decision which might be taken in the context of controlling nutrient inputs to lakes.

In contrast, operational decisions are made relatively frequently and can be readily and substantially reversed over time, even if at some significant cost. For such decision-making, some level of development is assumed to have resulted from developmental decisions taken in the past. Continuing the above example, the degree of phosphate removal by the plant over a given time would constitute an operational decision, which would depend substantially upon the characteristics of the plant installed.

Because the behavior of complex systems must be obtained numerically rather than analytically from their mathematical representation, it is convenient to assume that operational decision-making is a discrete process; that is, the decision taken remains in force for a specified period (the decision period), at the end of which a new decision is taken. However, there are often practical reasons for using the same decision structure. Then the operational management of the system involves a sequence of decisions which are to some extent interrelated through their effects on the system state. Furthermore, it is intuitively evident that each decision in the sequence should be related to the current system state, if the system's assimilative capacity is not to be impaired. Returning to the previous example, the degree of phosphate removal should be somewhat dependent upon algal biomass or other variables which denote the phosphorus-dominated state of the lake at a given time.

An operational policy is therefore defined as a procedure for selecting a sequence of operational decisions (one for each decision period) which takes due consideration of the system state.

This paper discusses the applicability of simulation and certain techniques from control theory to the derivation of optimal operational policies for controlling nutrient inputs to lakes, under the uncertainty associated with uncontrolled disturbances to the quality of the water impounded. The problems of uncertainty in the estimation of model parameters and the related uncertainty of field measurements are assumed to have been resolved previously. However, the effect of these uncertainties on the predictions will be considered in relation to the combined simulation/dynamic programming procedure developed in Section 5.

It may be argued that optimal operational policies for nutrient input control may be derived from operator experience or from simpler models which depend upon the availability of intensive time-series data for both the inputs and the lake. However, this is feasible only if the plant (or other developments) have already been installed. Such an approach is not applicable to the problem of determining optimal plant capacity (or scale of other developments), because no time-series data are available during the planning phase. Yet the solution to such a problem is in general strongly related to the operational policies to be adopted for the proposed development. The remainder of this paper is written with such problems in mind.

3 SIMULATION

Simulation is the most direct method of deriving optimal operational policies for renewable resource systems. System dynamics are represented by a set of differential or difference equations of state which may incorporate constraints on system behavior. Given the initial values of the state variables, methods for calculating inputs, and a plausible operating policy, the equations of state may be solved simultaneously to yield the system state over the long-term period of interest. Furthermore, the contribution of system operation towards some specified objective can be computed for the same period.

To obtain the optimal operational policy, another simulation run is performed which differs from the first only in that an alternative operational policy is implemented. The policy which results in the larger contribution toward the specified objective is denoted as currently optimal. Other plausible policies are similarly tested against that currently optimal policy until no further improvement can be found. Then the currently optimal policy is regarded as the true optimal operational policy for the system. The procedure is illustrated in Figure 2.

A major advantage of this approach is that complex models and trial operational policies may be tested. The only requirements are that the state equations and the contribution of system operation to the objective can be formulated algorithmically, and that computing facilities are adequate for the task. Simplicity of the optimization procedure is an added benefit. However, there are several inherent difficulties.

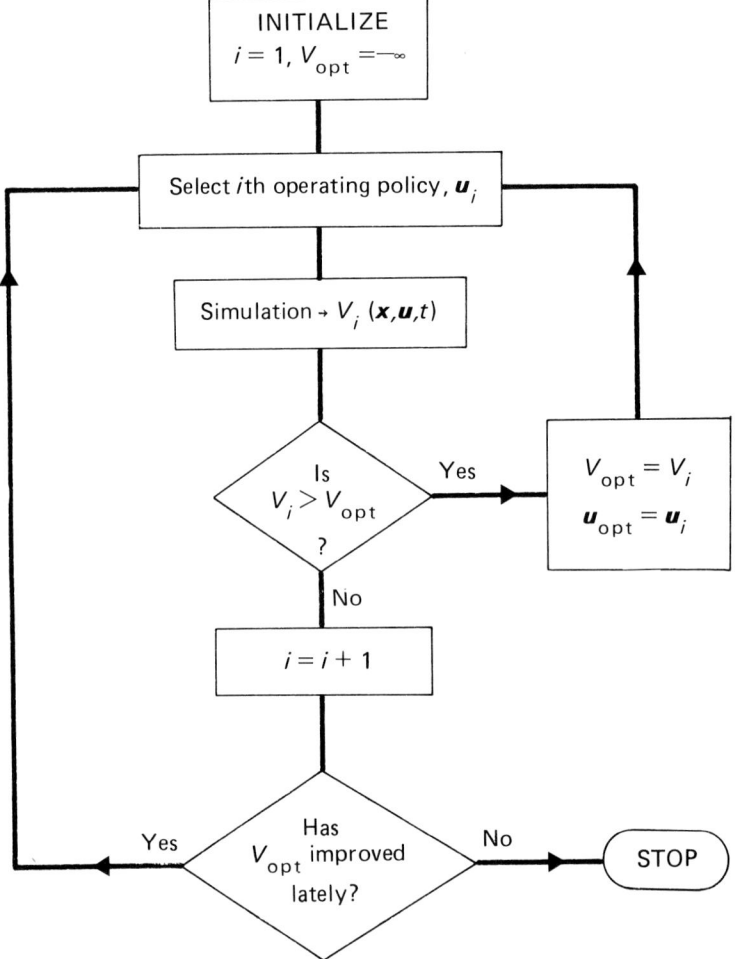

FIGURE 2 Procedure to obtain optimal operational policies by direct simulation. V_i is the value of the objective function resulting from applying the ith policy; V_{opt} is the current optimal value of the objective function; u_{opt} is the policy resulting in V_{opt}.

The first disadvantage of the simulation approach is that the currently optimal policy may yield only a locally optimal value of the objective function. The greater the complexity of the simulation, the more frequent is the occurrence of local optima, the more complex is the operating policy, and the less likely are search methods such as steepest descent or the "complex" method to find the globally optimal policy. When faced with this problem, Zuzman and Amiad (1965) used a partial factorial experimental design to determine the regions of the problem space which were to be examined by steepest descent. If the number of such regions is large, this may be unacceptable because of the computing time involved.

The second disadvantage of the application of simulation is a consequence of the simplicity of the optimizing procedure — the number of trial policies obtained by selecting, at the start of each of n decision periods, one element from a set of m admissible values of control is m^n — which can be extremely large even when m and n are quite small. To test even a substantial subset of these policies is unrealistic when the available computer time is limited or expensive. Furthermore, there may be important interactions between the initial state of the system and the proposed operating policies which require further simulation runs to elucidate.

The inputs to lakes within each decision period are not known with certainty. Frequently, they are assumed to conform to probability density functions whose parameters are possibly seasonally varying and are to be estimated from historical records. Simulation of such systems requires the generation of a sequence of inputs (one for each decision period) by random sampling from these density functions. If two different input sequences are so generated from the same set of density functions, significantly different optimal operational policies are obtained when the two different input sequences are used in an otherwise identical simulation procedure. That is, a different operational decision is chosen as optimal for the same (or comparable) decision period, when faced with the same system state in each case. Yet, such "mixed strategies" are suboptimal where the probability densities of the stochastic inputs are stationary or "cyclically stationary" in time (O'Loughlin, 1971). Any deterministic procedure applied to system models which incorporate stochastic elements will suffer from this contradiction.

In a dynamic programming study of optimal operational policies for simple water-resource systems, Hall and Howell (1963) suggested that the "pure strategy" which is expected to exist might be obtained by averaging the corresponding elements of the optimal policies which were derived under different sequences of the stochastic inputs. O'Loughlin (1971) found that such average policies were considerably less optimal than those obtained by the stochastic dynamic programming technique discussed in Section 4.4.

4 TECHNIQUES FROM CONTROL THEORY

4.1 Conventional Control Theory

Conventional control theory is taken to comprise those methods based on the use of transfer functions in the frequency domain (see, for example, Horowitz, 1963). The fundamental objection to the use of such techniques for the determination of policies for controlling nutrient inputs lies in the assumption of linear relationships between system inputs and outputs.

It is often argued that the assumption of linearity is reasonable for the treatment of small perturbations about some long-term (low-frequency) behavior of the observed variables (see, for example, Brewer, 1974). Even if this is true, the low-frequency behavior is usually assumed to be some equilibrium or simple trend, because this is the desired trajectory which the control policy aims to achieve. For the present problem, a nominal trajectory is not to be assumed for the reasons given in Section 1. As a consequence, nonlinear system dynamics must be assumed from the outset, and conventional control theory is then inappropriate. A further objection to these methods is the difficulty involved in incorporating constraints on state and control variables, which is vital in this case (Heidari et al., 1971).

4.2 Modern Control Theory

Modern control theory is taken to comprise those methods of control analysis carried out in terms of state space. In particular, the methods of state-variable feedback (Salmon and Young, 1979) have considerable appeal from the point of view of designing robust control policies. However, they suffer from the same deficiencies as conventional control theory when attempting to derive control policies for nutrient inputs to lakes.

In contrast, the technique of dynamic programming (Bellman, 1957) may be applied to general nonlinear systems in state-space terms. Constraints on both state and control variables actually reduce the amount of computation required and are readily implemented. It will become apparent from the next two sections that dynamic programming, by itself, is not without its difficulties when applied to the control of nutrient inputs.

4.3 Deterministic Dynamic Programming

Dynamic programming is based upon Bellman's (1957) principle of optimality which states that: "an optimal policy has the property that, whatever the initial system state and initial decision, the remaining decisions must constitute an optimal policy with respect to the state resulting from the first decision".

For the problem of control of nutrient inputs, dynamic programming may be reformulated as follows.

If $x(t)$ is a vector of n variables which define the state of the system at any time t, and $u(t)$ is a vector of m variables which define the operational policy, then the equations of state for the system may be written in discrete form as

$$x(t + \delta t) = x(t) + f[x(t), u(t), t]\delta t \qquad (1)$$

where f is some functional form defining the change of state from t to $t + \delta t$. In general, both x and u may be constrained differently at different times so that $x \in X(t)$ and $u \in U(x, t)$ are the sets of admissible states and controls, respectively, at time t. If the contribution of the controlled system to the objective function over the time t to $t + \delta t$ is

$$\Delta J = r[x(t), u(t), t]\delta t \qquad (2)$$

then direct application of Bellman's principle of optimality yields

$$I(x, t) = \min_{u \in U} \{r[x, u, t]\delta t + I[x + f(x, u, t)\delta t, t + \delta t]\} \qquad (3)$$

where $I(x, t)$ is the minimum cost which may be accumulated by proceeding from state

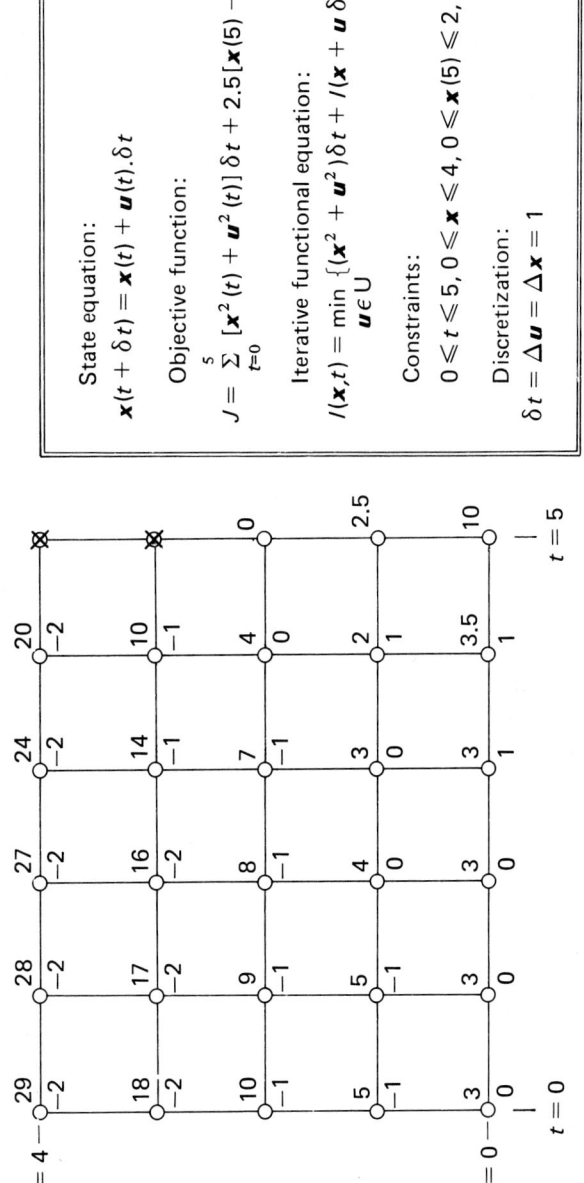

FIGURE 3 Dynamic programming solution for a simple problem (adapted from Larson, 1968). Note that at each discrete value of x and t, $I(x, t)$ is shown above the grid line and the optimal value of u producing I is shown below it.

x at time t to any admissible state at final time t_f, and $r[x, u, t]$ is the net cost accruing per unit time at time t.

The iterative functional equation (3) may be solved for discrete values of x and u to obtain the optimal long-term operating policy for the system by commencing with estimated values of $I[x, u, t_f]$ and proceeding backwards in discrete time-steps of δt until the initial time t_0 is reached. Figure 3 illustrates the form of the solution for a constrained problem in one state variable and one control variable. Although formulations which progress forward in time are possible, these require the inverse of the equations of state. If these equations are at all complex, finding their inverse is impossible, so that forward formulations are not feasible. In contrast, the backward formulation permits the use of an algorithmic rather than analytic specification of these equations, i.e., a simulation of the system over each period t to $t + \delta t$.

In the case of controlling nutrient inputs, the lake system must be considered stochastic, not only because of the uncertainty associated with inputs, but also because there are more variables exerting a significant influence on the state variables than can be incorporated in the dynamic programming formulation.

4.4 Stochastic Dynamic Programming

There are, fortunately, versions of dynamic programming which treat stochastic changes of system state over the decision period t to $t + \delta t$. The equations of state now include a set of random variables w_i, $i = 1, 2, \ldots, v$, which affect some or all of the state variables, so that

$$x(t + \delta t) = x(t) + f[x(t), u(t), w(t), t]\delta t \tag{4}$$

The parameters of the joint probability density function (PDF) of the random variables may vary with time, but the densities are assumed to be independent between successive decision periods. That is

$$P[w(1), w(2) \ldots w(T_f)] = P[w(1)]P[w(2)] \ldots P[w(T_f)] \tag{5}$$

where T_f is the number of decision periods and $P(y)$ denotes the probability of y.

Because future states are partly dependent on $w(t)$, only their probability of occurrence is known, even when the present state and the control policy are known. Hence a function of present and future states and controls can no longer be optimized. Instead, the expected value of the function is used. That is

$$J = \operatorname*{E}_{w(1), w(2) \ldots w(T_f)} \left\{ \sum_{t=1}^{T_f} r[x(t), u(t), w(t), t]\delta t \right\} \tag{6}$$

and

$$I(x, t) = \min_{u \in U} \{ \operatorname*{E}_{w} [r(x, u, w, t)\delta t + I(x + f(x, u, w, t)\delta t, t + \delta t)] \} \tag{7}$$

Computationally, the difference between deterministic and stochastic dynamic programming is the manner of evaluating the term inside the braces. In the latter case, for a given discretized state x and control u, each discretized value of w (denoted $w^{(a)}$; $a = 1, 2, \ldots, A$) is substituted into the square brackets of eqn. (7) and the result is multiplied by the probability of obtaining that $w^{(a)}$. The sum for all possible w is the required term. That is

$$E_w\{r(x, u, w, t)\delta t + I[x + f(x, u, w, t)\delta t, t + \delta t]\}$$
$$= \sum_{a=1}^{A} P[w^{(a)}]\{r(x, u, w^{(a)}, t)\delta t + I[x + f(x, u, w^{(a)}, t)\delta t, t + \delta t]\} \quad (8)$$

There are, however, several difficulties with such an approach. Firstly, by inspection of eqn. (8) it is apparent that a stochastic problem involves approximately A times the amount of computation associated with the deterministic counterpart. This can lead to prohibitive processing times for problems involving large numbers of discretized states — as is the case for the control of nutrient inputs to lakes.

There is a compensating factor, however. The formulation offers a direct means of imposing probabilistic constraints on nutrient concentrations (or other state variables) of the form

$$P[x > x_{max}] < P^* \quad (9)$$

where P^* is the probability with which the state may be outside the acceptable state, in a single decision period. These constraints, which were proposed in Section 1 as a major part of the operational objective, may be implemented as follows.

During execution of the computation implied by eqn. (8), $x(t + \delta t)$ for a given control u and for each $w^{(a)}$ are computed from eqn. (4) and checked to determine whether $x(t + \delta t) < x_{max}$. If any of the $x(t + \delta t)$ do not satisfy this condition, the associated values of $P[w^{(a)}]$ are cumulated. If this cumulation exceeds P^*, then the given control u is regarded as inadmissible. Of course, P^* may vary with time of year and is related to an annual probability via eqn. (5).

Secondly, for stochastic dynamic programming to be applied to a problem involving several stochastic variables, a technique is required for the evaluation of the $P[w^{(a)}]$, the probability of obtaining a particular set of discrete values of $w(t)$, the random variables, at any time t (see eqn. 8). In the case where the elements of $w^{(a)}$, w_i ($i = 1, 2, \ldots, v$), are uncorrelated, the product of the $P[w_i^{(a)}]$ yields the desired result, so that knowledge of the discrete univariate distribution for each w_i is sufficient for the purpose.

In general, some or all of the random variables will exhibit marked correlation with each other. Then the joint probability of obtaining each combination of discrete values of the w_i must be evaluated separately and stored for use when needed by the computational procedure. For reasons indicated in the next section, the joint PDF of the w_i may be specified in continuous form. Its conversion to the discrete form needed for dynamic programming may be accomplished by first assuming that all combinations of discrete values of the w_i are placed at the center of (hyper-)rectangles in v-dimensional space. Then the probability of obtaining a particular combination is given by the $(v + 1)$-dimensional (hyper-)volume between its (hyper-)rectangle and the projection onto the (hyper-)surface of the PDF.

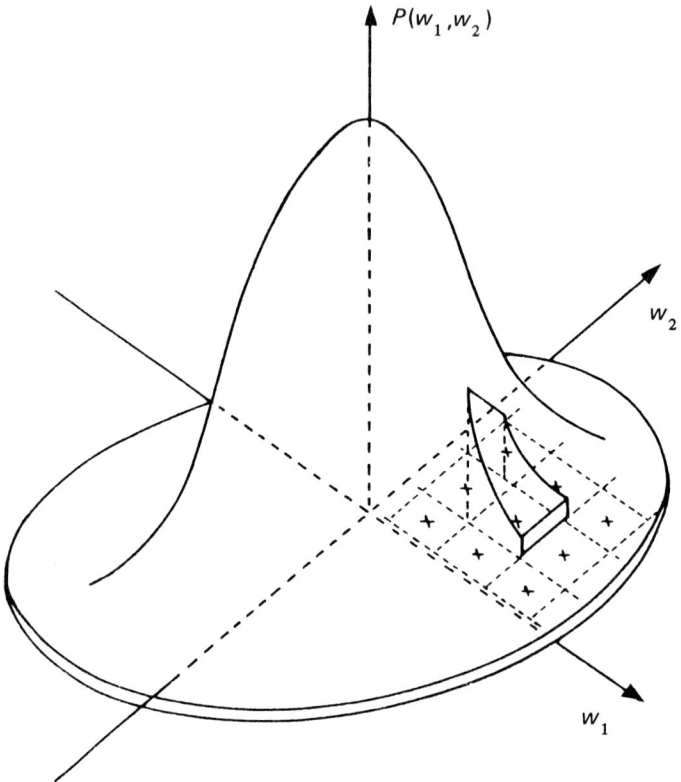

FIGURE 4 Joint probability associated with a particular combination of the stochastic variables (two-dimensional case).

The case of two random variables distributed joint-normally is illustrated in Figure 4. Doran (personal communication, 1972) developed an algorithm based on Gauss–Legendre quadrature to evaluate the probability elements from this distribution. Fisher (1974) generalized this algorithm to suit any multivariate normal distribution. It would, with minor changes, be suitable for any other multivariate distribution, so that the lack of comprehensive probability tables for multivariate distributions and the mismatching of element limits needed in the dynamic programming problem with those given in the available tables are no longer barriers to the application of stochastic dynamic programming.

Instead, the chief obstacle to such an application lies in the specification of the joint PDF for the random variables (w_i) so that it is appropriate to the particular system for which the operational policy is to be developed. It is in this respect that simulation plays a crucial role, as discussed in the next section.

5 COMBINED SIMULATION/STOCHASTIC DYNAMIC PROGRAMMING

There are two factors which prevent the direct application of stochastic dynamic programming as presented in Section 4.4 to lake inputs. Firstly it is difficult to specify a functional form for the state eqns. (4) which is an adequate representation of the changes which may occur in the lake state over a whole decision period. Even if this could be done, it will be virtually impossible to evaluate the joint probabilities $P[w_i^{(a)}]$ with an acceptable degree of statistical confidence due to lack of data, at least in the planning situation referred to in Section 2.

At this point, it is helpful to recognize that stochastic dynamic programming actually requires the probabilities of changes in the system state from each given discretized value $x(t)$ to any admissible discretized state $x(t + \delta t)$, during the decision period t to $t + \delta t$. Subsequently, these are referred to as "state transition probabilities". For simplicity of explanation here, these probabilities are assumed constant over all decision periods, but for environmental systems generally, different values for different times of the year would need to be considered.

In these circumstances, the state eqns. (4) may be replaced by an appropriate simulation of the internally descriptive type, so that the random variables w_i are replaced by the combined effects of uncertain (stochastic) inputs and endogenous variables which are themselves neither state variables nor control variables. In the phosphorus-removal example, if concentrations of algae, soluble reactive phosphorus, and particulate phosphorus were state variables for the lake system, a simulation incorporating phosphorus dynamics as influenced by endogenous variables such as uncontrolled inflows and wind-induced mixing could be substituted for eqns. (4), as indicated by eqns. (4a):

$$x(t + \delta t) = g[x(t), u] \tag{4a}$$

where g denotes the transformations in the system state produced by the simulation.

The right-hand side of eqn. (8) would then be

$$\sum_{b=1}^{B} P[x(t + \delta t)|x(t)]\{r[x, u, t] + I[x(t + \delta t), t + \delta t]\} \tag{8a}$$

where B is the number of alternative state transitions which may occur within the decision period in the dynamic programming formulation.

The frequency estimates of $P[x(t + \delta t)|x(t)]$ are obtained from a Monte Carlo simulation possessing the following properties: the initial condition is set at a discretized state $x^*(t)$, and the simulation period is one decision period, t to $t + \delta t$. Such a simulation procedure must be repeated for each admissible discretized value of initial state to obtain the full set of state transition probabilities required in the computation of eqns. (4a).

Because large numbers of discretized states are involved in a realistic dynamic programming formulation, it will generally require far less total simulation time to perform one large Monte Carlo experiment with the duration of each run still one decision period, but with the initial conditions randomly sampled from admissible state space. The state transition probabilities are then obtained from the fundamental multiplication theorem

$$P[x(t + \delta t)|x(t)] = P[x(t + \delta t), x(t)]/P[x(t)] \tag{10}$$

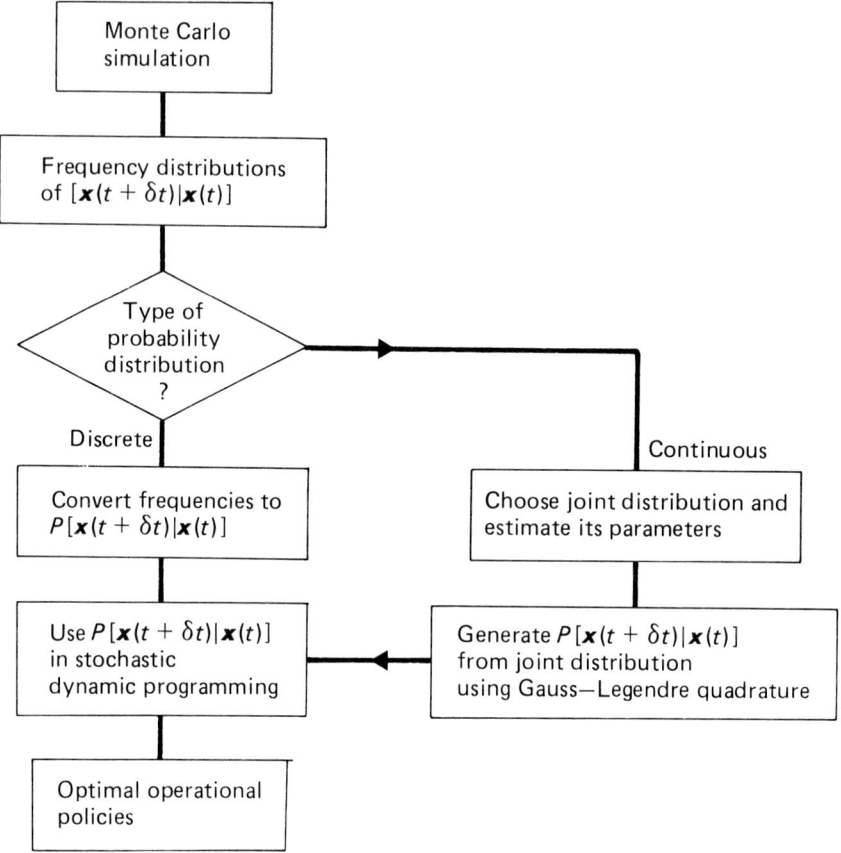

FIGURE 5 Combined simulation/stochastic dynamic programming procedure.

where the last term is the marginal probability which is derivable from the Monte Carlo estimates of the joint probabilities in the usual manner. The combined simulation/stochastic dynamic programming procedure is outlined in Figure 5.

Fisher (1974) argued that far less simulation might be needed to obtain state transition probabilities if some multivariate distribution were assumed for the joint probabilities $P[x(t + \delta t), x(t)]$, its parameters were estimated, and probabilities for stochastic dynamic programming were computed by Gauss–Legendre quadrature as described in the previous section. For a joint multivariate normal distribution, this proved efficient computationally, but it was not an adequate description of state transition in controlled arid grazing systems. For the case of nutrient inputs to lakes, it is unlikely to fare better because the simulations are of similar complexity. One of the methods involving direct frequency estimates of state transition probabilities is therefore to be preferred.

This discussion of the combined procedure contains an important feature concerning uncertainty. In another paper in this volume, Halfon and Maguire (1983) discuss the accumulation of uncertainty in predictions which result from prolonged operation of a simulation. There is very little such accumulation occurring in the combined simulation/stochastic dynamic programming procedure just described because no simulation is longer than one decision period.

In this section, the role of Monte Carlo simulation in the combined procedure has been emphasized. However, the optimal policies are obtained directly from stochastic dynamic programming. Hence, the problem does not arise that different policies are found to be optimal under different, but equally likely, sequences of stochastic inputs, as it did in the direct simulation approach of Section 3.

6 CONCLUSION

Any method for deriving management strategies for the control of nutrient inputs to lakes must recognize the uncertainties in their behavior induced by stochastic inputs and uncontrolled endogenous variables. In the particular case of deriving optimal operational policies, direct simulation adequately treats the complex dynamic nature of the problem, but it yields conflicting policies for equally likely sequences of the stochastic system state. Furthermore, even if systematic search techniques are employed, the simulation approach is computationally inefficient in comparison with techniques from control theory, because it does not take advantage of the structure of the sequential decision problem.

For ecological and philosophical reasons, methods from control theory are considered to be inappropriate if they require the specification of a target trajectory for the system state. Stochastic dynamic programming is acceptable from this viewpoint, and it exploits, to computational advantage, the structure of the sequential decision problem at hand. Furthermore, probabilistic constraints on the state variables may be incorporated; indeed, this can lead to improved computational efficiency. It is, however, limited in the number of state variables which can be comfortably included in the state equations. More importantly, some method is needed to obtain actual values for the state transition probabilities.

With suitable modification of the system state equations, Monte Carlo simulation may be used to provide state transition probabilities which reflect the uncertainties resulting from large stochastic inputs and uncontrolled endogenous variables. Furthermore, each run of the Monte Carlo experiment is only one decision period in duration, so that uncertainty in the predictions of system state does not have a chance to accumulate, as it does in the direct simulation approach.

REFERENCES

Bellman, R.E. (1957). Dynamic Programming. Princeton University Press, Princeton, New Jersey.
Brewer, J.W. (1974). Control Systems: Analysis, Design and Simulation. Prentice-Hall, Englewood Cliffs, New Jersey.
Fisher, I.H. (1974). Resource optimization in arid grazing systems. Ph.D. thesis, University of New South Wales.

Halfon, E. and Maguire, R.J. (1983). Distribution and transformation of fenitrothion sprayed on a pond: modeling under uncertainty. In M.B. Beck and G. van Straten (Editors), Uncertainty and Forecasting of Water Quality. This volume, pp. 117–128.

Hall, W.A. and Howell, D.T. (1963). The optimization of single-purpose reservoir design with the application of dynamic programming to synthetic hydrology samples. Journal of Hydrology, 1: 355–363.

Heidari, M., Chow, V.T., Kokotovic, P., and Meredith, D. (1971). Discrete differential dynamic programming approach to water resources systems optimization. Water Resources Research, 7: 273–282.

Holling, C.S. (Editor) (1978). Adaptive Environmental Assessment and Management. Wiley, Chichester.

Horowitz, I.M. (1963). Synthesis of Feedback Systems. Academic Press, New York.

Larson, R.F. (1968). State Increment Dynamic Programming. American Elsevier, New York.

O'Loughlin, G.G. (1971). Optimal reservoir operation. Ph.D. thesis, University of New South Wales.

Rosenbrock, H.H. and McMorran, P.D. (1971). Good, bad or optimal. IEEE Transactions on Automatic Control, AC 16: 552–557.

Salmon, M.H. and Young, P.C. (1979). Control methods and quantitative economic policy. In S. Holly, B. Rustem, and M.B. Zarrop (Editors), Optimal Control for Econometric Models. Macmillan, London, pp. 74–105.

Zuzman, P. and Amiad, A. (1965). Simulation: a tool for farm planning under conditions of weather uncertainty. Journal of Farm Economics, 47: 574–594.

Part Four

Commentary

UNCERTAINTY AND FORECASTING OF WATER QUALITY: REFLECTIONS OF AN IGNORANT BAYESIAN

Mark Sharefkin
Resources for the Future, Washington, D.C. (USA)

1 BACKGROUND

I was privileged to join in the Task Force meeting of November 12–14, 1979. Much of the terminology employed by the water quality modelers was new to me, or used in new ways: calibration, validation, model structure identification are notable examples. Because I am more at home in the jargon of decision theory and econometrics, I assumed at first that no more than the usual difficulties of translation were involved.

I am no longer so sure: in fact, I suspect that the two disciplines share more than terminological imprecision, and in fact, labor under some of the same substantive confusions. In this note I try to identify those confusions, and I will pose what are perhaps naive solutions.

2 ECONOMETRICS AND WATER QUALITY MODELING UNDER UNCERTAINTY: SIMILARITIES

Econometricians work almost entirely with nonexperimental data, whereas water quality modelers work with measurements they have made — with data that are the products of a conscious experimental design. However, since the Task Force meeting hardly touched on the experimental design problem — how best to spend a given budget in gathering data on a particular river or lake — I put this difference aside.

Much more important is the essential similarity: both econometricians and water quality modelers often work with data that are quite "poor" relative to the rich and complicated systems they wish to study. In at least this sense, most of the systems that econometricians study are, in the terminology heavily favored by several Task Force meeting participants, "badly defined systems". What some of those participants call "model structure identification", econometricians call "specification". Many would agree that, given a "model structure" or, in econometrics, a "specification", the logically secondary step of estimation is relatively straightforward and less interesting.

What to do? Econometricians have generally left the specification problem to the imagination, or worse. One notoriously common practice is to run all plausible regressions, choose the one with the highest R^2 value, and report it as if it was the first regression run.

That something is seriously wrong with this kind of practice has long been recognized by econometricians, but a consensus about just *what* is wrong, and about what would be better, is only now developing (see, for example, Leamer, 1978). I cannot report with authority about the prevailing situation in water quality modeling, but the discussions in Vienna left me with the impression that something similar may be happening there.

3 "MODEL STRUCTURE IDENTIFICATION" OR "SPECIFICATION": WHAT IS THE ISSUE?

Because both econometrics and water quality modeling operate in data-poor (i.e., data-expensive) regimes, both must make the best possible use of all available information. Some information lies buried in the data, and we call it "data information"; sometimes that data information can be efficiently summarized by a few sample statistics. But some relevant information is prior information, information held by the investigator before even looking at the data. It seems to be a fact that there are many varieties of prior information: Bayesians learn, or try to learn, to live with that fact and its implications. One important implication: different investigators will interpret the data evidence in different ways, because their interpretation of that evidence — their *a posteriori* estimated model structure — is the result of an exercise in which prior information and data information are combined. Bayesians can boast a systematic way of combining the two types of information; non-Bayesians often make that combination implicitly, and sometimes opportunistically. Bayesians feel obligated to report fully their prior information and the map of that prior information into *a posteriori* information. Non-Bayesians see themselves as under no such obligation: frequently their prior information, and the use they have made of it, can only be seen dimly, in their final, reported estimates.

"Model structure identification" and "specification" are processes in which prior and data information are combined. At the Task Force meeting I sensed underlying disagreement about what prior information was available on the systems being modeled and about what that prior information was worth. There also seemed to be open disagreement about how to combine prior and data information. The first such disagreement is understandable, and even predictable; the second is neither.

4 WHAT KIND OF PRIOR INFORMATION IS AVAILABLE FOR WATER QUALITY MODELING?

Individuals will differ in their prior information, and from one perspective much of the discussion at the Task Force meeting consisted of articulation of those disagreements. Two papers, both concerned with specification and estimation of models for the control of algal blooms in lakes, can be contrasted to illustrate the range of prior information deemed relevant by at least some water quality modelers.

Paper A (Chahuneau et al., 1983) describes a diffusion equation based hydrodynamical model of a lake: that model will be one module of a larger management-oriented model of the lake. All parameters of the underlying partial-differential equation were assigned a certain prior value, except for two; nonlinear programming was used to fit those two remaining parameters by minimizing a quadratic distance between observed and predicted temperatures at two depths. Paper B (Hornberger and Spear, 1983) aims at determining a best strategy for control of algal blooms in a lake, but the approach is entirely different. A simple dynamic model of the system is constructed – with relatively little *a priori* structure imposed. That model, like the model of Paper A, has many more parameters than can be "confidently estimated" from data available on the lake. The authors of Paper B raid the literature for decent values of many of the relevant parameters which have been measured elsewhere, and then fit the remaining undetermined parameters by imposing the requirement that the fully-fitted equation reproduce a few rough, qualitative features of the lake, including of course the spectacular and objectionable algal blooms, the control of which motivates the whole exercise.

At one level, the essential difference between the two approaches is a difference in the interpretation and treatment of prior information. The authors of Paper A bring to their work two kinds of prior information: measured characteristics of the lake, including the temperature distribution, and the "prior information" embodied in the diffusion equation employed, and in the coefficients of that equation assigned fixed values before the optimization. The authors of Paper B, on the other hand, import prior information on those physical parameters measured in similar systems elsewhere and on the gross qualitative features of the system they are trying to control.

Who is right? That question is meaningless. A more instructive question is: which judgment is closer to the mark regarding the kind of prior information we now have, and might buy at reasonable cost, on lake ecosystems? Below I offer some speculation: here, I look at the problem in the following perspective. Because of my ignorance of the subject, I would be very unwilling to let either prior or data information dominate the mapping of prior into posterior for water quality modeling: this much I learned at the Task Force meeting. For that reason, I would choose to cast prior information in a form in which it might be substantially modified by the kind of sample information (data information) readily available on water quality systems. My suspicion is that only the approach taken in Paper B will do. The only way in which the key piece of prior information exploited in Paper A – the partial-differential equation – can be "discredited" by the data information is if the optimization assigns the two estimated coefficients values that are wildly implausible.

But it is a poor critic indeed who cannot find fault with everything. Turning to the way in which prior information is mapped into posterior information, I can find fault with both Papers A and B, and can even find a way to make some possibly constructive suggestions.

5 MAPPING PRIORS INTO POSTERIORS

Paper B devises a method for mapping priors into posteriors and is explicit about that method. Prior information in this case is of two kinds: the four nonlinear dynamic equations chosen to represent the system and rectangular prior probability distributions on the parameters entering into those equations. Those intervals defining the prior probability intervals are taken from the literature on other, presumably similar, water bodies.

The "data" information enters in a way that is novel and is *the* novel feature of the paper. "Data" means certain gross structural features of the system, in particular the algal blooms. These are data in the sense that they are the only observed features of the dynamic system, and they are also data in the sense that they are to be tested against prior information.

Paper B sets up that confrontation as follows. Each prior parameter set implies a specific nonlinear deterministic equation set. If that equation set defines a system trajectory consistent with the "data" — the inequality constraints imposed upon the real-time behavior of the system — then the corresponding parameter-space point is assigned to one region (behavior), and to another region (nonbehavior) if the inequality constraints are violated. Monte Carlo methods allow exploration of the parameter-space hypercube, and — through run-by-run comparison of the output with the inequality constraints — partition off the original parameter-space hypercube into two (presumably connected) regions, "behavior" and "nonbehavior". If the behavior subregion is "large", then the prior information is concluded to be reasonable and reasonably compatible with the data evidence. The implication, for management and control, is that selection of a point from the behavior region, and use of the corresponding equation, may be a sensible procedure.

Much as this procedure cheers a Bayesian, I lodge the following complaint: the way in which the procedure uses data evidence to map priors into posteriors is, to say the least, less than subtle, and may seriously distort the data evidence. In particular, posterior probability distributions are always uniform distributions on a subregion of the support hypercube of the original prior distribution. Put another way, the data evidence can never be strong enough to drag us out of the original prior parameter region. The very binary nature of the inequality constraints on system trajectories — those trajectories are either consistent with the inequality constraints or they are not — rules out any finer discrimination between prior points. (See the example in the Appendix.)

One of the stronger arguments, if not for Bayesian econometrics then for a style of presentation of econometric results with a Bayesian flavor, goes as follows: the best way to get a sense of the power and persuasiveness of the data evidence is to report the way in which the data evidence maps priors into posteriors. In Paper B, only one such mapping is presented. In the next section I suggest alternative methods — for computation and reporting — in the spirit of Paper B.

Paper A can be similarly dissected. Prior information in this case is the partial-differential (diffusion) equation which is the basis for the hydrodynamical model of the lake, taken together with those coefficient values which are assigned certain point values, typically from the literature on other, similar water bodies. The data information consists of the relatively few measurements made on the lake, typically temperatures at specific points and levels, and several inflow values.

The approach in Paper A is classical (as compared to Bayesian): the remaining (unassigned) model parameters are determined by nonlinear optimization, with the criterion being the sum of squared deviations between data and model prediction points.

For purposes of comparison with Paper B, however, we choose to imagine that the authors of Paper A were to rework their prior information and data information, this time in a Bayesian spirit. This could be done as follows. First assign prior probabilities to the two model parameters that are not assigned certain point values. Those prior probabilities on model parameters imply corresponding prior probabilities on the physical variables for which some data exist, obtained by solving the model repetitively over the prior intervals.

Now, beginning from the two optimized model parameters obtained in the non-linear optimization, compute equal likelihood contours in the space of the unassigned model parameters. Taken together, the equal prior probability surfaces and the equal likelihood surfaces define a set of best compromises between prior and data information. They also indicate how decisive or persuasive the data evidence is against the sample evidence. If, for example, the data likelihood surface embraces all reasonable priors, then the data evidence is, for this way of comparing data and prior evidence, relatively weak.

My own suspicion is that, for a large, over-parameterized model, this will usually be the case. A key criterion in the choice of a model should be the following: that the model permits a constructive and instructive confrontation between prior and data evidence. Larger, over-parameterized models typically will fare badly under this criterion.

6 SOME CONSTRUCTIVE SUGGESTIONS

I have argued, above and elsewhere, that both the model order criterion — or the information criterion from which it derives — and least-squares estimation have serious deficiencies (see Sawa, 1978; Leamer, 1979). The first is simply meaningless, and the second is, except for a few very special cases, strictly dominated. Here is a proposal for reform, followed by a proposal for a fair test of the relative merits of the modeling approaches exemplified by Papers A and B.

My proposals for reform are couched in the form of a set of statements. I hope these statements will meet with general agreement. First, in water quality modeling for water quality management, both prior and data information are exploited. This is a triviality worth repeating: there is data information, typically too sparse to say anything on its own, and there is prior information, again often too sparse to be decisive on its own. The reasons for the second situation are different than the reasons for the first. Some kinds of prior information are relatively weak without certain particular pieces of information that would be very expensive to obtain. I have in mind here information of the following kind: the boundary information which, in principle, is needed to apply a diffusion equation to a large water body.

Second, an ideal reporting style for analyses of a situation in which neither prior nor data information is decisive is reporting the mapping of priors into posteriors: that is, in showing how several priors, somehow chosen over a reasonable range, are mapped into posteriors by the data.

Third, an implied criterion for choosing among water quality modeling approaches is their convenience and ease in assisting us in this exercise. This criterion will, in practice, heavily favor models with small numbers of parameters. That is, the entire procedure of specifying and adjusting priors becomes cumbersome once more than two or three parameters are involved.

Finally, a promising suggested test of the approach of Papers A and B. The test should be made in the context of a particular management problem. Begin with the assumption that the "true" system is that described by the approach taken in Paper A. Use the hydrodynamical and ecosystem models of that approach to generate pseudodata on the behavior of the system: each data set should be generated so as to permit application of either the Paper A or Paper B approaches. Now approach that data in two ways. For the Paper B approach assign some parameters from other water bodies, and fit the rest with our proposed modification of the approach of that paper. Do the same for Paper A, this time fitting the unassigned parameters with the technique of that paper. Finally, assign a loss function on the effect to be controlled, and compare — at each prior probability/data probability point — the losses imposed by the use of the two approaches.

REFERENCES

Chahuneau, F., des Clers, S., and Meyer, J.A. (1983). Analysis of prediction uncertainty: Monte Carlo simulation and nonlinear least-squares estimation of a vertical transport submodel for Lake Nantua. In M.B. Beck and G. van Straten (Editors), Uncertainty and Forecasting of Water Quality. This volume, pp. 183–203.

Hornberger, G.M. and Spear, R.C. (1983). An approach to the analysis of behavior and sensitivity in environmental systems. In M.B. Beck and G. van Straten (Editors), Uncertainty and Forecasting of Water Quality. This volume, pp. 101–116.

Leamer, E. (1978). Specification Searches. Wiley, New York.

Leamer, E. (1979). Information criteria for choice of regression models: a comment. Econometrica, March.

Sawa, T. (1978). Information criteria for discriminating among alternative regression models. Econometrica, November.

APPENDIX: A SIMPLE EXAMPLE

Suppose that we are studying a dynamical system known to be described by eqn. (1) and initial condition (2):

$$\dot{x}(t) = (a_1 + a_2 \cos t) x(t) \qquad (1)$$

$$x(0) = c \qquad (2)$$

We are asked to estimate the two scalar parameters a_1, a_2; the only data we are given is in the form of one inequality restriction (3) on $x(t)$:

$$x(t) \geqslant L, t \in [t_1, t_2], 0 < t_1 \leqslant t_2 \qquad (3)$$

Because the solution to (1) is:

$$x(t) = c \exp(a_1 t) \exp(a_2 \sin t) \tag{4}$$

condition (3) implies the inequality

$$a_1 t + a_2 \sin t \geq L/c, \quad t \in [t_1, t_2] \tag{5}$$

For each value of t in the interval $[t_1, t_2]$, inequality (5) defines a closed half plane of the (a_1, a_2) plane, so that the set of restrictions (5) defined as t varies over that interval defines the intersection of a set of closed half planes. For convenience (and without loss of generality) choose units so that our prior hypercube in (a_1, a_2) space is the unit square $[0, 1] \times [0, 1]$, and consider the trivial case in which the t-interval reduces to one point: $t_1 = t_2 = \bar{t}$. Then the method of Paper B tells us to summarize the "data" $x(t) \geq L$ by constructing the half plane

$$H = \{(a_1, a_2) / a_1 \bar{t} + a_2 \sin \bar{t} \geq L/c\} \tag{6}$$

and by restricting our attention, in (a_1, a_2) space, to

$$[0, 1] \times [0, 1] \cap H \tag{7}$$

Compare this with an analog of the more familiar form of data summary. Consider the loci of points in the (a_1, a_2) plane defined by constant values of the loss function

$$Q(a_1, a_2) = (a_1 \bar{t} + a_2 \sin \bar{t} - L/c)^2 \tag{8}$$

Those loci are lines parallel to the boundary of the region (7). Clearly the data summaries provided by (7) and (8) are very different.

AUTHOR INDEX

Beck, M.B., 3

Chahuneau, F., 183
Chapra, S.C., 293

des Clers, S., 183

Fedra, K., 259
Fisher, I.H., 357

Gardner, R.H., 245

Halfon, E., 117
Hornberger, G.M., 101

Ikeda, S., 205
Itakura, H., 205

Jørgensen, L., 173

Koivo, H.N., 339
Kondo, T., 225

Maguire, R.J., 117
McLaughlin, D.B., 305
Mejer, H., 173
Meyer, J.A., 183

O'Neill, R.V., 245

Reckhow, K.H., 293

Sharefkin, M., 373
Somlyódy, L., 129
Spear, R.C., 101

Tamura, H., 225
Tanttu, J.T., 339

van Straten, G., 157

Whitehead, P.G., 321

Young, P., 69

SUBJECT INDEX

activated sludge process, 14, 15
adaptive prediction, 339–356
aggregation, 119, 120, 126
  error, 122, 123, 269
  spatial and temporal, 293
Akaike's information criterion (AIC), 226, 229, 230
algae (phytoplankton)
  bloom, 13, 33, 92, 103, 109, 112, 183, 374, 375
  carbon content, 158
  dead material, 31, 32
  death rate constant, 34, 35
  growth and photosynthetic properties, 34, 178, 225, 331–333
  interference with BOD test, 31, 37
  model, 157–171
  nutrient, zooplankton system, 52–59
  zooplankton model, 248–252
ammonia (see nitrogen, ammonia)
Anglian Water Authority, 33, 325, 330
Attersee lake (Austria), 274–287
automatic water quality monitoring, 321, 329
  telemetry for, 43, 321, 329
autoregressive moving average (ARMA) model, 339

Bayesian viewpoint, 72, 75, 95, 315, 316, 373–379
Bedford-Ouse river (UK), 33–40, 48, 52, 56, 321–336
behavioral pattern
  behavior, nonbehavior sets, 77, 103, 105, 111, 119, 122–124, 125, 263–266, 271, 274, 276–278, 376
  problem-defining, 262, 287
Bellman's principle of optimality, 362
bias error, 247, 248, 255, 301
biochemical oxygen demand (BOD), 14, 30, 31, 33, 326
  cumulative predicted distribution, 329
  decay rate constant, 27, 30, 34, 36, 37
  interaction with DO, 226–229, 232–240, 323–325, 328, 343–354
  sources and sinks, 31, 38, 40
black box models, 4, 19, 44, 78, 94, 229, 261, 294, 301, 302
Bormida river (Italy), 226, 231–240
bottom friction coefficient, 129, 132, 137, 138

calibration, 3, 9, 136, 190–193, 373
  relationship with prediction, 5, 45, 48, 52, 288

catastrophe theory, 13, 14
chemical oxygen demand (COD), 205, 208, 210, 211, 216–218
Chézy coefficient, 138
chloride concentration, 326
chlorophyll-a, 33, 39, 113, 165, 179, 205, 207–210, 213, 216–218, 264–266, 333
  –carbon ratio, 269
  estimated rate of addition, 38
  sources and sinks, 38
Cladophora alga, 82–86, 102–115
cluster analysis, 205, 214
coefficient of variation, 53, 59, 197, 201, 249, 284, 286
continuously stirred tank reactor (CSTR) idealization, 93, 323, 324
correlated parameter errors, 56, 59, 161–163, 190–192, 253, 264, 269
correlation coefficient
  simple, 248
  partial, 248–250
correlated structure of parameter space, 278, 279
  associated frequency distribution, 282
Cramér–Rao inequality, 161
Crank–Nicolson numerical scheme, 185

data information, 374, 376
deterministic approach, 7, 9, 260, 261
detritus, 268, 269
dilemma of prediction, 45, 59–62
dispersion
  atmospheric plume model, 252
  coefficient of, 11, 12, 36, 93, 94, 316
  coefficient of vertical, 185, 186, 190, 191, 195, 197, 272
  in groundwater systems, 306
  mechanisms, 11, 12, 21, 34, 35, 37, 92, 324
  model, 12, 35, 36, 48, 93–95, 375, 376
dissolved oxygen (DO), 33, 208, 216, 217, 326
  associated with sediments, 190
  cumulative predicted distribution, 329
  interaction with ammonia-nitrogen, 321, 330–333
  interaction with BOD, 21, 27, 28, 30, 226–229, 232–240, 323–325, 328, 343–354
  photosynthetic production, 225, 226
  reaeration rate, 28, 31, 34–36
  sources and sinks, 31, 38, 39
distributed-parameter models, 50, 76, 93, 174, 175, 181, 184, 305–308, 324

dominant modes of behavior, 35, 60, 75, 77–79, 86
dominant-mode model structure, 80, 110
dynamic programming, 359, 361–364
  combined with simulation, 366–369
  difference between deterministic and stochastic, 366–369
  stochastic, 361, 364–366

econometrics, 373, 374
eigenvalue–eigenvector analysis, 77, 214
epilimnion, 158, 165, 186, 194, 272
error analysis (see also sensitivity), 50, 246, 294, 301
  comparison with sensitivity analysis, 247–255
errors (uncertainty)
  boundary conditions, 313, 315
  distribution (see probability density functions)
  initial conditions, 170, 267, 313, 315
  input measurements, 129, 144–151, 188
  in source rate, 316
  measurements, 159, 165, 309, 313, 315, 316, 324
  model structure, 159
  of sampling, 266, 322
  parameters, 164, 245–256, 296, 299, 316
  spatial, 158, 159
  temporal correlation in, 311, 312, 314
error-loss function, 12, 23, 122, 136, 163, 164, 168–170, 227, 228, 340, 344
  gradient of surface of, 12
  linearization about minimum, 191
  sum of squares, 72, 122, 157, 191, 193, 215
error variance–covariances
  correlated state-parameter, 51
  current state, 51
  future inputs, 51, 53
  initial state, 53, 124
  measurements, 174
  minimum variance prediction, 340
  parameters, 12, 46–48, 51, 79, 80, 124, 157, 164, 183, 344
  predicted state, 51, 183, 312, 313
  residual errors, 51, 165
estuaries
  algal productivity, 111
  flushing dynamics, 87–92
eutrophication, 8, 102, 130, 183, 205, 274, 294
expected value of a function, 247
experiment, 52, 53
  active, 10–12, 15, 69
  natural, 10, 12, 13, 15
  passive, 10, 13, 14, 69
  planned, 10, 11, 15, 70, 77, 93, 373
  relationship with uncertainty, 152, 153, 259, 260, 301
  sampling frequency, 10, 51, 176, 197
  tracers, 11, 92, 93
exponential weighting of past data (forgetting factor), 344, 347–349, 352, 355, 356

extended Kalman filter (EKF), 28, 42, 51, 246, 322, 331
  gain matrix, 53

feedback control, 362
fenitrothion, 117–127
  hydrolysis, photolysis, volatilization, 117
Fisher information matrix, 161
forecasting (prediction) errors
  distributions of, 302
  due to parameter uncertainty, 163, 194, 245–256
  first-order analysis of, 308
  k-step ahead, 340
  prediction ensembles, 282–287
  propagation, 45, 50–59, 157, 162, 170, 245, 293–303, 305–319
frequency distributions of errors, 246, 247, 252–255
frequency response analysis, 10, 11, 361

Gauss–Newton algorithm, 174, 175, 191
groundwater flow, 105
group method of data handling (GMDH) algorithm, 225, 339
  partial polynomials in, 225, 230
  revised, 225–240

heat budget, 185, 187, 188
homoscedacity, 295
hydrodynamical model, 129–154
  1-D equations of motion and continuity, 130
  2-D and 3-D representations, 151, 152
hydrological sciences, 4, 12
hypolimnion, 158, 165, 185, 186, 190, 191, 194, 272
  aeration, 183
hypotheses
  failure of, 26, 27, 34, 35, 41–44
  falsification of, 29, 31, 33–35, 40, 69, 81
  generation of, 266–273, 287
  speculation about, 29, 31, 37–44, 114

identifiability, 12, 35, 56, 72, 80
  over-parameterization, 79, 80, 180, 377
  spurious content, 61
  surplus content, 16, 44, 56, 61, 72, 82
identification
  system, 3–5, 9, 324
  time-series model, 78
implicit finite difference scheme, 132
innovations process, 79, 92
in situ field observations, 6, 70
  "experiment" reconstruction from, 43
instrumental variable algorithm, 80, 316
invalidation, 26

Kolmogorov–Gabor polynomial, 230
Kolmogorov–Smirnov statistic, 77, 109, 125, 328

Kriging algorithm for interpolation, 315

lake Balaton (Hungary), 129–154
lake Biwa (Japan), 206–221
lake Charlevoix, Michigan (USA), 298
lake ecosystem, 274–287
lake Esrom (Denmark), 181
lake Glumsø (Denmark), 178
lake Nantua (France), 184–201
lake Ontario (USA, Canada), 16, 44, 157–171
lake–watershed systems, 293
least squares
  algorithm, 72, 341
  estimation, 169, 191–194, 246, 316, 344, 355, 356, 377
Leslie matrix model, 254, 255
light
  extinction coefficient, 84, 106, 158, 272, 277
  growth-related parameters, 105, 106, 180, 267
  shading effects, 112, 267, 272
likelihood function, 159
  equal likelihood contours, 377
  maximum (see maximum likelihood estimation)
longitudinal flow velocity, 130

marine ecosystem, 264–270
Marquardt's algorithm, 174, 175, 191, 194
marsh hydrology model, 248–252
matrix sweep technique, 132, 133
maximum likelihood estimation, 157, 160, 193, 316
metalimnion, 192
Michaelis–Menten coefficient, 163, 179, 180, 272
mixing length, 186, 190
models
  input/output, 12, 229, 358
  internally descriptive, 78, 226, 230
  linked set of, 7
  mechanistic, 78, 104, 226
  steady-state gain of, 86
  time-constant of, 86
  time-variable coefficients, 87
model-building procedure, 74
model discrimination, 21
model order estimation, 19, 20, 42
model structure identification, 9, 15–31, 52, 69, 75, 81, 262, 373, 374
  definition of, 19, 20
  of time-series model, 79
monitoring networks (see automatic water quality monitoring and experiments)
Monod kinetics, 21, 35, 104, 267
Monte Carlo simulation, 43, 71, 76, 77, 84, 103, 111, 123–125, 130, 134, 136, 146–151, 170, 191, 194–197, 245–256, 259–289, 302, 327–329, 367–369, 376
multidimensional scaling (MDS) approach, 205–222

natural variability, 246, 247, 252, 255
nitrates
  effects on potable supply, 321, 325, 327
  in river Thames and river Lea, 321
nitrification, 331–333
nitrogen
  ammonia, 103, 158, 165, 208, 321, 326, 333
  cycle, 163
  inorganic, 113
  intracellular (algae), 179
  limiting nutrient, 104
  nitrate (see also nitrates), 103, 158, 165, 208, 210, 212, 216–218, 321, 333
  organic, 158, 179, 208
  scenario, 115
  total Kjeldahl, 158, 165
  total load, 92
nonlinear programming, 375
nutrients
  budget, 91
  control of inputs to lakes, 357–369
  phytoplankton, zooplankton system, 52–59
  transformation processes, 8, 205
  uptake and release, 17

off-line estimation scheme, 23, 24, 43
on-line estimation scheme (see recursive estimation)

parameter estimation, 9, 23, 69, 75, 81, 119, 157, 227, 228, 274, 324
  in partial differential equations, 37, 190
Peel–Harvey estuary (Australia), 82–87, 102–115
pelagic food web, 267–270
pesticides, 126
pH, 126, 208, 216–218, 330
phosphorus
  available, 84, 107, 108, 276, 280
  cycle, 163
  loading, 277, 294–300
  organic, 158, 179
  particulate, 272, 276, 281, 367
  phosphate, 111, 112, 158, 165, 178, 190, 208, 265–270, 272, 274, 278, 358
  scenario, 83, 103
  sediment, 83, 84, 107, 108, 112, 181
  soluble, 83, 104, 107, 108, 158, 178, 367
  total, 158, 165, 208, 216–218, 275, 294–297
plug-flow reactor (PFR) idealization, 93
posterior information, 375
potable water supply, 327, 329–331
Prandtl number, 186
prey–predator oscillations, 269
primary production, 180, 267, 269, 272, 276
  relationship with nutrient loading, 284, 286, 357
principal components analysis, 77, 214–221
prior information, 374, 375, 377

probability density function
  discrete, 146, 150
  lognormal, 253, 254
  normal (Gaussian), 126, 134, 146, 150, 159, 187, 194, 251, 253, 254, 296, 366, 368
  rectangular (uniform), 106, 123, 124, 126, 134, 146, 150, 151, 187, 278, 279, 376
  skewed, 253
  triangular, 124, 126, 253

rainfall–runoff models, 4, 13, 95
random search algorithm, 122
real-time forecasting, 329, 339
recursive estimation, 23, 24, 26, 52, 72, 79, 87, 91
  adaptation of parameter values, 25, 28, 40
  constant parameters, 30, 34, 37
  filtering theory, 51, 170, 316
  instrumental variable, 80
  least squares, 72, 341, 342, 344, 355, 356
  smoothing, 80
  states, 30
  time-varying parameters, 23–25, 32, 33, 38–40, 79, 80, 333–353
reductionist approach, 6, 71, 75, 82
regression analysis, 134, 179, 222, 230, 294, 295, 374
residuals, 51, 79, 91, 193
  probability distribution of, 192
  (model) errors, 296, 301, 325, 326
Richardson number, 186, 190, 191
river Cam (UK), 12, 26–32, 52, 343–354
river Rhine (Netherlands), 12, 48

Saginaw bay, lake Huron (USA), 15
salinity, 87, 89, 90
sanitary engineering, 4
scientific method, 26, 69, 70, 73, 81, 82, 95, 262
sediments, 118, 119, 127
  adsorption, 124
  interstitial water, 118
  model, 181
  reactions in, 190
  sampling, 190
seiche event, 133, 152, 185
sensitivity (see also sensitivity analysis)
  classification, 110–112
  coefficients, 50, 52, 166, 248, 249
  (Jacobian) matrix, 174, 311, 315
  to parameters, 129, 130, 136–139, 190, 191
sensitivity analysis, 10, 12, 48, 53, 61, 71, 111, 112, 115, 123, 125, 136, 288, 305
  comparison with error analysis, 247–255
  first-order, 50, 248, 302, 310–315
  generalized, 114
  second-order, 310, 311
  statistical, 50

small perturbations analysis, 88, 248, 309, 361
southern North Sea, 264–270
spatial variability, 252
specification, 373, 374
speculative simulation modeling, 84, 85, 96
spline fitting, 175–178, 181, 315
spruce budworm, 117
square-root algorithm, 344, 355
steepest descent algorithm, 174, 175, 191, 360
streamflow models, 4
Streeter–Phelps model, 22
striped-bass population, 254, 255
suspended solids, 14, 33, 117, 210, 216–218
system
  environment, 10, 11, 17, 18, 28
  external description, 17
  internal description, 17
systems ecology, 4

Theil's inequality function, 122
thermocline, 185, 188, 190, 191, 277
time-series analysis, 5, 43, 44, 323
  recursive methods of, 73
toxic substances, 126, 127
transparency, 205, 208
trial and error fitting, 8, 43, 130, 136, 259
trophic states, 294
  eutrophic, 296, 297, 299, 300
  hypereutrophic, 296
  mesotrophic, 296, 297, 299
  oligotrophic, 207, 296–298

univariate tests, 109, 110
unobserved state variables, 40

validation, 8, 69, 81, 130, 139, 140, 194, 233, 234, 245, 326, 373
  holistic, 71
verification, 3, 101, 111
vertical transport model, 183–201
  eddy diffusion process, 183, 185

wastewater treatment, 4, 14
weighted least squares estimation, 160, 162, 169, 170
white noise sequence, 79
wind
  drag coefficient, 129, 132, 138, 139, 186
  -induced water motion, 129, 185, 357, 367
  shear stress vector, 151, 188
  velocity vector error, 129

zooplankton
  carnivorous, herbivorous, and carbon content, 158, 163
  grazing function, 179
  nutrient, phytoplankton system, 52–59, 166, 209, 266–270

U. Förstner, G.T.W. Wittmann

Metal Pollution in the Aquatic Environment

With contributions by numerous experts
2nd revised edition. 1981. 102 figures, 94 tables.
XVIII, 486 pages
ISBN 3-540-10724-X

Contents: Introduction. – Toxic Metals. – Metal Concentrations in River, Lake, and Ocean Waters. – Metal Pollution Assessment from Sediment Analysis. – Metal Transfer Between Solid and Aqueous Phases. – Heavy Metals in Aquatic Organisms. – Trace Metals in Water Purification Processes. – Concluding Remarks. – Appendix. – References. – Subject Index.

The contiunued demand for **Metal Pollution in the Aquatic Environment,** combined with rapid advcances in this field, have necessitated the preparation of this new edition which includes coverage of all important developments made since 1978.

From the reviews: "This is an impressive and useful compilation of a large amount of data on heavy metal in the natural environment. The book contains nearly 500 pages and 70 of these are taken up by the references, of which the majority were published in the last ten years. I found the book lucidly written and on the whole easily readable. …
…In conclusion this is a very valuable book that has been well produced. It will be, as I expect, consulted regularly by many scientists interested in the contamination of the natural environment by metals. …" *Hydrobiologia*

"… the authors present what is probably the first succesful compilation on a world-wide basis, including a critical inventory and evaluation of the prevailing investigations in water, sediments and organisms…
…The book contains an extensive list of references (ca. 2000), is well written in a lively and critical style…
…The book is more than a necessary inventory; it should stimulate further research and be helpful in solving problems. I highly recommend this book to everyone involved in aquatic trace metal research, wheter biologist, chemist or geologist." *Marine Chemistry*

Springer-Verlag
Berlin
Heidelberg
New York
Tokyo

Seawater and Desalting

Volume 1
By A. Delyannis, E.-E. Delyannis
1980. 189 pages. ISBN 3-540-10206-X

From the contents: The water problem. – Desalting processes. – Raw material seawater. – Distillation. – Components of distillation plants. – Combined distillation plants. – Dual purpose plants. – Waste heat as energy source. – Nuclear energy as heat source. – Geothermal, energy as heat source. – Solar energy as heat source. – Scale formation and prevention. – Treatment of feed water. – Corrosion. – Materials of construction. – Ion exchange. – Electrodialysis. – Rev. osmosis and ultrafiltration. – Rev. osmosis membranes. – Reverse osmosis process. – Economic consideration. – Iceberg utilization.

Literature on water and desalinization, covering both basic research and engineering aspects, has up to now been spread throughout a large number of journals. This, together with the vast patent literature, has made the reference work of scientists and engineers a time-consuming affair.
To provide the scientific community with a systematic survey of the relevant literature, the authors have compiled in this book over 2100 citations, providing abstracts of the more important articles. New developments in various desalinization processes are reviewed in detail, and data on recently erected desalinization plants given. Particular attention is paid to the patent literature in view of the new technologies involved. Over 650 patents from the recent literature are in cited separately in each section and summarized in the Patent Index.
With its conclusion of literature from related fields, such as analytic chemistry and corrosion research, this book will be a valuable reference not only to scientists and engineers active in desalting research, but also to geoscientists, oceanologists, ecologists, and corrosion engineers.

Springer-Verlag
Berlin
Heidelberg
New York
Tokyo